Hunters and Hunted of the Savannah

World of Wildlife: AFRICA

Hunters and Hunted of the Savannah

ORBIS PUBLISHING·LONDON

From the original text by Dr Félix Rodríguez de la Fuente
Scientific staff: P. de Andres, J. Castroviejo, M. Delibes, C. Morillo, C. G. Vallecillo
English language version by John Gilbert
Consultant editor: Dr Maurice Burton

Contents

Acknowledgments

Des Bartlett/Bruce Coleman: 110, 111, 131, 161, 211, 223
F. Bel/Jacana: 21, 151
R. Bousquet/Jacana: 106
J. Brownlie/Jacana: 27
J. Burton/Bruce Coleman: 23
Camera Press/Zardoya: 89
Bob Campbell/Bruce Coleman: 271
M. A. Castaños: 88
E. Cerra: 287, 296, 297
N. Cirani: 189
A. Coleman: 147, 149, 152
Dubios/Jacana: 14
Edistudio: 7, 8, 15, 19, 33, 42, 62, 66, 82, 84, 127, 153, 159, 175, 245, 250, 259, 261, 262, 269, 276, 288
J. M. Fievet/Jacana: 135, 150, 270
C. A. W. Guggisberg: 257
Alfonso Gutiérrez/Edistudio: 1, 4, 29, 36, 47, 49, 53, 57, 61, 65, 67, 73, 74, 102, 107, 109, 120, 185, 191, 200, 219, 221, 227, 228, 230, 231, 235, 243, 275, 283
Ray Halin/Edistudio: 156
Hoa-Qui Editions: 13, 30, 58, 79, 128, 168, 186, 194, 224, 225
David Hughes/Bruce Coleman: 137
P. Jackson/Bruce Coleman: 195, 196
Klages/Atlas: 29
Armelle Kerneis/Jacana: 37, 116
C. de Klemm/Jacana: 40
J. Lalanda: 26, 63, 77, 80, 93, 141, 146, 148, 201, 225, 233, 254, 274
Baron Hugo van Lawick/Camera Press: 292
A. Margiocco: 180
E. Münch/APSA: 165

Norman Myers/Afrique Photo: 24, 142, 199, 203, 218, 268
Norman Myers/Black Star: 82, 117, 124, 130
Norman Myers/Camera Press: 70, 71
Norman Myers/Bruce Coleman: 10, 11, 37, 50, 51, 59, 66, 138, 205, 264, 276, 280
Norman Myers/Salmer: 262, 290
J. Nuñez: 238
Jaime Pato/Prensa Española: 78, 92, 114, 115, 133, 145, 154, 167, 294
D. M. Paterson/Bruce Coleman: 176, 246, 253
L. Pellegrini: 163
Photo Researchers: 123, 173
André Picou/Afrique Photo: 199, 249, 291
André Picou/Fotogram: 202
G. Pizzey/Photo Researchers: 160
Masood Quarishy/Bruce Coleman: 97, 236, 278
Robert/Jacana: 48
Félix Rodríguez de la Fuente: 42, 52, 68, 100, 104, 169, 171, 178, 187, 206, 239, 251, 253, 255, 300
John Rushmer/Carl Östman: 15
Alan Root/Okapia: 32, 39, 240, 247, 249
James Simon/Bruce Coleman: 135
M. Socias: 84, 86, 88, 118, 122, 165, 208, 215
Time-Life: 17, 28, 46, 54, 90, 101, 105, 174, 193
Simon Trevor/Bruce Coleman: 35, 60, 99, 165, 181
J. P. Varin/Jacana: 19, 95, 113, 154, 171
Vasselet/Jacana: 14
G. Vienne/Jacana: 151
Verzier/Jacana: 212
Albert Visage/Jacana: 3, 82, 98, 103, 138, 160, 170, 176, 182, 188, 190

Foreword

Naturalists and zoologists from all over the world travel thousands of miles to visit the magnificent grasslands of East Africa. The fauna which is to be found there is unrivalled anywhere else in the world. It is no coincidence therefore that many of the greatest game reserves and national parks can also be found in this area. They provide an ideal place for the detailed study of animals in their natural environment. Here, the natural biological balance which had existed for millions of years before the coming of the white man has been preserved and thus a very wide variety of wildlife can be observed and studied. It is fitting therefore that we begin our survey in this ecological haven.

The hunters of the savannah like the cheetah and the lion are often considered to be villains, but in fact it is necessary for hunter and hunted to live together. In this way each species has to fight to survive. The hunted like the gazelle, the gnu, the zebra and the buffalo have, over the years, developed many different behaviour patterns which help them to survive. This is nature's way of keeping only the strong and the healthy alive. The vulture too, has its part to play. Once battle has taken place it comes along and cleans up the carcass, thus preventing the spread of desease.

The fascinating story of all these animals as seen on location is told in this the first volume of *World of Wildlife*.

World of Wildlife

Africa:
Hunters and Hunted of the Savannah

INTRODUCTION

We have the good fortune to live on a planet where life has developed in myriad different forms to produce that wonderful world of animals, of which we ourselves form such a tiny part. If it were possible to transport ourselves to one of the great national parks of East Africa, we should be overwhelmed by the variety and magnificence of the animal life there. It is a world of which we know hardly anything beyond what the cinema and television have shown us; and the traveller with an amateur interest in zoology would find it impossible to sort out the variety of creatures ranging the wide savannah before him. Where could he begin? The many species of tiny graceful gazelles, the larger antelopes with their majestic horns, the gnus and the zebras—the multitude of different forms and colours would throw him into confusion.

More than one million living species

Zoologists devote themselves to the scientific classification of this incredible variety of animal life. On the basis of anatomical and physiological similarities, as well as on their more obvious resemblances, animals can be grouped in *genera*, and identified as individual *species*.

Facing page : The white pelican (*Pelecanus onocrotalus*) is one of Africa's largest and most common of water birds. Pelicans live on fishes which are stored in the pouch suspended from their huge beaks. Here they are seen perched on a dead tree in the middle of a lake.

Survival in the wild means 'to eat and not be eaten'. Conserving energy whilst keeping a watchful eye open for predators or potential victims comes naturally to every animal, an attitude perfectly conveyed by this group of cheetahs in Serengeti National Park.

INSECTS: more than 1,000,000 species

INVERTEBRATES (excluding insects): 232,000 species

FISHES: 20,000 species

BIRDS: 8,600 species

REPTILES: 6,000 species

MAMMALS: 5,000 species

AMPHIBIANS: 1,500 species

THE ANIMAL KINGDOM

Already more than one million species of animal have been described and classified. Most of these are invertebrates, and above all insects, of which there are a million or more species. The vertebrates comprise little more than 40,000: 20,000 species of fish, 8,600 species of bird, 6,000 reptiles, 5,000 mammals and about 1,500 amphibians. All these figures are of course approximate and provisional, as new species are constantly being discovered, particularly among the fishes and the invertebrates; and especially among the insects, where new species are being described and named almost every day.

The many species, as we have said, can be grouped into genera. In this way lions, tigers, leopards, pumas and other similar animals belong to the genus *Panthera*; while dogs, wolves, jackals and coyotes belong to the genus *Canis*.

There are numerous distinct species of animals which do not possess particular names in everyday speech, because their separateness as a species is only recognised by scientists. This is the reason why every species is given its own 'Linnaean' name in Latin, according to the principles first established by the great Swedish scientist Linné. Each species name consists of two parts: the first indicates the genus and the second the species, so that the lion is *Panthera leo*, the tiger is *Panthera tigris*, the leopard *Panthera pardus*, and so on.

The genera are grouped into Families, the Families into Orders, and the Orders into Classes. In this way the genus *Panthera* belongs to the family Felidae, while the genus *Canis* belongs to the Canidae. But both these families are included in the order Carnivora, since they are meat-eaters, and the Carnivora are classified, with other orders, among the class Mammalia, or mammals; which, in their turn, belong, together with the birds, reptiles, amphibians and fishes, to the Vertebrata. And it is with the vertebrates that *World of Wildlife* will be particularly concerned.

The distribution of species

This total of a million or more distinct species of animal is not distributed evenly throughout the countries and seas of the planet. Most species are adapted to a particular environment, and cannot live in different surroundings: we do not find gorillas or ostriches in Alaska, nor polar bears and seals on the coasts of Africa.

This unequal distribution of species is due, above all, to the very different environmental conditions (climate, soil, vegetation, the presence of other animal species, food sources, etc) of the various regions of the globe, as well as to the many changes undergone by the earth's surface during the course of time. The modifications of the continents have caused certain species to remain isolated for millions of years, so that they have retained their original characteristics. A typical example is the echidna, a species cut off on the Australian continent since the early days of evolution: although it is a mammal it reveals much of its reptilian ancestry, lays eggs and looks rather like a hedgehog or a sea urchin.

The lion is not always the fierce hunter described by so many writers. The male lion, in fact, is a rather lazy creature who lets the female kill the prey and then steps in to claim the largest portion. When his appetite is satisfied he spends the greater part of the day basking in full view of the herbivorous animals which provide his sustenance.

Animals in their surroundings

If we were to study animals solely according to their systematic classification, then we would have to consider together those animals which belonged to the same genus. Since lions, tigers and jaguars are all closely related, we would study the lions of Africa, the tigers of Asia and the jaguars of America all together, although none of these species depends at all directly upon the others.

Within the African animal community, for instance, lions depend for their existence upon zebras and antelopes, but not directly upon their nearest relatives, the leopards. In fact lions, which are relatively heavy and slow-moving animals, could not survive in an environment that was not populated with large herds of herbivores. And at the same time, the zebras and antelopes of the savannahs depend upon the existence of lions and other predatory animals, since without the population control that these hunters exercise they would soon multiply to such an extent that they would destroy their grazing and eventually die out from starvation.

For a long time, zoologists concerned themselves almost exclusively with the classification of species and with the description of their individual anatomical characteristics. In this study they worked only with skeletons or with dissected dead animals. Recently, however, this concern with systematic classification and anatomical description has been relegated to second place. Naturalists have begun to study animals in their natural environment, and to observe their behaviour in relation to the many other species with which they co-exist.

In *World of Wildlife* we shall not present animals according to their classification, nor limit ourselves to anatomical description. We shall study together those animals that live together, being particularly concerned with the ways in which they behave, their relations with other animals, and the manner in which they have, through evolution over many millions of years, adapted themselves to their surroundings.

The major part of each chapter is devoted to a group of animals, how they live, love and reproduce themselves; the organisation of their family and their community; how they attack or defend themselves from their enemies; how they form a part of the environment in which they are established, and how they hold their own in the daily battle for survival. But systematic classification will not be ignored: the place of each species in its genus, family, order and class will be summarised in special sections at the end of each chapter, together with a description of the general anatomical characteristics of each category and the way in which it is related anatomically and physiologically, and in its behaviour, with others.

The uneven distribution of wild life around the world is due in part to the nature of the environment, which varies from one region to another, to climatic conditions and – to an even greater extent – to links in prehistoric times between continents that have since drifted apart. Zoogeographers, studying the world distribution of mammals, have divided the map into five main regions of fauna. Islands have separate characteristics, their plant and animal populations showing distinct, often unique, patterns of evolution; those of islands formerly linked to continents differ from those of islands showing volcanic or coral origin.

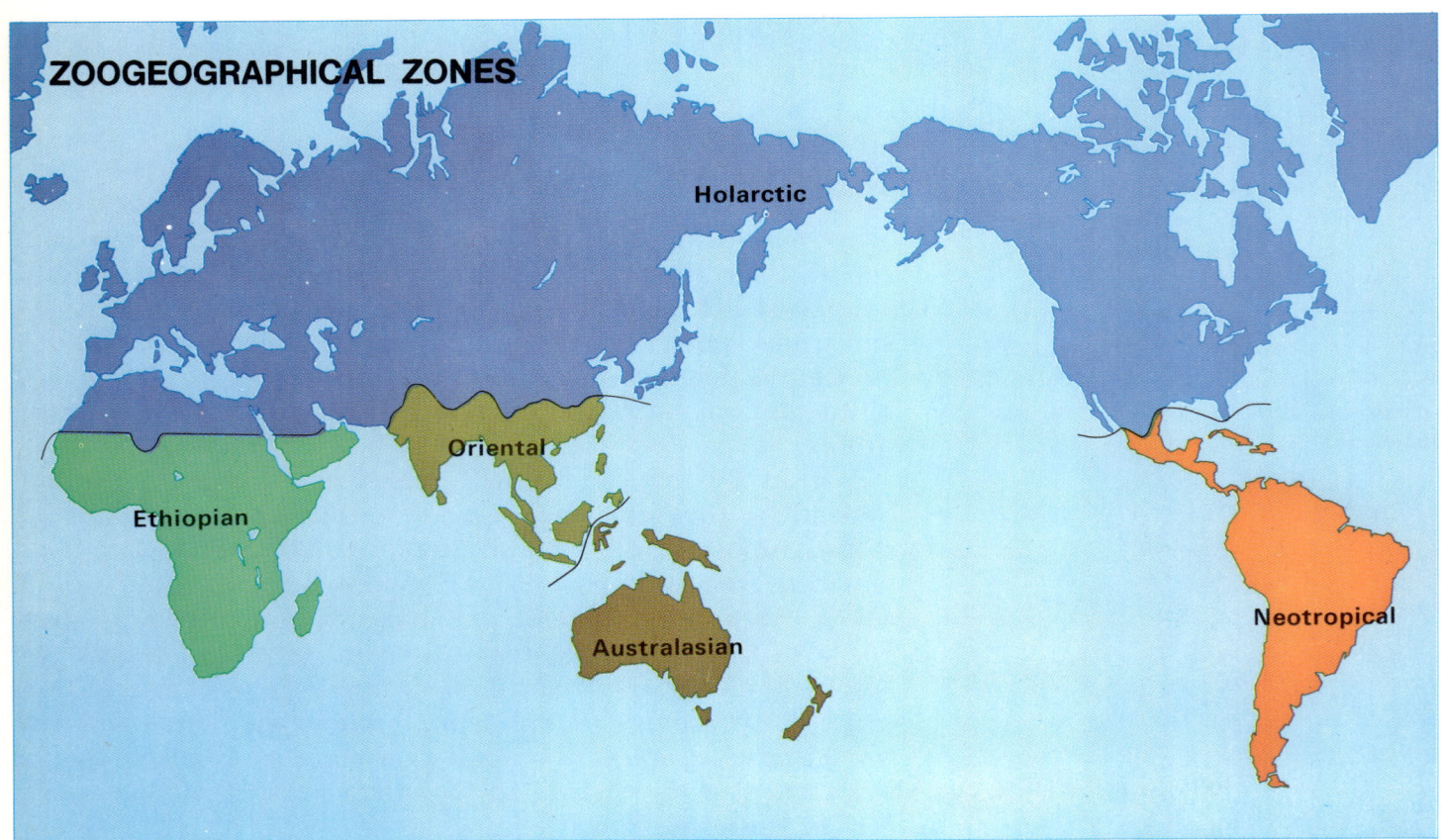

ZOOGEOGRAPHICAL ZONES

Holarctic

Oriental

Ethiopian

Australasian

Neotropical

The zoogeographical zones

World of Wildlife has been organised according to zoogeographical and ecological principles. The world has been divided into areas of animal distribution known as zoogeographical zones, which coincide, in the broad sense, with the continents. To begin with, we shall deal with the animals of the Ethiopian zone, which comprises the whole of Africa south of the Sahara; and after this, with those of the Holarctic zone, comprising Europe, North America, north Africa and most of Asia. Following these, we shall deal with the animals of the Neotropical zone (South America), the Oriental zone (southern Asia), and the Australasian zone; and finally with the oceans.

Within each of these zones we shall look at the different kinds of natural habitat: desert, forest, mountain country, etc. To begin our study of African wildlife, therefore, we shall consider the wide savannahs of East Africa, one of the few remaining areas where animals still survive in conditions unchanged for thousands of years.

This will be the beginning of a great photographic and scientific safari around the world: we shall force our way through the thickest jungle, climb the highest peaks and plunge into the ocean depths—to capture all the immediacy and spontaneity of the wonderful world of wildlife.

The African savannah is an ideal place for the detailed study of animals in their natural surroundings. This particular expedition, whose tents are pitched in the shade of acacias in Serengeti National Park, is studying the behaviour of the little known *Lycaon* or African hunting dog.

CHAPTER 1

The savannahs and steppes of Africa

The 19th-century explorers and hunters who first probed the interior of East Africa had to make the long sea voyage to Mombasa in Kenya and then strike inland on foot with a team of porters to carry their baggage. Their expeditions were fraught with difficulty and danger. Today, though the route is similar, the risks are fewer and progress is more speedy and comfortable, thanks to aeroplanes and sturdy vehicles such as trucks and Landrovers, as well as an array of modern equipment, adaptable to all types of terrain and weather. Yet the first impression is not unlike that which must have struck the early explorers and colonists—and it is somewhat discouraging. The port of Mombasa itself lies heavy under a sticky, suffocating canopy of heat. The lowlands are stark and bleak, covered only by thorny scrub and dwarf acacias. The dusty, arid expanses of the immense Rift Valley, flanked by steep slopes, are broken by the occasional brackish, salty lake; and the moment the weary traveller finds a shady spot by a stream he is assailed by an army of mosquitoes.

Eventually, however, the scrub and the stunted bushes are left behind and the landscape opens out into high plateau country, unforgettably spectacular. The tall green grasses of the level plains are ruffled by fresh breezes and welcome shade is provided by high, umbrella-like acacias. The rich pastures stretch to the distant horizon.

Animals can be seen grazing peacefully everywhere—especially antelopes of many sizes, hues and patterns, some with unwieldy bodies in striking contrast to their elegantly curved horns. They include the splendid Grant's gazelle, with its plain sand-coloured coat, the agile little Thomson's gazelle, its flanks striped with black bands, the ungainly hartebeest and the top-heavy brindled gnu. Herds of such creatures mingle in their thousands with

Facing page : A group of Thomson's gazelles, distinguishable by the black stripes on their flanks, seek the welcome shade of the tall savannah acacias. Should one of them scent the presence of a predator, the violent trembling of his body immediately alerts the others and sends them scampering to safety.

Previous pages : A group of African elephants in typical savannah surroundings. One of them smells and probes with his trunk at the acacia branches, prior to breaking off a choice morsel. Elephants habitually collect much more food than they can eat, leaving piles of debris behind them when they move on.

countless numbers of zebras. It is a sight which, in modern times, is unique to this region, a moving testimony to the richness of animal life as it once existed in other places, before man began to destroy it with his urban and industrial development schemes – or more simply out of simple delight in killing.

It is not by accident, therefore, that naturalists and zoologists flock from Europe and America to visit the magnificent grasslands of East Africa, nor is it coincidental that the world's most extensive natural game reserves are situated in these parts. This is also why we have chosen to begin our survey of the world's fauna in a part of Africa which is ecologically unrivalled, with a large and varied animal community scarcely touched by man's depredations. In fact, the perfect biological balance which existed for millions of years prior to the arrival of the first white colonists has been preserved, thanks to the protection afforded by the great national parks and game reserves.

The savannahs

We must take a closer look at the grasslands that are so characteristic of most of the African landscape, briefly explaining how they originated and how they were fortunate enough to be so richly endowed by nature.

Some 25 million years ago, in the Miocene period, the earth's climate was much wetter than it is today and there were no Sahara and Arabian Deserts. Africa was almost entirely covered by huge, impenetrable forests, which also extended from the Atlantic to India and South-east Asia, giving the whole area an almost uniform type of fauna. But in time changes took place so that the inhabitants of these regions no longer enjoyed conditions favouring their continued development. The climate became gradually drier so that the area previously occupied by forest dwindled, and the difference in terrain between Africa and Asia became more marked. Animal life tended to split into two groups, each proceeding along a separate evolutionary path.

Irregular and infrequent periods of rain alternated with long dry seasons, encouraging the growth of enormous grassy expanses, punctuated here and there by shrubs and bushes. Nowadays trees are found in greatest number on the banks of rivers and lakes. Though generally of modest height, they possess a highly developed root system, enabling them to absorb a maximum amount of moisture from the soil. Some, such as the acacias, have umbrella-like foliage; others, such as the enormous baobabs, have tremendously thick trunks, capable of storing large quantities of water. The soil also stimulates the growth of a number of different grasses – *Sorghum, Panicum, Digitaria, Setaria, Themeda triandra, Pennisetum,* etc – as well as legumes and other herbaceous plants. This type of vegetation, perfectly adapted to periods of irregular rain-fall, makes up the savannahs.

The savannahs stretch in a broad fringe from Senegal to Ethiopia, bounded to the north by the sub-desert steppes and to the south by the equatorial rain forest bordering the Gulf of Guinea. At the level of the equator they are intersected by the vast expanse of forest forming a belt across the African continent

Facing page : The giraffe is a marvellous example of anatomical adaptation for survival, its long neck enabling it to nibble the shoots and leaves of very tall trees. Here a group is seen in characteristic guard position, each looking in a different direction so that between them they scan the widest possible area.

from the Atlantic to the Mitumba Mountains. Beyond that point, along the frontiers of the Congo Republic and covering the major parts of Uganda, Kenya and Tanzania, the savannahs extend without a break southwards into Zambia, Rhodesia, Mozambique and South Africa (where they are known as the High Veld and Middle Veld).

Ecologists have divided the world into a series of regions where soil and climate tend to sustain a dominant form of vegetation and a homogeneous fauna. These major zones include tundra, taiga, temperate forest, Mediterranean-type maquis (former woodland, now covered by dry scrub), savannah, tropical rain forest, steppe and desert. These regions are known as vegetation zones and are recognised by the majority of ecologists.

Nowhere are the savannahs so extensive as on the African continent, although South America, India and Australia have similar landscapes. Furthermore, the great North American prairie and the steppes of Asia have much in common with the African savannahs and, despite the fact that climate and terrain vary considerably, ecologists classify them together as grassland vegetation zones.

Immense areas of grassland in East Africa are of volcanic origin, with a thin layer of soil covering a thick layer of hardened lava. This type of terrain impedes the growth of trees and bushes with extended roots and is therefore devoid of forest.

It has been reckoned that the average annual rainfall in a desert area is between 0 and 12 inches. Steppe regions receive between 12 and 30 inches per year, while rainfall ranging from 30 to 60 inches is necessary for the development of vegetation in savannah and other grassland and woodland areas, provided the precipitation is evenly distributed and that it occurs in a single season, followed by a long dry period. It is therefore obvious that the alternating dry and rainy seasons characteristic of tropical and sub-tropical Africa provide highly favourable conditions for savannah flora and fauna.

The areas devoted to savannah are periodically extended by fires which burn down adjacent woodland and forest. Even before man learned to create fire, dry grass might ignite spontaneously under the penetrating rays of the tropical sun, as a result of a lightning flash or in the wake of a volcanic eruption. But once man taught himself to strike sparks from flint, fire raged even more freely and destructively through forest, savannah and steppe. Such conflagrations can only be halted – and then not with any degree of certainty – in places where the forest growth is exceptionally thick or where a river forms a natural barrier. Torrential rain or absence of vegetation are equally effective extinguishers of the flames.

Plants and plant eaters

The plants that are best adapted to the harsh conditions of a dry climate and stand up both to fire and to the continual assaults of animal life, are the various forms of grasses. We know from our own experience how a smooth lawn in a garden or park, patchy from being trampled underfoot, begins to grow again after a

An observation post overlooking Ngorongoro Crater, deliberately built near a water hole where both carnivores and herbivores congregate. These high platforms are ideal for studying the region's wild life without causing any disturbance.

Facing page : The savannah grasslands are frequently devastated by fires (*upper picture*) which may rage for days on end. Once the fire is extinguished (*below*) vast areas are left scarred and blackened, only the trees proving resistant to flame and heat as a result of their thick bark.

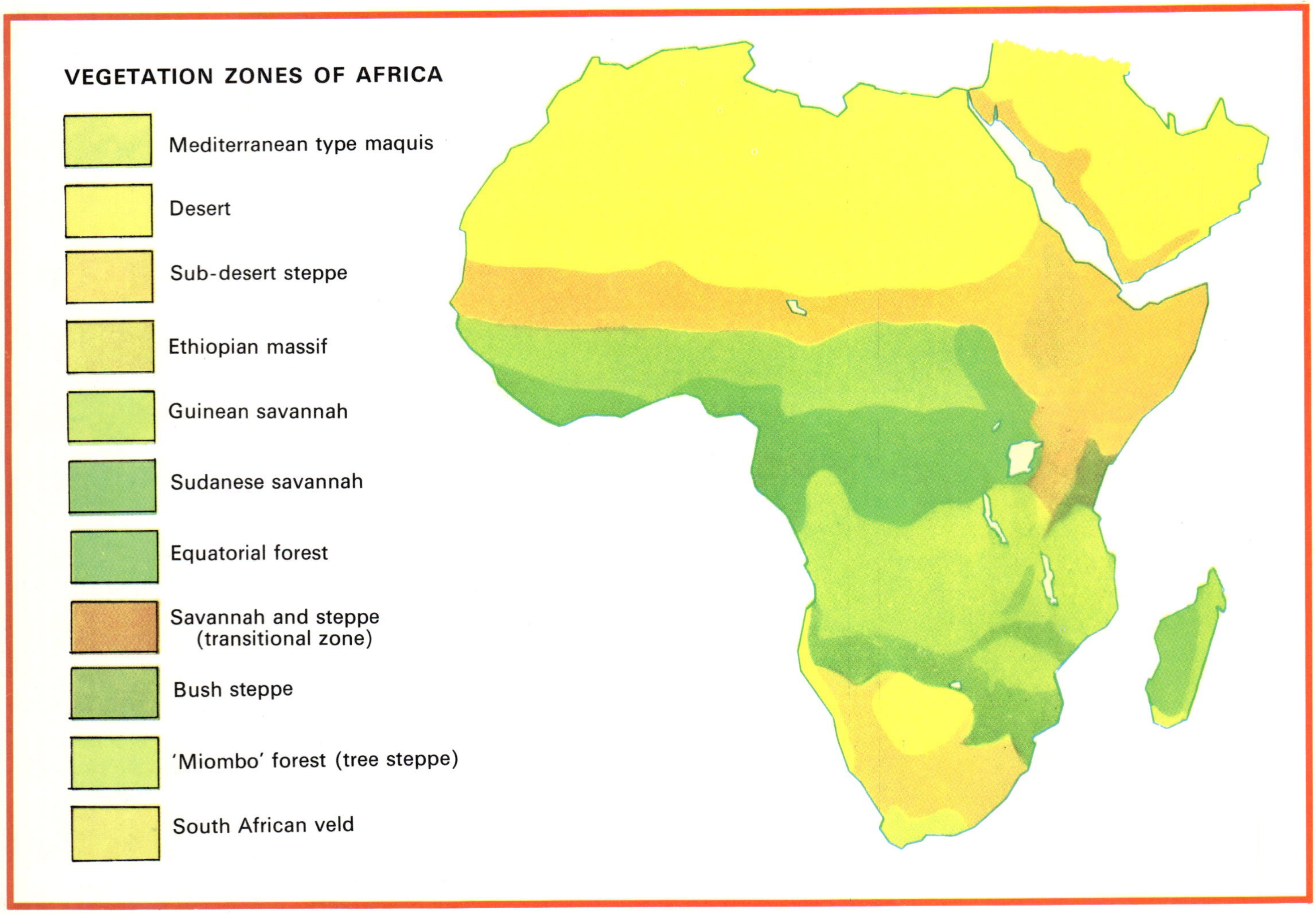

VEGETATION ZONES OF AFRICA

- Mediterranean type maquis
- Desert
- Sub-desert steppe
- Ethiopian massif
- Guinean savannah
- Sudanese savannah
- Equatorial forest
- Savannah and steppe (transitional zone)
- Bush steppe
- 'Miombo' forest (tree steppe)
- South African veld

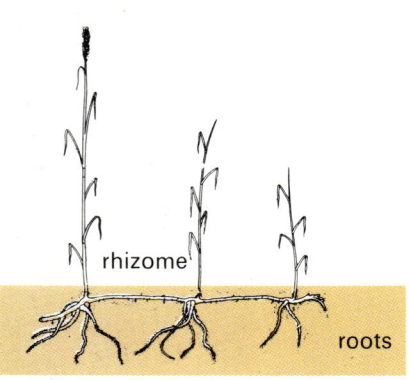

rhizome

roots

Many herbaceous plants have a specialised type of growth in the form of widely branching rhizomes, or underground stems, which throw-up stems above ground and roots below. This enables them to send up new growth after being chewed by animals or burnt by fire.

generous dousing with water. What few of us realise is that we owe this property of recovery to innumerable generations of plant-eating creatures roaming the world's grasslands: over millions of years the vegetation has adapted itself to the continual destructive actions of teeth and hooves. In fact, herbaceous plants and herbivores have developed side by side.

We do not know exactly when such plants first appeared on earth, but examination of the dental structure of primitive members of the horse family (Equidae) has led scientists to conclude that the ancestors of modern horses were already grazing on grassland in the Miocene period. The crowns of their incisors and molars would seem to have been admirably formed for snapping off and chewing grass. Such a discovery appears to confirm the theory that plants and herbivores are inseparable and that they have evolved together for at least 25 million years.

The growth region of herbaceous plants is situated at the base of the stem and not at the tip (or apical meristem), as is the case with trees. It is this important characteristic that enables grass which has been trampled by animals or damaged by fire to shoot up anew. Furthermore, many plants possess a special underground structure, consisting of vigorous, branching rhizomes, only the upper parts of which protrude above ground level. Such species are capable of spontaneous regrowth after being destroyed, whereas other annual forms are dependent upon alternating rainy and dry seasons for their development.

Some grasses, such as *Hyparrhenia*, withstand fire thanks to their specialised seed structure. The seeds are encased in a bristly capsule of stiff hairs, spirally arranged in such a way that, under the combined influence of the heat of the fire and the coolness of the night air, they expand and contract alternately, burying themselves deeply in the ground. Thus they are protected from fire and drought until the next rainy season, when a new plant appears.

Reproduction of grasses and many other kinds of plants takes place without the assistance of insects, occurring either by wind-activated cross-pollination or by multiplication of underground rhizomes. This is another reason why they spread so freely over immense areas.

Grasses are not only ideally equipped by nature for survival but in addition possess high nutritive value. They contain silica, a mineral substance which, in the opinion of some biologists, accounts for the tall stature of the herbivores. So this unique type of vegetation provides food for the enormous herds of ungulates (hoofed animals) that are so characteristic of the East African landscape. In Kenya's Masai Mara savannahs and in Tanzania's Serengeti grasslands, it is estimated that there are more than 150 animals to every acre.

Hooves, toes and speed

Anyone visiting the high plateaux of East Africa for the first time must be immediately impressed by the sheer weight of numbers of the animal population. Equally astonishing is the extraordinary diversity of shapes and sizes. The hordes of zebras and gnus

In the cool of the late afternoon a pack of fierce African hunting dogs routs a herd of zebras. The chosen victim is almost always the straggler, exhaustedly falling far behind its companions. The savage dogs will bring it to the ground, tear it apart with claws and fangs, and leave only a pile of bones for scavenging birds and animals.

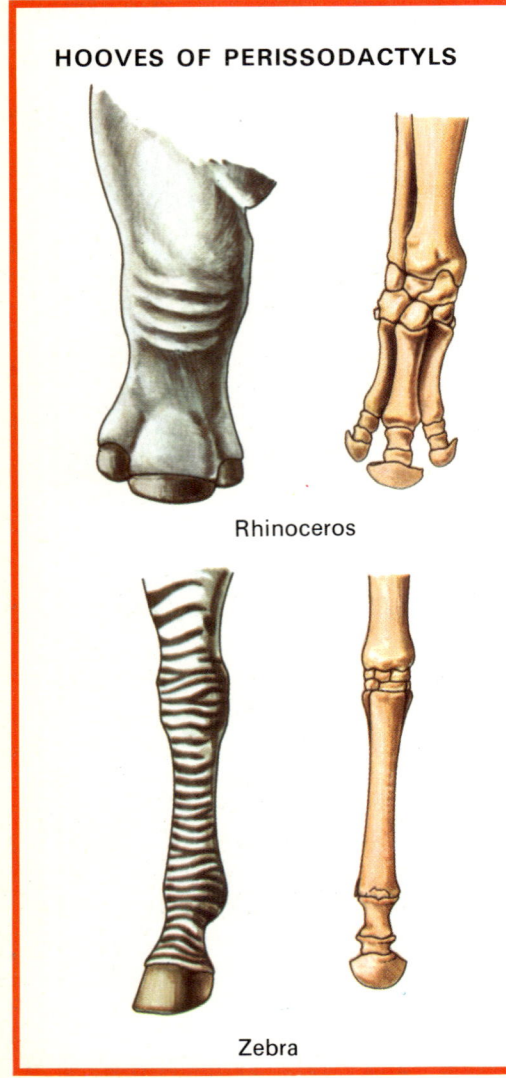

HOOVES OF PERISSODACTYLS

Rhinoceros

Zebra

HOOVES OF ARTIODACTYLS

Hippopotamus   African buffalo

Facing page: Lengthening of the limbs and the reduction in the number of toes has enabled certain species to become fast runners. Impalas (*above*) are artiodactyls – even-toed ungulates – with remarkable jumping ability. The zebra (*below*) is a perissodactyl or odd-toed ungulate, who also relies on speed to escape pursuers.

cover the plains like a living sea, while agile gazelles run and leap in every direction. The hartebeestes can be distinguished from the other antelopes by their slender silhouettes and elongated heads. The massive elands, weighing up to 1,200 lbs, are in extreme contrast to the delicate little dik-diks, scarcely larger than hares. All these animals certainly appear to differ from one another as regards size, form and colour, but closer investigation reveals an important resemblance – they all have long muscular legs terminating in hooves. These are perfectly adapted to hard terrain, enabling the animals to travel long distances in search of rainy regions or carrying them as swiftly as possible out of the reach of predators.

This means of movement common to all these vegetable and plant eaters has led scientists to group these animals together as ungulates. The characteristic feature of these animals is the horny sheath or hoof which extends over and protects the last joint of the toe, reducing to a minimum the area of contact between foot and ground. All evidence shows that this specialised structure is an adaptation facilitating rapid movement and perfectly suited to life on prairie and savannah.

One toe, three toes . . .

If we were to take an even closer look at the ungulates we would immediately notice that many of them have unusually long legs, and in fact that some of the species balance themselves on a single digit or toe – a unique anatomical feature. Most of us know the hoof shape of a horse, the culmination of an evolutionary process lasting millions of years, in the course of which it has adapted itself to a life spent galloping across level grasslands. The Equidae – horses, zebras and the like – have actually lost their original four toes during their lengthy evolution, and are now reduced to one toe on each foot. Their bulkier cousin, the rhinoceros, supports himself with three toes on all four feet. The tapir has four toes on the front feet, three on the back feet. All these ungulates are classified in the order Perissodactyla – hoofed quadripeds with an odd number of toes.

. . . two toes, four toes

Other animals, including the gazelles – who are among the fastest of all runners – have solved their locomotive problem in a less spectacular way. Over the ages they have retained either two or four toes, similarly protected by hooves. The animals possessing an even number of toes are classified in the order Artiodactyla and are capable of far greater physical dexterity and manoeuvrability than the Equidae. Witness, for example, the astonishing agility of wild goats in scaling precipitous rocks, the confidence of antelopes among the hazards of swampy terrain or the wonderful suppleness of impalas and their high-jumping skills.

The artiodactyls also include enormous creatures such as the hippopotamus who, having greater need of a solid supporting base than limbs designed for swift movement, has likewise retained four toes on each foot. Yet the Suidae (pigs and swine),

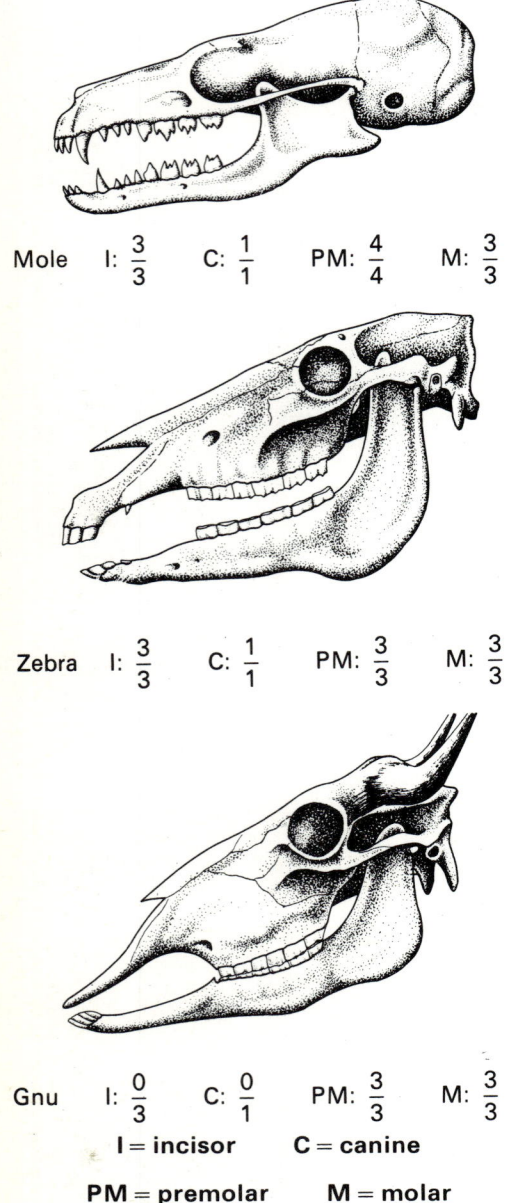

Mole I: $\frac{3}{3}$ C: $\frac{1}{1}$ PM: $\frac{4}{4}$ M: $\frac{3}{3}$

Zebra I: $\frac{3}{3}$ C: $\frac{1}{1}$ PM: $\frac{3}{3}$ M: $\frac{3}{3}$

Gnu I: $\frac{0}{3}$ C: $\frac{0}{1}$ PM: $\frac{3}{3}$ M: $\frac{3}{3}$

I = incisor   **C = canine**

PM = premolar   **M = molar**

Skulls and dental formulae of animals adapted to different kinds of diet: the figures give the number of teeth each side of the upper and lower jaw. The mole, one of the insectivores, has a complete set of teeth, conical-shaped incisors, well developed canines and sharp molars, enabling it to pierce and chew the hard body coverings of insects. The zebra and gnu, both herbivores, have large, four-cornered, flat teeth in the area of the cheek-bone, ideal for crushing grass stems; the canine teeth, having no real function, are reduced or absent. In the Equidae (such as zebras) the incisors act like pincers, whereas in the Ruminantia (such as gnus) the upper incisors are missing. Instead there is a hard pad in the front of the upper jaw for the lower incisors to bite against.

including athletic creatures such as the wild boar, warthog and barbirusa, also have four toes on each limb, all of which can be used. In fact, they have a double advantage, making do either with all four at once or with just the two inner digits, according to the nature of the terrain and their manner of moving. In addition to this efficient machinery of motion, all the artio-dactyls, with the exception of the hippopotamus family and the wild swine, possess a specialised digestive system which classifies them as ruminants or cud-chewing animals (for a detailed description of ruminatory digestion see pages 22-24). Many of these artiodactyls, particularly the males, are also equipped with horns.

The importance of teeth

The limbs of ungulates are perfectly suited for easy and rapid progress across steppe, prairie and savannah, providing the necessary stamina and speed for long-distance travel, especially the return after the rainy season to familiar, favourite pastures. However, the evolutionary process was by no means confined to the animals' motive ability; it applies equally to the structures and organs concerned with chewing and digesting – teeth, jaws and stomach.

Anyone who has tried to tear up a clump of wild grass, fibrous and often razor-sharp, can confirm that it is no easy task. Farmers cope with the problem by using tools specially designed for such a job. Nature has provided the ungulates with similar tools – a system of dentition which has been progressively transformed over the centuries – enabling them to cut and crush vegetable matter without effort. In the most specialised cases some of the incisors have disappeared and those that remain have grown in size so that they are eminently suitable for cropping herbaceous plants. The canine teeth have failed to develop properly or have also vanished. In other instances incisors or canines have become over-developed and been transformed into defensive tusks, as with the hippopotamus and its relatives and the wild swine. The premolars have grown into the same shape as the molars so that both forms of dentition are sufficiently strong to be used jointly for crushing and grinding. The teeth of antelopes and zebras are very similar in appearance, with large, high crowns, the vertical layers of enamel sandwiched between two harder areas of ivory and cement. This composition keeps the crowns of the teeth sharp and keen, preventing wear and tear which might have a rounding effect and render them unsuitable for chewing. This would certainly occur if all the teeth were uniformly smooth and if the enamel acted as a protective covering for the ivory, as is the case with the meat-eating carnivores and mixed-diet omnivores.

The efficacy of the dental structure has been further refined by complex modifications of the maxilla or upper jaw and of the masticatory muscles. This enables ruminants, in the process of grazing on grasses and other plants, to carry out a variety of complicated chewing movements (up and down, side to side, back to front) which make their feeding smooth and effortless.

How herbivores digest their food

Almost all living creatures are able to digest the food substances on which they depend for survival. This is made possible by the action of certain glands secreting digestive juices. No vertebrate, however, is capable of wholly assimilating cellulose, the essential component of plants. How then have the herbivores been able to obtain the nourishment they need? Once again nature has provided its own remarkable solution. At the time when the phytophages or plant-eating animals first appeared on earth, an association evolved—known scientifically as symbiosis—between the animals themselves and micro-organisms capable of digesting the plants' cellulose content. Insects feeding on vegetable matter harbour in their digestive tracts colonies of protozoa performing this function. Such micro-organisms were also found in the intestines of the reptiles of the Mesozoic period. Thanks, therefore, to the millions of bacteria and protozoa living in their digestive tracts, the ungulates are able to derive the maximum nutritional benefit from the vegetation on which they habitually feed.

At the best of times, however, the nutritive value of these foods is so poor that the animals must consume them in very large

Zebras devote a major part of their time to grazing, but since they digest grass incompletely they are compelled to ingest large quantities in order to derive adequate nutritive value.

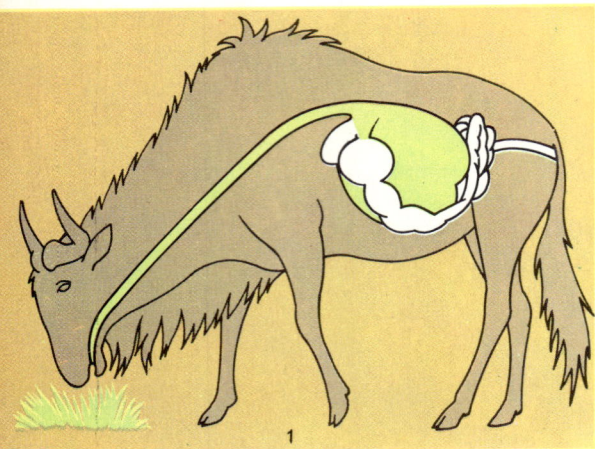

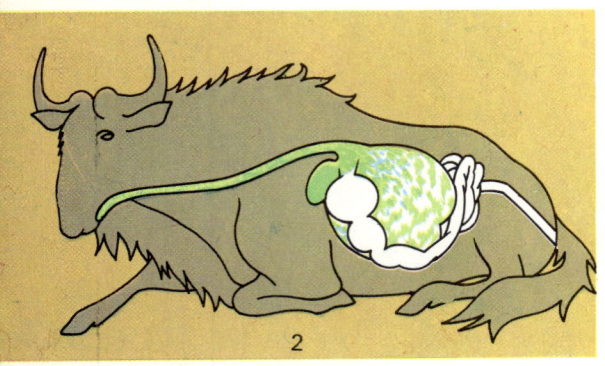

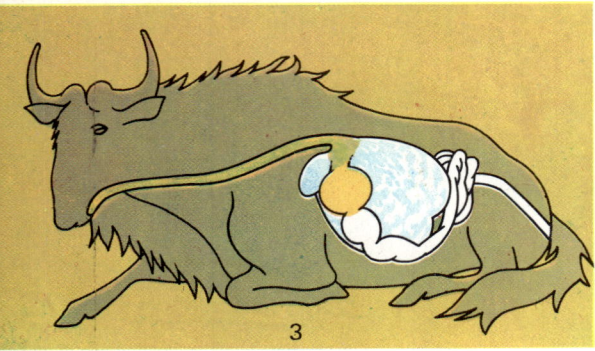

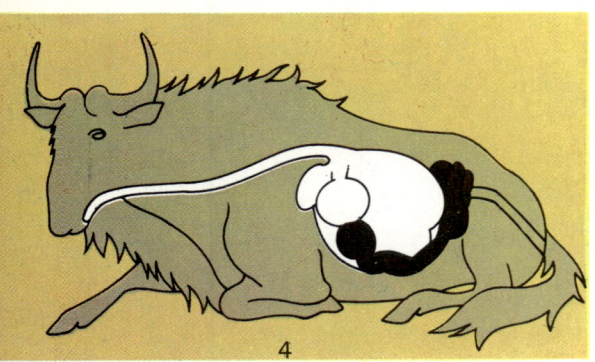

1. The gnu, a typical ruminant, swallows grass which enters the rumen or paunch. 2. The food bolus is regurgitated and masticated in the reticulum. 3. The food is reswallowed and passes into the omasum where the water is absorbed. 4. The digestive process is completed in the abomasum and intestines.

quantities to obtain the necessary amount of calories. This is why the digestive system of the herbivores has been extended and undergone, in the course of time, unusual modification.

When a zebra feeds, it cuts the grass with its incisors, which are strong enough to break up the toughest and most fibrous stems. Chewing is then done by the premolars and molars, with the aid of the powerful masseter muscles; the same process is employed by all species of horses. The resultant pulped grass enters the stomach, where the gastric juices go to work, and reaches the small intestine. As in all herbivores, this is of considerable length and it is here that all the usable protein material is absorbed. The residue passes into the caecum (exceptionally large and capable of containing up to about 17 pints), where it remains long enough for the cellulose content to be fermented by bacterial action and so rendered assimilable. After this the digesting mass reaches the large intestine where the water content is absorbed into the bloodstream.

This complex digestive process is not completed. A large proportion of the ingested matter leaves the body as excrement, and the fact that grasses are partially resistant to natural digestive processes helps to disperse them. Some grass seeds are inevitably unaffected by the action of the gastric juices and micro-organisms in the herbivores' digestive tracts. Mixed with excreta, they find their way back into the soil, sometimes many miles from the spot where they were consumed by the constantly moving animals, there to germinate and develop anew.

Animals that chew the cud

For thousands of years sheep have been the backbone of the agricultural economy in many of the world's temperate regions. These creatures are the best known of all so-called ruminants, a familiar feature of many a rural scene. One of the most immediately striking aspects of their behaviour is their seemingly unlimited capacity for food, verging indeed on gluttony. They manage to devour grass at an incredible rate, as if even a momentary distraction might have disastrous consequences. Ruminants in the wild display similar concentration and single-minded passion, but with more reason. Take, for example, a gnu, grazing near a thicket of acacias. The animal feeds with its nose buried in the grass, eyes lowered, its ability to hear extraneous sounds severely handicapped by reason of its incessant chewing activity. At such times all sensory faculties are considerably diminished, leaving it at the complete mercy of ever-watchful predators. The lion, concealed in the tall grass, or the leopard, motionless among the branches of a tree, wait for just such moments of inattention to pounce. So it is clearly a matter of life and death for the gnu to take its meal as quickly as possible.

It is at this point that the mechanism known as rumination comes into operation—an important stage in the evolutionary process of mammals. The reason for the gnu's apparent greed is that it is storing up the largest possible quantity of vegetable matter. Only later, having retreated to an area where no danger lurks, will it begin chewing at leisure and digesting the food it

The gerenuk is an antelope that feeds mainly on leaves. Its well developed, muscular limbs make it possible for the animal to stand upright on its hind legs.

Previous page : The animals of the African savannah gather peaceably together at water holes, whether they feed on grass, as do the zebras, or on leaves, like the giraffes.

has already swallowed. Now, however, its head is held high and all the senses are alert to the potential presence of an enemy.

This specialised digestive mechanism is of course quite different from that of the zebra, and it goes without saying that the process of rumination or chewing the cud was not evolved overnight. Modern ruminants have a stomach containing four separate chambers—the paunch or rumen, the reticulum, the omasum or manyplies, and the abomasum. Some authors have also claimed that the specialised two-chambered stomach of the non-ruminant pigs and hippopotami displays the initial stages of a similar evolutionary process, that of the hippopotamus showing the more highly developed stage.

In the ruminants proper, the grass, mixed with alkaline saliva, is hurriedly swallowed and passes into the rumen where it is moistened by gastric fluids containing bacteria. This vegetable mass is then regurgitated in the form of a rounded bolus for a lengthy period of mastication. The regurgitation is the result of antiperistaltic (upward-moving) contractions of the esophagus, paunch and diaphragm, which force the food gradually back into the mouth. The spasms are similar to a series of hiccoughs. The masticated food bolus is then swallowed again, passing this time into the omasum, where the water is reabsorbed. Now, practically solid in form, it reaches the abomasum or digestive chamber, the glands of which secrete the digestive juices. Here the last stages of chemical digestion take place and the food passes through to the intestine. This complex modification of the digestive system not only enables the animal to take in its food supply very rapidly but also permits complete assimilation of the vegetable diet.

Grass and leaf eaters

Many of the non-carnivorous animals of the savannah are grouped together and are described as phytophages, or plant-eaters. But not all of these animals enjoy an identical diet. Some of them, for example, feed on grass (herbivores), others on leaves (phyllophages), and yet others on seeds, roots, fruits etc. This diversity results quite naturally from the fact that not every part of the African savannah has a uniform covering of vegetation. Even the vast expanses of grassland usually give way sooner or later to a different type of terrain—punctuated perhaps by a gallery forest along the course of a river, bounded by forest proper or merging with tree- and shrub-strewn steppe. Much depends of course on whether it is a moist or dry region. The savannah plays host equally to animals that feed exclusively on grass, to those that depend on a leafy diet and to those quite happy with a mixture of both forms of nourishment.

The herbivores and phyllophages never stray too far from the areas where their natural food supply abounds. Ecologists, who make a study of animals and their environment, call such areas 'biotopes'. The tropical savannahs display three distinct types of biotope—grass, forest gallery and shrub. Typical representatives of each of these specialised vegetational regions are the zebra, the waterbuck and the dik-dik respectively. All are plant eaters but have their varied and individual preferences.

Nature's crop rotation

It seems astonishing that the grasslands are capable of supporting such a dense concentration of herbivores without exhausting their natural resources; yet it is quite logical when one bears in mind that the grass and the grass eaters have evolved side by side for millions of years.

The herbaceous plants of the savannah are naturally able to withstand both the continuous trampling of the ungulates' hooves and the non-stop onslaught of their teeth. What is more, each different species of grass seems to attract—as if by selection—its own species of herbivore. This ensures that the different types of grass preserve a natural balance, neither spreading too rapidly (because ignored by plant-eating animals) nor becoming too sparse as a result of their excessive demands. It is a wonderful fact that no wild African herbivore risks exhausting its own natural food supply. This is in large measure due to the herds' custom of ranging far and wide over their entire habitat, observing instinctively a timetable which farmers, in more scientific fashion, know as crop rotation. Enormous numbers of animals regularly move from one region of savannah to another so that the grass is never completely used up.

When the Masai shepherds first brought their domestic cattle to graze on the enormous stretches of pasture hitherto roamed only by wild ungulates, they managed within a few years to reduce the greenest of grassland to steppe and semi-desert. The cattle had lost their natural instinct of rotation and, by devouring the grasses to which they were partial down to the roots, left the hardier, less popular, species untouched. Consequently, the tough plants, poor in nutritive content, multiplied at the expense of the edible vegetation.

The vast areas of waste land where small flocks and herds have taken over from larger wild animal communities should serve as a warning of the danger we run in trying to alter a complex but harmonious natural relationship which has existed for innumerable centuries. To introduce relatively small numbers of domesticated species into a region which is promptly abandoned by its traditional fauna is to ensure the speedy destruction of the ecological balance. Ironically, man's efforts to improve and develop his environment may have unexpected side effects. The noxious tsetse fly, for example, is part of the basic zoological community of the African savannah. Previously it helped to check the unrestricted growth of the wild cattle population. Its wholesale destruction over vast areas may well result in a rate of breeding among cattle which, if not sensibly controlled, could have disastrous consequences.

Because herbivores are not attracted by the same types of grass, there is enough food for all of them. Four typical herbivores are—from left to right—the zebra, gnu, gazelle and hartebeeste. Zebras show a preference for fairly tall and fibrous grasses, gnus for the somewhat shorter species. Gazelles feed mainly on low grass while hartebeestes have a liking for dryish stalks, rejected by other herbivores. This economic use of natural resources ensures that plant and animal life alike is healthy and abundant, and that several species of herbivore can occupy the same habitat without competing for food.

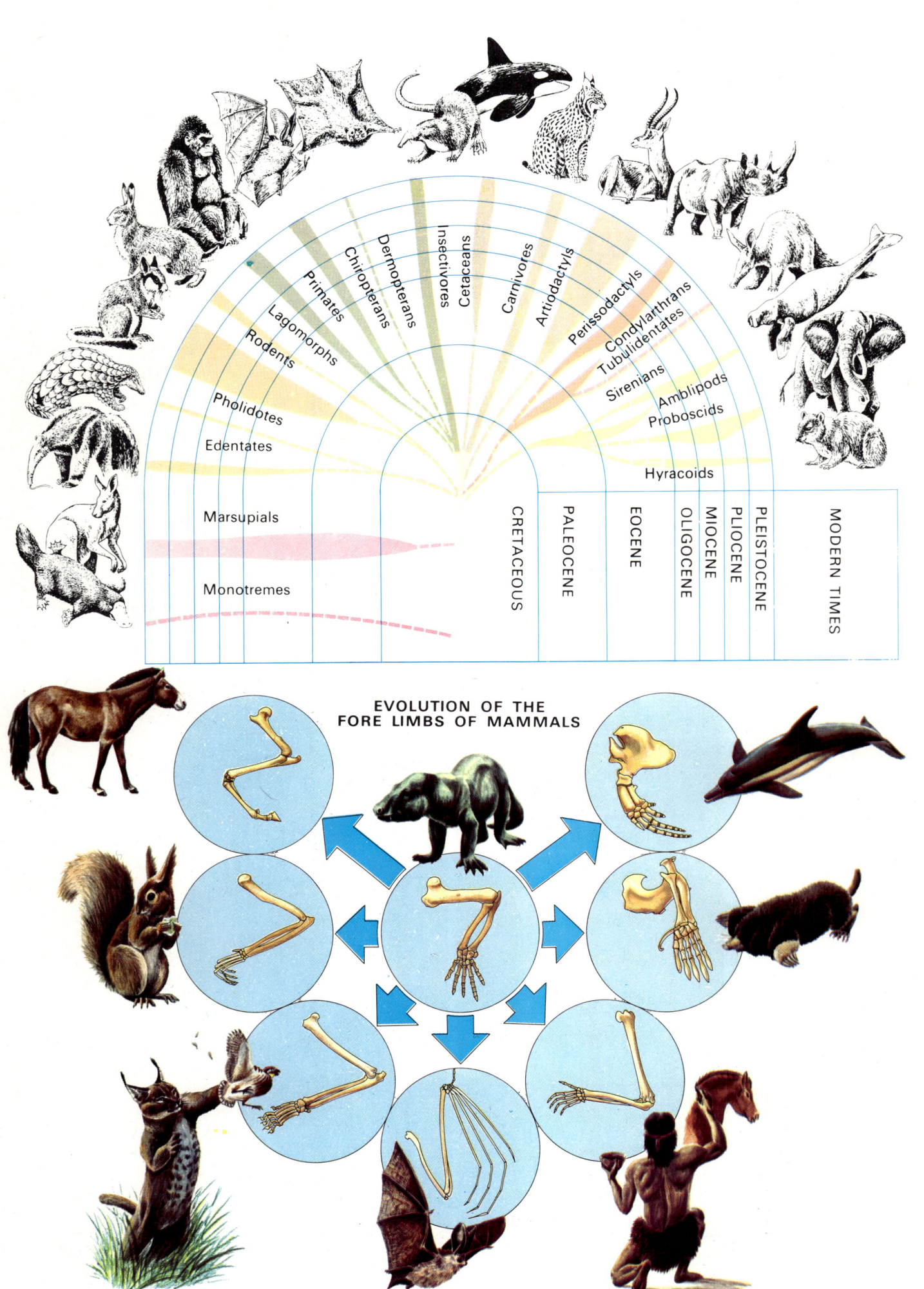

Chiropterans
Primates
Lagomorphs
Rodents

Pholidotes

Edentates

Marsupials

Monotremes

Dermopterans
Insectivores
Cetaceans
Carnivores
Artiodactyls
Perissodactyls
Condylarthrans
Tubulidentates
Sirenians
Amblipods
Proboscids

Hyracoids

CRETACEOUS

PALEOCENE

EOCENE

OLIGOCENE

MIOCENE

PLIOCENE

PLEISTOCENE

MODERN TIMES

**EVOLUTION OF THE
FORE LIMBS OF MAMMALS**

CLASS: Mammalia

Animals have not always been as they are today, of course, but have evolved over long periods of time. The most complicated of them are in fact descended from much simpler organisms. All living creatures are now known to have a common origin. Paleontologists have attempted, and are still trying, to determine and define the successive, related stages linking the various groups of the animal kingdom, and to trace the development of each group from more primitive branches.

About 180 million years ago the earth was inhabited by creatures which defy precise scientific classification either under the heading of reptiles or mammals, simply because the characteristic traits of these two classes of animal appeared very slowly and gradually, at one stage virtually overlapping each other. It is therefore quite impossible to fix a rigid dividing line between the first mammal and its nearest ancestor.

The mammals (class Mammalia) are derived from the Therapsida order of reptiles. These were carnivorous creatures, with skull, teeth and limb structure so similar to mammals that they are sometimes referred to as mammal-like reptiles. Fossil remains so far excavated have provided no information about organs and tissues, so that we shall probably never know whether they had a hairy skin covering or warm blood. But it is quite possible that some of them had reached that stage of evolution.

We are just as much in the dark regarding the appearance of the first mammals, but we are able to make an intelligent guess based on the reconstruction of a creature called *Melanodon* which lived in North America some 160 million years ago. This animal was small—about the size of a rat—with a hair-covered body, was nocturnal, tree dwelling and fed on insects. It was mammals of this type which, because of their biological advantages, were able to fill the void left by the large reptiles. During the Cenozoic era—also known as the Age of Mammals—which extends back from the present day about 65 million years, these creatures evolved into their modern forms.

The reason that the mammals outlasted their reptile predecessors and eventually became the dominant zoological group, to which man also belongs, was due above all to the development of their brain—and especially to the formation of the cerebral cortex or neopallium. This is the seat of the intelligence, defined by some authors as the adaptive faculty enabling animals to react correctly to unaccustomed circumstances. Thanks to modifications of the skull, which became simpler in structure as it expanded, the mammal brain, proportionately larger in relation to body weight than that of the reptiles, developed rapidly.

The predominance of mammals over reptiles was not due solely to their mental evolution but also to their acquisition of a heat-regulating mechanism enabling them to maintain a constant body temperature regardless of external temperature variations. For this reason they are termed homoiotherms, together with birds, which have the same peculiarity. It is a natural advantage which allows them to lead uninterruptedly active lives, whereas the amphibians and reptiles, popularly though unscientifically known as 'cold-blooded' animals, become lethargic in winter.

A vital role in the homoiothermic process is played by the coat (or pelage) of the mammals, which may be compared to the scales of reptiles. The hairs covering the skin may be of different types—long hairs for insulation, short, stiff hairs for protection or sensitive hairs (vibrissae) acting as organs of touch; and other hairy appendages include manes, tails and quills. But not all mammals are completely covered in hair. Elephants, for example, have only a few hairs, unevenly distributed. The pelage of certain kinds of whale consists of tactile hairs on the lips while in other cetaceans such hairs are only present in the developing embryo.

This strange creature with webbed feet is the platypus (*Ornithorhynchus anatinus*), one of the most primitive of living mammals, order Monotremata. It has a duck-like beak and lays eggs, yet suckles its young.

Facing page (above): A chart showing how mammals have evolved from mammal-like reptiles. Each branch represents a different mammal order, its width varying according to the paucity or abundance of the species it contains. Inter-related orders are shown similarly coloured. (*Below*) a diagram illustrating one aspect of mammalian evolution, the development and adaptation of front limbs (originally reptile-like) in various species.

The kangaroo is a marsupial, whose reproductive mechanism is unlike that of the placental mammals. The young are not fully developed at birth and grow in the mother's marsupium or pouch, in which the milk-yielding teats are situated.

The mammalian coat also has another function – camouflage. It helps its owner to blend with the natural surroundings in such a way that it is not immediately visible to predators or potential victims. Nature endows many animals, especially young ones, with spots, stripes and similar patterns, as well as a periodic change of colour. Thus the ermine and the Arctic fox turn white in winter so that they are practically invisible in the snow. But not all animals are compelled to hide; many of them bear clearly defined marks or appendages which serve as recognisable signals for those of their own kind and even for other species. Thus the lion's mane gives a distant warning of his presence to his companions, while the white patch on the hind quarters of the roe deer shows up clearly to those behind when night falls or as he plunges into a thicket.

Each hair has at its base a sebaceous gland, secreting an oily substance that is both a waterproofing and lubricating agent, preventing it from breaking or falling off. But some animals periodically lose their hairy coats, which are then automatically replaced by new ones. This process of moulting may take place in the course of a fixed season or once or twice a year; with certain species it is a continuous occurrence.

Among all the representatives of the class Mammalia, only the primates – among whom is man himself – possess vision capable of distinguishing a wide range of colours. The eyes of other mammals either lack this faculty or have incomplete colour vision, nor is their sense of sight as keen as that of birds. To compensate for this, their sense of hearing is often especially well developed, as a result of the presence in the ear of a chain of tiny bones. Their sense of taste resides in the cells of the tongue known as the taste buds. As for their sense of smell, this is particularly important among the more primitive insect-eating mammals as well as in the Canidae, who depend on it to locate their prey.

The essential characteristic of the mammals – distinguishing them from all other vertebrates – is their secretion of milk. The milk is produced by the females' mammary glands and passes through ducts to the teats or nipples situated at the tips of the breasts or on udders, from which the young suckle. The most primitive types of mammals are the Monotremata (platypus, echidna etc) whose mammary glands are not joined but have their separate outlets in the fur of the abdominal wall, along which the milk runs to be licked up by the young. The egg-laying Monotremata, and the marsupials, differ from the rest of the mammals in the manner in which they give birth.

Most mammals are viviparous – that is, they bring forth live young. In marsupials, such as kangaroos and opossums, the process is incomplete, for the young, when they come into the world, are still in an embryonic stage, being formed in the pouch or marsupium situated on the mother's belly. But in the majority of mammals, the fetus develops to maturity inside the mother's body. The period between conception and birth is known as pregnancy or gestation, its duration varying according to species. This process represents a significant leap forward in the evolutionary pattern. The developing fetus in the mother's womb is attached by the umbilical cord to a disc-shaped organ called the placenta. This is linked to the mother's circulatory system and permits an interchange of food and waste so that the fetus derives the nourishment and oxygen vital for its development within a constant environment. Reptile embryos require similarly stable conditions for their development, but this is much more a matter of chance; for instance, some species bury their eggs in sand warmed by the sun. It was because the mammals were able to reproduce freely, regardless of environmental variations, that the numerous species succeeded in dispersing themselves so rapidly around the world.

The natural protection enjoyed by the developing embryo up to the moment of birth is thereafter provided by the parents. The care of mammals for their young is indeed remarkable, unequalled by any other group. In fact, the newly-born young find themselves quite incapable of coping with life without the active assistance of the adults who give them the necessary training for survival. The period of infancy for these animals lasts all the longer because vital skills taught by example take precedence over normal instinctive behaviour patterns. Young rodents have relatively little family experience, their life cycle being very short and their food easy to find. On the other hand, young foxes, who feed on these same rodents, need to perfect their hunting techniques by watching their parents, and consequently their family ties are closer and more enduring.

Some mammals, thanks to new adaptations, particularly to their limbs, succeeded in conquering and populating the limitless expanses of air and the unplumbed ocean depths. Here they found their sustenance and sought refuge from their enemies. The bat is able to fly as expertly as a bird, thanks to wings consisting of a thin membrane of skin supported by the elongated fingers and the long forearms. The cetaceans (whales, dolphins and porpoises) are able to swim, having developed a body shape similar to that of fishes. Their hind limbs have atrophied and the fore limbs have been transformed into flippers.

The majority of mammals, however, were and are terrestrial tetrapods — creatures with four feet or limbs — able to move freely by reason of their modified joints. Their limbs are longer and more upright in relation to the body than the sprawled limbs of reptiles. Such animals move in various ways. The plantigrades support their bodies on the palms and soles of the feet and are relatively slow moving. Digitigrades are able to stand and walk on their toes and onguligrades on their hooves — both types being capable of sustained and high-speed locomotion.

This wide distribution has only been achieved by animals that have managed to adapt themselves to their surroundings in a number of remarkable ways. It is not simply a matter of seeking and finding food, but equally of deriving the maximum benefit from it. That is why the dentition of mammals has been perfected in the course of evolution in order to perform a variety of practical functions. A typical mammal has three types of teeth. The incisors are particularly well developed in rodents, who use them for cutting and fragmenting vegetable matter. The canines are especially large and powerful in the carnivores, who use their tusks and fangs as weapons, while the ruminants, having no need of them because of their plant-based diet, are not generously provided with them. Finally there are the premolars and molars which are adapted for grinding food, like a millstone. In some mammals a first set of milk teeth, lacking molars, is succeeded by a second complete set.

The number of teeth, their size and their shape, are decisive features of mammal classification since they provide valuable clues to the relationship between different species. They are of even greater significance for the paleontologist who, by examining fossil remains of teeth, can deduce not only the form of diet but indirectly much of the pattern of life of a long-extinct animal species.

There are other less spectacular but equally significant characteristics that distinguish mammals from other classes. Almost all mammals, for example, breathe with the assistance of a muscular diaphragm which separates the lungs and four-chambered heart from the other vital organs. Most of them possess an auricle, or fleshy external ear. The occipital condyles (rounded ends of bone) connect the base of the skull to the vertebral column. The pelvic bones have dwindled and fused to form a cavity in which the fetus resides during the gestation period; and there is a secondary palate separating the mouth from the nasal cavity which allows the animal to breathe while it is eating.

The mammals are divided into three sub-classes, according to their method of reproduction. Lowest of all are the prototherians or egg-laying mammals that suckle their young. The metatherians or marsupials are the pouched mammals and the eutherians are the placental mammals, the most numerous and advanced of all.

The dolphin is a mammal which has adapted itself perfectly to its aquatic environment. Intelligent and playful, it is forced to come to the surface at intervals in order to breathe.

CLASSIFICATION OF MAMMALS

Sub-classes	Orders
Prototheria	Monotremata
Metatheria	Marsupialia
	Insectivora
	Dermoptera
	Chiroptera
	Edentata
	Pholidota
	Lagomorpha
	Rodentia
Eutheria	Cetacea
	Carnivora
	Tubulidentata
	Proboscidea
	Hyracoidea
	Sirenia
	Perissodactyla
	Artiodactyla
	Primates

CHAPTER 2

The grim game of life and death

Many film directors and writers of romantic adventure stories, allowing their imaginations to run riot, have painted a misleading picture of Africa, the dark continent. If we are to believe them, the savage carnivores of the mysterious, impenetrable jungle wage incessant war on the peaceful herbivores, even, if driven by dire need, eating one another. The action is relentless and continuous. The predators lie in ambush for their unwary prey. Bloodthirsty lions roar incessantly; maddened elephants, trumpeting their fury, trample all in their path; panic-stricken antelopes and zebras flee headlong across the immense plains. In the heart of the virgin forest Tarzan swings from his lianas and catapults himself through the air, to the background accompaniment of monkey chatter and a miscellany of animal grunts, howls and shrieks. It is all very blood-curdling and enjoyable but it is sheer fantasy.

More responsible observers, filming and writing from first-hand experience, have done much to correct such fanciful versions. Documentary films in the cinema and on television have helped to consign Tarzan and his animal friends to their proper place in the pages of pure fiction. Yet it still comes as a surprise to anyone visiting East Africa for the first time to find that the dominant impression is one of calm and tranquillity. The landscape of the steppes and savannahs is pleasantly undisturbed and it is in these open spaces, not in the forest depths where animals rarely show themselves, that a profusion of wild life can be seen and studied at leisure. The casual visitor, camera at the ready for instant action, may in fact find it all a trifle boring. The lions, far from roaring and baring their fangs, are more likely to be dozing lazily in family groups in the shade of the acacias, looking benevolent rather than ferocious. Patience may be

Facing page : This peaceful scene of a lion and its cubs is not uncommon in the African savannahs. Despite its fearsome reputation the 'king of beasts' conserves its energy, as do all mammals. The lioness weans her cubs during the rainy season and then prepares them for the life and death struggle of the wild in which they, as the strongest, will call the tune.

This leopard, lying at ease in the fork of an acacia, hardly looks like a terrible killer, but his fangs and claws make him one of Africa's most formidable predators.

Facing page : A group of zebras grazing tranquilly in Serengeti National Park. The presence of nearby gnus is reassuring, for the carnivores seem to prefer them and will often attack them first, giving the slower moving zebras time to escape.

rewarded, however, by the glimpse of an engaging young cub, not much larger than a kitten, teasing its drowsy father by pawing the tufted tip of his tail. The only response of the tolerant parent is to wave the tail indolently from side to side to keep off the flies.

Even more unexpected is the sight of antelopes and zebras grazing peacefully only a hundred yards or so away from the sleeping lions. Now and then they may raise their heads to glance at their carnivorous neighbours, as if to reassure themselves that they are still there. And should an elderly lioness, glutted with a recently digested meal of freshly killed meat, pad heavily through the grass, the hartebeestes and gnus will barely take the trouble to watch her amble past and disappear in the distance before resuming their endless ruminating.

Another animal hardly seeming to live up to expectation and reputation is the leopard. This handsome feline with his spotted coat is generally considered to be the fiercest and most daring of all wild beasts. The truth is that he is a profoundly lazy animal, basking contentedly while propped between the branches of an acacia tree. No noise can disturb his after-dinner siesta; not even the click of the camera shutter will cause him to flutter his eyelids.

A stay of several months in savannah country and a guided tour of all the important nature reserves in Kenya, Uganda and Tanzania would provide our visitor with thousands of feet of fascinating film material; but he will be fortunate if he comes back with even a fifteen-minute sequence of lions attacking zebras, leopards chasing gazelles or elephants charging through a forest glade. Happily for them, the herbivores face their daily dangers with as much cool indifference as we do ours. Statistically speaking, the zebra runs about as much risk of being eaten by a lion as a city dweller does of being knocked down and killed by a car. Just as we drive around in traffic without consciously thinking of the attendant dangers, so the zebra grazes peaceably alongside the lion. After all, the nervous energy of all living creatures is limited. A permanent state of tension and strain between hunters and hunted would be impossibly wasteful and quite at variance with the natural laws of economy and moderation.

As long as the great carnivores are occupied with the remains of their previous kill and their hunger satisfied, they do not indulge in any unnecessary, superfluous movement; nor do they go through the strenuous motions of hunting for the sheer love of killing. So it is not the lions who are responsible, as has usually been claimed, for reducing the numbers of the herds. Only the extreme necessity of survival will drive them to the hunt. More-over, the animals on which they prey are so perfectly and natural-ly attuned to the movements and intentions of the carnivores that their complex and highly efficient alarm systems only have to be brought into action at the last moment. But then the peaceful scene is amazingly transformed. A herd of gnus, for example, which only a few minutes ago was scattered over a wide area, will now converge into semi-circular formation, heads up, sniffing the air, bellowing nervously and gazing fixedly at the spot where the lions are probably lying concealed.

It is when they decide to hide that the lions become a threat to their victims. At such times the least suspicious movement, the

faintest scent—no matter how remote—is sufficient to unleash terror and confusion, a sense of atavistic panic, culminating in wild, undirected flight, which the hunters are so quick to exploit. As night falls, there may be a fleeting glimpse of something black in the golden grass—the dark spots behind the lioness' ears as she twitches them nervously. Shattering roars echo across the empty plain. Now there is not a single gazelle, zebra or hartebeeste to be seen in movement. The grim game of life and death is about to commence.

The lions' banquet

Few regions in the world can compare in beauty and lushness with Tanzania's Ngorongoro Crater and plains of Serengeti. These are natural parklands which have been converted into a huge game reserve, roughly half the size of Wales, where immense animal communities live in harmony. Conditions are so ideal here that scientists travel from all over the world to study the ecology and observe the behaviour of the great carnivores, monkeys and ungulates in their natural surroundings.

The following vivid account of one savage aspect of the region's wild life is taken from a naturalist's diary.

'One rainy night in February we had pitched our camp on the edge of the reddish track leading from the crater of Ngorongoro to the village of Seronera, in the heart of Serengeti. Suddenly a strange clamour split the silence. Short, shrill cries mingled with the thunderous pounding of hundreds of animals' hooves. The uproar came from the south, from the direction of the crater. A few seconds later our encampment was struck by what appeared to be a cyclone and the air was filled with the powerful stench of the stable. In the darkness we got the momentary impression of a herd of runaway horses being pursued by a devilish pack of maddened dogs. In fact, the animals were zebras, with a bunch of lions hot on their heels. They had passed so close that they had only just missed flattening ourselves and our tents. The piercing neighs of the zebras, like short, sharp, rhythmical barks—astonishing to those of us who had never heard them—lasted for some minutes as the thunder of hooves became fainter and was then heard no more.

'All of a sudden a deep, long roar broke the silence once more. A lion was signalling its companions that a kill had been made, the hunt successful. Its powerful voice, noble and authoritative, echoed in the darkness—an exciting sound, for we knew it held out the promise of a remarkable and rewarding spectacle, a lion's banquet.

'It was midnight. In these parts it would not be day for seven hours. We knew that this particular pride of lions was a large one. It was more than likely that by dawn no trace of zebra would still remain. But there was always the possibility that some of the hunters might have branched off in another direction or even, as often happens, that the perpetrators of the killing might have been a couple of lionesses, accompanied by their relatively inexperienced cubs, with less ravenous appetites than their parents. In any event we were in our Landrover before the sun

Facing page : A straggling zebra has been killed by a band of lionesses and the cubs are allowed their share of the tasty flesh. Once the body is stripped, the scanty remains are devoured by the scavengers of the savannah.

Vultures take to the skies in mid-morning on rising warm air currents, scanning the broad plains for food. They are a sure guide to the presence of dead animals, attracting other carrion eaters to the site of a recent kill.

Facing page (above) : A lioness sinks her fangs into the neck of a zebra. She will then drag the heavy carcase to a convenient spot where her entire family will devour it at leisure. (*Below*) despite its fleetness of foot, this small Thomson's gazelle has been unable to escape from its relentless enemy, the cheetah.

was up, zig-zagging to and fro across the broad, acacia-studded plain, hoping to stumble on our lion family.

'Grant's gazelles and isolated groups of zebras were grazing peacefully on the dew-freshened grass as if the events of a few hours ago had never occurred. Each side had resumed its normal everyday round. The herbivores grazed and ruminated, the drowsy carnivores gathered new strength. The game of life and death had been sudden and brief.

'Our task was not easy for in the darkness we had been unable to pinpoint the exact direction from which the roars of the wild animals had come. Nor could we rely on being led towards the lions' banquet by those most trustworthy of all savannah guides, the vultures. For these great winged carrion eaters only take to the skies after the first hour of sun sends warm currents of air coursing upwards from ground level. Then they soar up to a height of several hundreds of feet, scanning the plateau with their keen eyes for food.

'By mid-morning we had still found nothing. Finally the vultures took wing, circling and finally swooping downwards, claws extended, wings not fully unfurled, like weird parachutists. Their objective was a hollow, covered with tall grass, which we would certainly not have been able to find on our own. A few hundred yards away we caught sight of some large yellowish shapes, contrasting sharply with the emerald green of the grass. A dozen lions were stretched out asleep. Nearby a confused brown mass resolved itself into a clamorous throng of spotted hyenas, jackals and vultures, fighting one another for the last morsels of the zebra killed ten hours previously.

'While the lions dozed, two cubs, apparently about six months old, toyed with a zebra hoof, now brown and polished from rubbing against the ground. We were able to film the hyenas at close range. They crushed the bones left by the lions with their tremendously powerful jaws and then sucked out the marrow. The more agile jackals gnawed scraps of carcase in the long grass while three white-backed vultures fought one another for a last bleeding hunk of flesh.'

Natural cycle, ecological pyramid

Descriptions such as these understandably touch our hearts and stir our sympathies. It seems unjust that this lively and attractive member of the horse family, with its cheerfully striped coat, should meet a painful and grisly end at the claws of a lion. And as if this were not sufficiently unpleasant, there is the additional ugly spectacle of hyenas and vultures gorging themselves on the remains of the kill. Scientists, however, take a more objective and long-range view of such events. For them the death of an animal is not such a terrible drama as might at first glance appear. It is one inevitable stage in the natural cycle of matter, known to ecologists as the food chain, which begins with sunlight, continues with the hyena and the vulture and is completed by bacteria—only to begin all over again.

What part would the zebra have played had he not been killed? In simple terms, he would have continued to transform energy

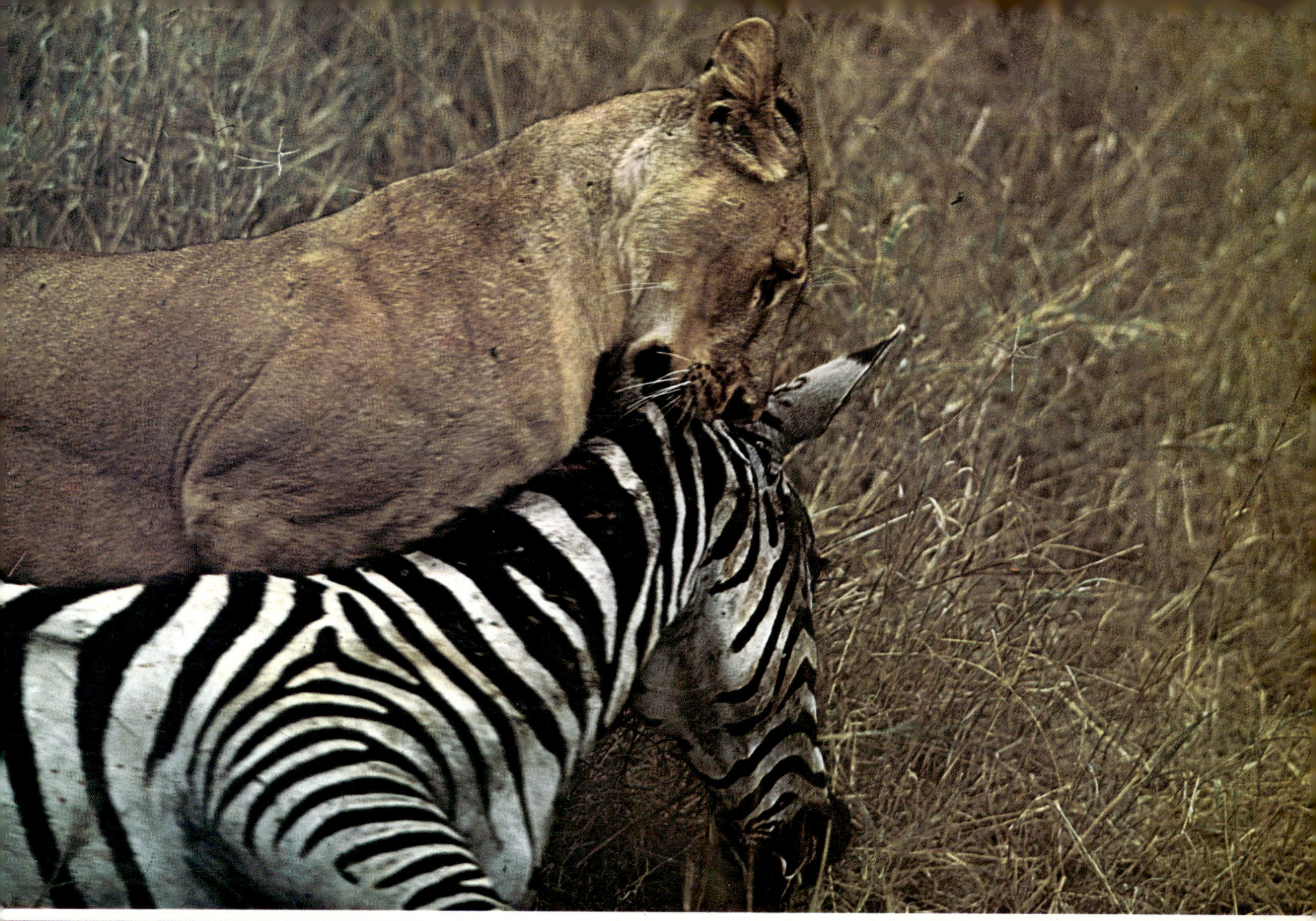

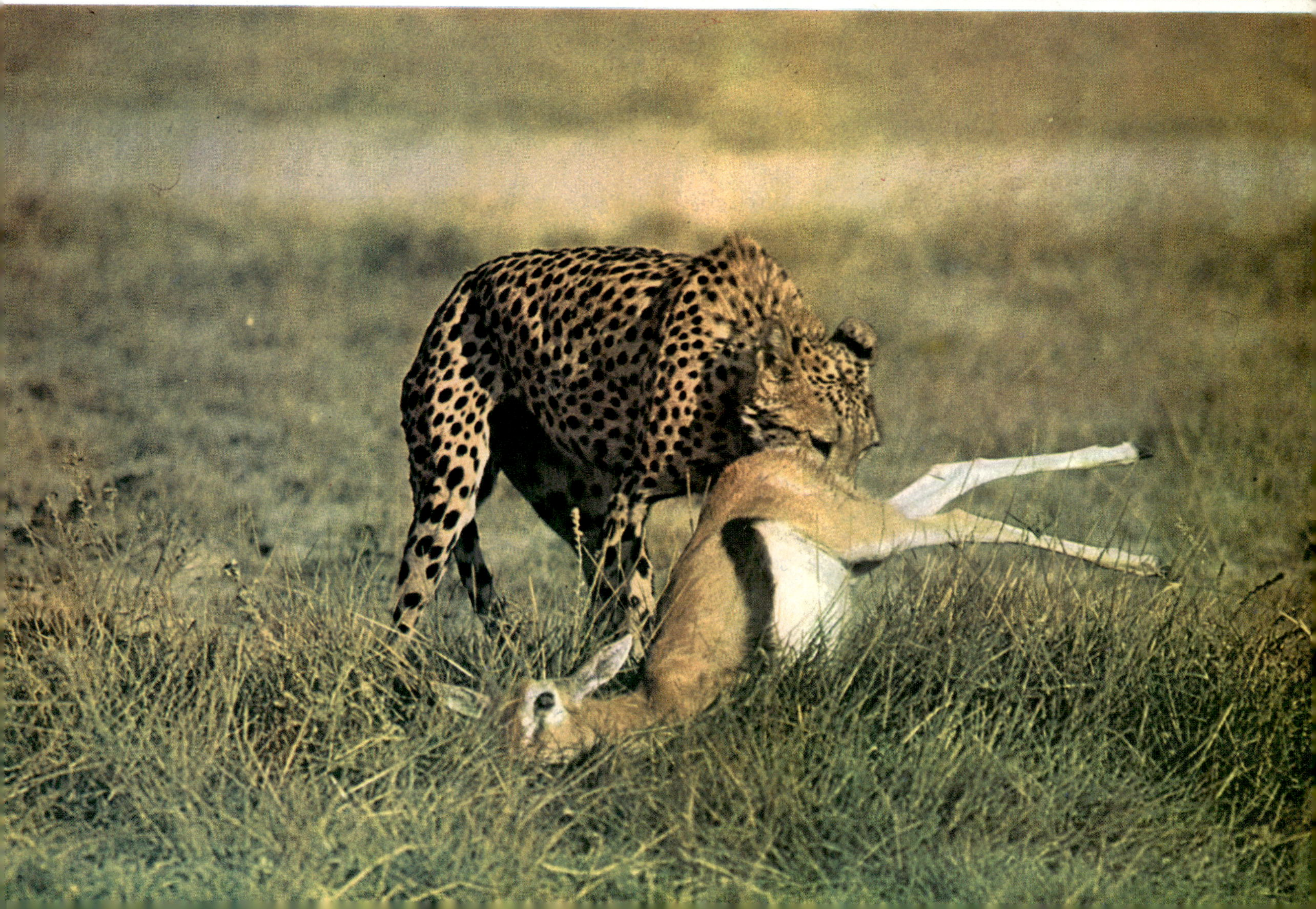

SUPER-PREDATORS

PREDATORS

PHYTOPHAGES

PRODUCERS

A simplified diagram of an ecological pyramid, showing the various stages of the food chain in a savannah-type plant and animal community. The plants or producers serve as food for the phytophages or plant eaters, who are in turn eaten by the predators. The latter may themselves become the prey of the super-predators. A large amount of accumulated energy is utilised at every level, with the numbers of species decreasing at each successive stage.

accumulated by the grasses on which he fed into a different form of energy—his own. And had he reached adult status he might well have reproduced his kind, creating in his turn individuals capable of transforming vegetable protein into animal protein. As for the lions, who do not take direct advantage of vegetable substances, they are compelled to kill in order to find the energy indispensable for their survival and the perpetuation of their own species. They then give birth to other lions who in due course are destined to play a positive role in preserving the natural balance between the living and the dead, which has existed since the beginning of time. Finally, so that no energy should go to waste, still complying with nature's fundamental laws, hyenas, jackals and vultures devour what the lions leave behind.

Here we have the framework for a system which biologists call an ecological pyramid. At the base of the pyramid are the producers—green plants (including grasses and trees) capable of transforming inorganic salts, water and carbon dioxide into organic material. Then come the phytophages or plant eaters, who are able to transform the energy accumulated by the vegetable substances into animal energy. The next stage is that of the predators who are obliged to kill to keep themselves alive and in order to reproduce. Finally there are the scavengers who feed on the carnivores' leavings. This is of course a very simplified version of a complex system, omitting many intermediate stages. The pattern is rarely so neat and convenient. The zebra, for example, may fall prey to the leopard who may himself be the victim of a lion. So one has to include the additional category of super-predators, in which the lion finds its rightful place. Thus we have a food chain—only one of many—comprising grass, zebra, leopard, lion, hyena, jackal and vulture.

To eat and not be eaten

This natural law which governs life on the African plains is by no means confined to regions inhabited by wild animals—it is the characteristic pattern of all living things. This game of life and death began many millions of years ago, when the first creatures appeared on earth. Apart from certain micro-organisms, all such creatures depend on solar energy for their growth and development. But few of them are able to make direct use of it; only the green plants are capable of absorbing the water and carbon dioxide provided by the sun and atmosphere and transforming it into the organic foodstuffs which are the source of energy, by the process known as photosynthesis. Animals cannot make much use of nutrients in their inorganic form (apart from water) and rely on intermediaries. Some are able to absorb organic foods in the form of vegetable matter; others are unable to ingest or digest plants, and therefore kill the vegetarians in order to derive the benefit of the substances accumulated in their tissues, while the scavengers in turn obtain their sustenance from the victims' remains. Finally all the organisms return after the animals' deaths to the soil, where they are acted upon by decomposing agents such as bacteria.

The cycle is self-renewing and endless. Minerals are absorbed by the roots of green plants, then transformed and stored up in

The leopard, perfectly camouflaged among the leaves, has an indolent air but is alert to the slightest hint of weakness or incaution when stalking its herbivore victims.

Herbivores such as antelopes risk being attacked from the air as well as on the ground. The martial eagle (*Polemaetus bellicosus*), unlike the carrion-eating vulture, hunts live prey.

their tissues, then taken in at the various levels by every member of the animal kingdom. The destiny of each living thing is to eat and be eaten so that the great energy cycle can be completed and renewed. Nevertheless, at every successive level of the pyramid, an animal will do its utmost to avoid being killed. This instinct of self-preservation is vital for the perpetuation and development of the species. So the basic law which governs animal life can be formulated as 'To eat and not be eaten'.

Animal heroes and villains

Until modern times, society, understandably concerned chiefly with man and his own affairs, was inclined to attribute human qualities, virtues and faults alike, to animals. Myths, fables, fiction—and even scientific treatises—habitually represented the tiger, the lion, the wolf and the eagle as cold, merciless, blood-thirsty killers, while deer, sheep, gazelles and doves were symbols of gentleness and innocence. Such deeply-rooted beliefs and prejudices were largely responsible for the campaigns of destruction launched against wild animals, widely held to be harmful and therefore outside the law. To exterminate predatory meat eaters, proclaimed the self-righteous killers, was to protect the harmless, useful herbivores. It seemed quite reasonable to kill as many lions as possible to allow the zebras to survive. Logic also dictated that all you had to do to ensure that deer lived in peace was to hunt down the wolves who made a practice of eating them.

In the early part of this century a group of well-intentioned Americans decided to preserve the deer population of the Kaibab Plateau in Arizona by eliminating the creatures that preyed on them. Pumas, wolves and coyotes were systematically done to death and entry to the preserve (later to form part of the Grand Canyon National Park) was forbidden to all hunters.

The early effects of this conservation programme were astounding. During the first five years the animal population grew steadily. The deer were strong and handsome, feeding on the lush pastures without fear of lurking carnivores. But this Garden of Eden was doomed. Soon the deer multiplied to such an extent that the vegetation, subjected to unendurable pressures, was completely transformed. The grass, nibbled down to the roots, had no time to grow again; the trees, stripped of their bark, slowly died. The areas most exposed to soil erosion rapidly degenerated into desert. Among the deer themselves, increasing numbers of weak, deformed individuals were born. This was partly due to a natural food deficiency and partly to the fact that their parents were old, unhealthy animals who would normally have been killed by predators. Long before the deadline fixed by the well-meaning animal lovers for the successful outcome of their enterprise, the once-proud herd had vanished, leaving behind isolated groups of emaciated beasts scarcely able to stay alive on their diet of dry wood and thin grass. In a few years their numbers had been decimated. The shadow of death engulfed what had traditionally been an animal paradise, proving conclusively that the wild carnivores, the dreaded killers, have a vital part to play in nature's balanced pattern, even if her laws appear to be implacable.

Killers who ensure life

The predators of the savannah who feed mainly on herbivores are in a way the natural allies of their victims – protectors rather than destroyers! This may seem a surprising and somewhat contradictory statement but easy to understand once one has grasped the fundamental principles of ecology.

Throughout nature there is and always has been an incessant, relentless struggle for the food which provides energy. The

In Kenya's Masai Amboseli reserve, near Mt Kilimanjaro, vultures gather to share the remains of a lions' banquet. Although not the pleasantest of sights this is an inevitable stage in the continuous cycle of life and death on the African savannah, helping to maintain the natural equilibrium of the regional fauna.

plants, representing the producers in the food chain, have for millions of years served as sustenance for the phytophages (plant eaters). It is not fanciful to suggest that even among plants natural defence mechanisms slowly evolved which enabled them to stand up to the depredations of the creatures who devoured them. Silicates in the tissues of grasses, for example, would have helped to minimise the damage inflicted by the herbivores' champing teeth (although nature, ever impartial, helped the herbivores to surmount this problem by providing them with teeth so adapted as to offset the silica's abrasive effects). Similarly, the sharp spines of some acacias and mimosas might be envisaged as a form of natural protection against the tough mouths of rhinoceroses and the long, prehensile tongues of giraffes. Consequently such animals rejected the unappetising branchlets in favour of the smoother, more succulent stems and leaves, and did less damage to the plant as a whole.

The interesting point here is that the plants played an important part in the evolution of those animals which were compelled to adapt themselves to a new diet and thus attain greater efficiency in bodily structure and function. Equally influential was the group immediately above the plant eaters in the food chain—the carnivores who preyed on them. In finding it easier to catch sick, feeble or elderly victims, the predators systematically weeded out the weaker herbivores, permitting only the fittest to survive—another significant contribution to the evolutionary process.

An animal stalking its prey expends a large amount of energy, whether coaxing it from a burrow, springing on it from the branches, chasing it across a plain, throwing it to the ground, pursuing it into deep water or swooping down on it from the sky. The victims too use up energy in evading would-be captors, employing a variety of defensive methods—colours and patterns for camouflage, limbs for running and jumping, claws for burrowing and climbing, wings for soaring out of reach. Over the ages their protective armament has become more formidable, their weapons more deadly.

According to the scientists who have formulated the theory of evolution, such adaptations occurred very gradually in the course of successive geological ages and were the joint result of mutation and natural selection. Characteristics and abilities which are *acquired* by individuals during their lives are not passed on to their descendants. Modifications in form, growth and behaviour which may appear in successive generations of a family are caused by structural changes in the genes and chromosomes—known as 'mutations'—and these are hereditary.

The appearance of mutations does not explain entirely why certain species evolved in particular ways to adapt themselves successfully to new environments. For mutations come about by accident and many have a harmful rather than beneficial effect on the individual. So we must consider them in conjunction with the process of natural selection—the survival of the fittest—in which the predators play a role.

Let us take another look at the hoofed animals. A series of mutations was responsible for the progressive lengthening of

Facing page (above) : From the top of an anthill a female cheetah and her cub scan the horizon in the hope of spotting prey. *(Below)* a freshly-killed Thomson's gazelle is dragged back to the lair. The cheetah's hunting technique is in perfect accord with its physical capacities—a silent, stealthy approach and then a headlong pursuit of the victim at speeds of up to 70 miles per hour.

Although it bears a strong resemblance to the wild boar of temperate forest zones, the warthog is a creature of the African steppes and savannahs. Its canine teeth, jutting out from the mouth in the form of tusks, are fearsome weapons, as many a predator has discovered to its cost.

their limbs, enabling them to stand firmly and run rapidly on hard ground. The outer toes were gradually reduced in size until they served no useful purpose whatsoever. Among the horses (Equidae), for example, the two outer toes were lost early in evolution and two more, the second and fourth, became much reduced in size, leaving only one functional toe, the central toe encased in a hoof. It was now far better equipped to escape from predators. Those members of the Equidae whose limbs were not so modified were decimated by carnivores who needed to expend far less energy attacking them than in chasing futilely after the single-toed, fleet-footed horses. The latter became not only faster but larger and heavier, while the more primitive equids eventually became extinct. With the disappearance of the older species the carnivores' hunting and feeding habits changed once more so that many of the surviving equids were again at risk.

This relentless process of natural selection is not, however, confined to an animal's physical features; it also influences behaviour—the way it lies in ambush, hunts, fights and defends territory. In the unending battle between carnivore and herbivore all the senses must function at the highest level of efficiency. Animals must be able to see without being seen, hear and interpret sounds whilst remaining silent and motionless, scent prey without themselves offering tell-tale clues to an enemy. Once the victim is located, the implacable business of tracking it down begins, with the hunter finally coming to grips with the hunted. Apart from man himself, who has refined the art of killing at a distance, the only creatures endowed with similar deadly ability are certain poisonous snakes and the small archer fish *Toxotes jaculator*, which shoots a jet of water from its mouth to bring down an insect in flight. The carnivores, however, have to tackle their victims physically in order to kill them.

The odds do not necessarily favour the attacker. Should the hunter carry a heavier weight of muscle than its prey it will tire more quickly, and then superior speed and stamina will tell: the lighter, more agile zebra or antelope will often outpace a pursuing leopard or lion. Nor need victory inevitably go to the predator once the contenders are locked in combat. It is essential for the carnivore to kill its desperate victim without suffering disabling wounds. The horns of an oryx, for instance, can prove dangerous, even fatal. To avoid death is not enough, for a deep wound or a bad fracture may handicap hunting capacity, immobilising the carnivore and exposing it in turn to other predators. So energy must be conserved, the fatal encounter kept short and free of risk.

The carnivores and herbivores, hunters and hunted, have therefore evolved side by side. As the plant eaters became faster, stronger and more intelligent, so too did the meat eaters react to changing conditions through the double mechanism of mutation and natural selection. The heavy, clumsy predators of 60 million years ago would have found it impossible to catch the modern herbivores, let alone compete with today's carnivores. There has been a striking example of this evolutionary contrast in recent times. The Australian dingo, introduced by the early

Solar energy

Scavengers

Herbivores

Plant-eating insects

Insectivores

Predators

The energy cycle. Plants use sunlight to manufacture organic substances from inorganic material by photosynthesis. Plant tissues are ingested by insects (to be eaten by insectivores) and by herbivores. The food chain continues with the predators and scavengers. Finally bacteria break down the waste materials and dead bodies of animals, the resultant inorganic salts returning to the soil to be absorbed once more.

With its powerfully muscled body, precise movements and marvellously co-ordinated reflexes, the leopard is one of the most efficient of hunters, a specialist in sudden, lightning attack. It has adapted both to the dampness of tropical forests and the extreme dryness of deserts. Such flexibility usually denotes an advanced stage of animal evolution.

Facing page : Stretched out in comfort along a branch, a leopard feeds on a gazelle it has just killed. After enjoying the tastiest morsels on the spot, the leopard frequently denies a share of the spoils to scavenging jackals, hyenas—even lions—by dragging the carcase out of reach up a tree.

primitive settlers, was able to revert to its wild state. Not only did it feed on the kangaroo and a variety of herbivores but it also attacked many of the continent's traditional predators—the marsupial wolf (*Thylacinus cynocephalus*), the tiger cat (*Dasyurus*) and the Tasmanian devil (*Sarcophilus ursinus*). These latter animals, unable to contend with wild creatures far more fierce and powerful, have almost disappeared where they had previously been abundant.

It is clear that in the natural selection of carnivores there are two influencing factors, one determined by the nature of the prey, the other depending on competition with similar predatory species—as occurred in Australia between the dingo and the tiger cat. A third, less immediately obvious, factor relates to competition between members of the same species, dictated by their need for a particular type of food. This too has encouraged the development of highly specialised faculties, enabling the peregrine falcon, for example, to power-dive on its prey at over 150 miles per hour or the cheetah to reach up to 70 miles per hour across level grassland in pursuit of a gazelle.

The phytophages, in order to avoid being killed and eaten, have developed their own protective techniques and behaviour patterns. When hunter and hunted come to grips, provided both are strong and healthy, there is a fair balance of forces and only the greatest effort on the part of the carnivore will bring about the death of its prey. Energy is therefore conserved by refraining from tackling the fittest members of the herd. Signs of age or infirmity can be sensed intuitively and ambushes laid for stragglers. In this way predators perform a positive function by restricting the spread of infectious disease, preventing the births of abnormal individuals and keeping a general check on excessive population growth.

So we have the paradoxical situation of the 'villains', grim purveyors of death traditionally considered to be harmful, actually assuming a heroic guise as the most certain and reliable guarantors of continuing life for the species. What we have to guard against is the temptation to apply human moral standards to the activities of animals. In their world the terms 'good' and 'bad' have no validity.

The savannah's living mosaic

We return now to the Serengeti, selecting a vantage point on top of one of the rocky hills or *kopjes* that from time to time break the flatness of the landscape. From here we can look down on immense expanses of green grassland and thousands of grazing animals. A casual glance will reveal the presence of large herds of gnus and zebras as well as scattered families of ostriches. Thomson's gazelles are there in quantity while groups of Grant's gazelles, giraffes and impalas browse among the acacia thickets. Hyraxes, squat little animals resembling marmots, dart about the rocks beneath our feet. On the higher slopes we may catch a glimpse of a party of shrieking baboons or, if we are exceptionally lucky, stumble across the nest of a Verreaux eagle, a huge, handsome bird of prey with black plumage.

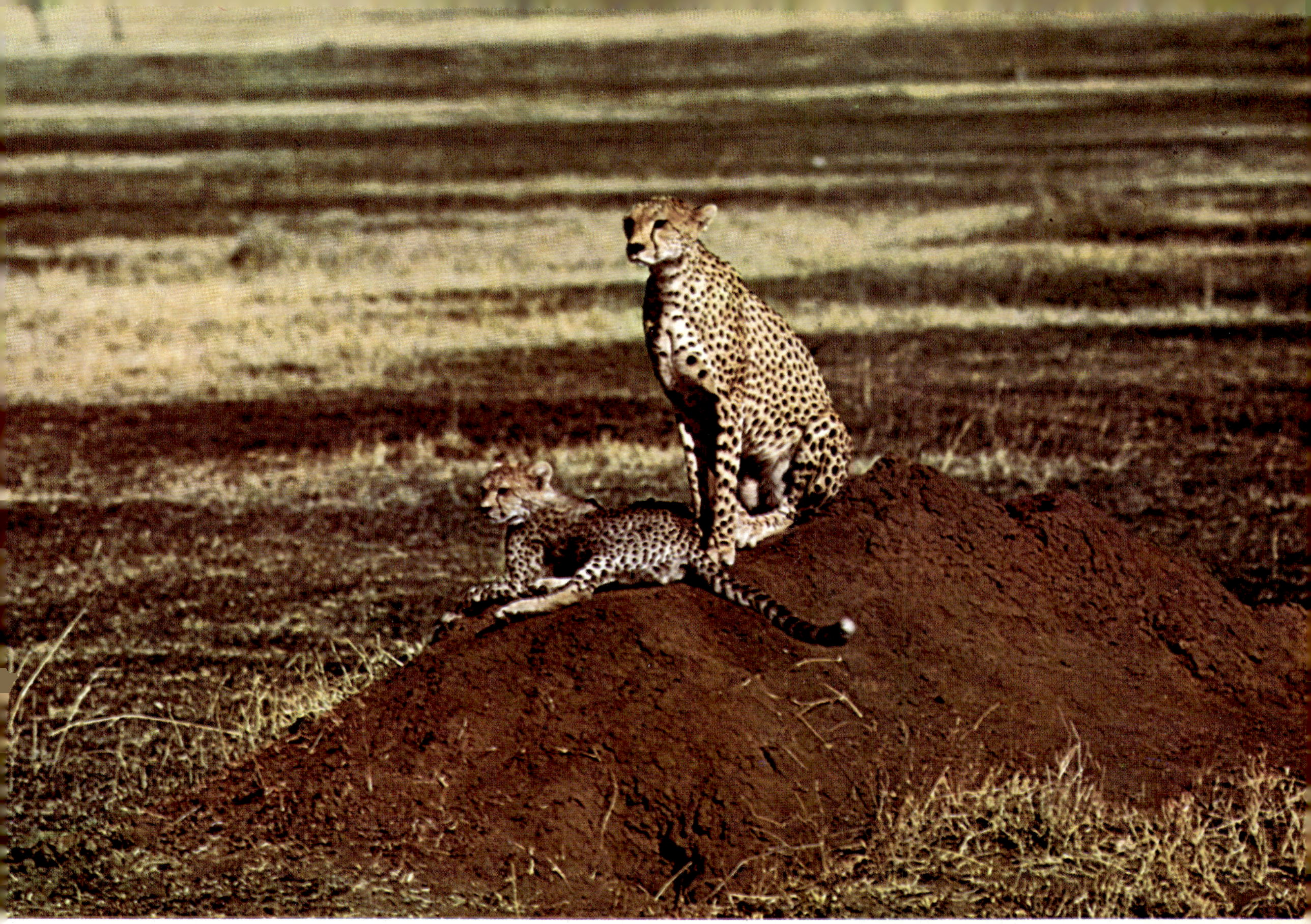

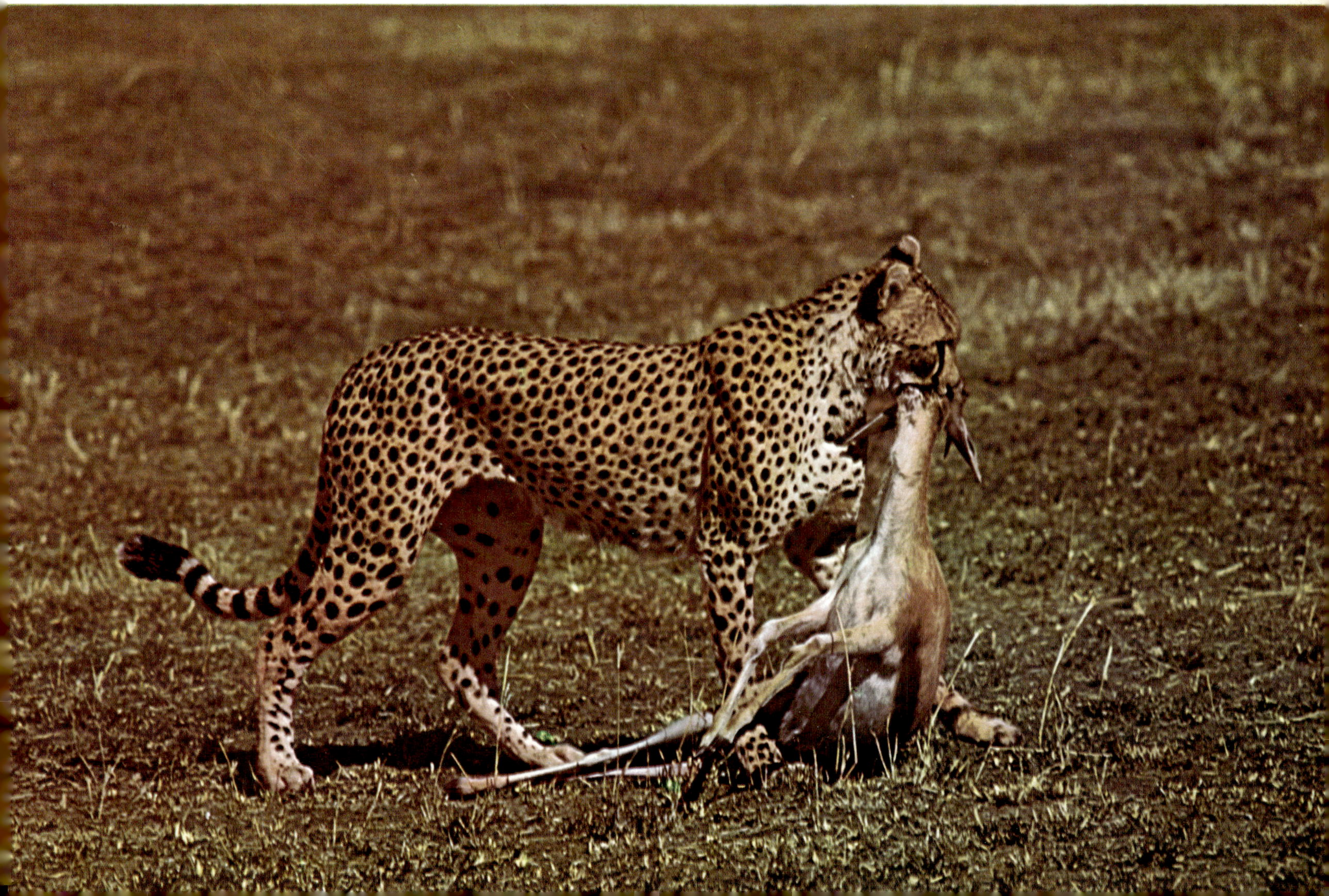

The long-legged, long-necked ostrich is one of the most unusual of savannah animals. Not only is this bird unable to fly but it actually prefers the company of gnus, zebras and hartebeeste, sharing the same kind of food. Together they band in common defence against predators, the far-ranging vision of the ostriches complementing the remarkable hearing of the ungulates.

These and many more species are of course widely distributed over the savannah, moving from one clearly defined area to another, where they can be certain of finding food, water and shelter; and if we take a closer look at the habitats or biotopes of the various animal communities, we soon notice that the pattern is much more complex than at first appeared.

In fact mammals, birds, reptiles, fishes and amphibians all live here together, so that the image of a vast, colourful mosaic of wildlife is an apt one. It is a paradise for naturalists, who can enjoy unparalleled opportunities for studying an enormous range of animal species. The specialist in mammals will find ungulates, gazelles, lions and cheetahs; the ornithologist a spectacular array of bustards, Guinea fowl, sand-grouse, francolins and fierce birds of prey. The herpetologist is surrounded by lizards (from chameleons to monitors), tortoises and snakes. The fishes that teem in the rivers and lakes—especially the strange lungfishes of the order Dipnoi—delight the most experienced of ichthyologists, and the huge crocodiles and other amphibians lining the banks make up another group demanding separate study. As for the entomologist, he will hardly know where to turn first, with such a multitude of termites, driver ants, dung-beetles, praying mantises and other wonderful species at his disposal.

One practical way of surveying the huge zoological population of the African savannah is to divide it into smaller groups or communities. We must remember, however, that even these divisions are artificial inasmuch as every animal has ecological links with others. The mammal community, for example, might seem to be entirely separate from those of birds, reptiles, fishes, amphibians and insects; after all, the herbivores derive their nourishment and energy from the grass on which they feed, the carnivores consume the herbivores, and the carrion eaters feast on the remains. Yet the mammals are not completely independent. The mongoose feeds on reptiles and the otter on fishes, crustaceans and molluscs. Other mammals are closely dependent on insects and birds. The aardvark eats termites while the ratel gorges itself on the honeycombs of wild bees—just two examples of mammal-insect relationship.

Links between mammals and birds are equally important. Most mammals harbour irritating insects on their skin or hide. Rhinoceroses and buffaloes provide regular hospitality to small birds called oxpeckers, which perch on their backs and remove the small ticks that plague them. A similar service is provided by the buff-backed heron or cattle egret. In addition both species of bird warn their hosts, by flying up, of potential danger. Ostriches too act as sentinels for the grazing herbivores with whom they mingle peaceably. With their long necks and remarkable eyesight, they are quick to spot lurking predators and their alarm signal is a loud, angry hissing. Ostriches, although they have lost the use of their wings, are able to run as rapidly as their ungulate companions, feed on the same type of vegetation and are threatened by the same predators. Here is a perfect example of the interdependence of two communities; and other instances of close mammal-bird links are provided by the predatory eagle, which feeds chiefly on rodents, and the carrion-eating vulture.

The different species all transform the natural resources of their environments into energy-giving food but they cannot do this without engaging in a continual struggle against the other members of their community. This is a condition of survival for hunters and hunted, plants and animals alike. If the world consisted of one vast stretch of savannah from which carnivores such as lions and leopards had been banished and the only herbivores were zebras, the latter would, in theory, lead an idyllic existence. There would be plenty of grass, no competition from other herbivores, no threat from prowling predators. Inevitably, however, their numbers would multiply to excess and their pastures, unable to cope with the greatly increased demand, would wither beyond hope of recovery. The species would be doomed to extinction.

Alternatively, imagine an earth containing only zebras and lions. The lion population, unmenaced by other carnivores, would expand without limit and the result would be the disappearance first of the zebras, then of the lions. Pushing the argument even further, suppose the grasses were somehow to develop abrasive substances which ruined the teeth of the herbivores, denying them food. The result would be the same — a savannah empty of animal life.

Ever since life began this bitter battle for survival among the many species of our planet has been waged. They have fought to eat and to avoid being eaten, each in their different way dependent on the life-giving energy radiated by the sun. Out of this competitive activity new forms of life have evolved, and species that were once dominant have been superseded by others—the only testimony to their existence being fossils.

Life marches relentlessly on, adapting to changing circumstances and surroundings, working towards an increasingly efficient functioning of body and mind. Although the long-extinct dinosaurs and primitive mammals may not appear to fit into the general pattern, progress has been steady and uninterrupted. In terms of strength, agility, mobility, skill and intelligence, predators and prey alike have shown continuous development over the centuries. Unless man puts an end to it all by meddling with and upsetting nature's equilibrium (quite apart from the unthinkable possibility of destroying his world in even more dramatic fashion) there is no reason why this trend should not continue indefinitely.

It is not only the larger creatures which are engaged in this fight for existence. The microbes attacking the human organism and the flowers that attract insects with their glowing colours and heady perfumes are just as much involved in the struggle. It is merely that predators such as the lion resolve the problem in a more clear-cut and recognisable manner. The carnivores do not derive their nourishment from sunlight, nor do they quietly nibble grass. They kill and they devour. Nature has equipped them with weapons specifically for this purpose, whether they are deadly underground hunters such as the weasel, organised bands of killers like hunting dogs or solitary

Hyraxes, roughly the size of hares, live in groups of thirty individuals or more among the *kopjes* – rocks that jut out like islands in the African savannahs and deserts. They are lively little creatures, usually agile enough to evade the clutches of eagles and other predators.

Following pages : A hyena and a group of vultures feeding on the carcase of a gnu. This is a common scene on the high plateau where countless thousands of grass-eating animals congregate. It is a striking example of the close relationship that exists between the variegated communities of mammals and birds.

The *lycaon* or African hunting dog is one of the most savage of predators on the savannah. Living and hunting in packs of about a dozen, they spell certain doom for any herbivore rash or unfortunate enough to be separated from its fellows.

prowlers such as the leopard. Methods vary but the result is always predictable. Every plant eater has its predator, each with an allotted part in the community's life pattern, the perpetuation of species by natural selection—assuring life by inflicting death. It is a law of which neither predator nor victim is consciously aware, but which both have faithfully observed throughout the ages.

The hunters and their world

The world's many carnivores use widely differing methods to capture their prey. The large group of Canidae, for example, have perfected elaborate techniques of pursuit and attack. Wolves, hunting dogs, jackals and coyotes are tireless creatures with tremendous staying powers, their bodies admirably suited to running long distances. They possess a long, deep thoracic cavity, supple limbs with claws, and hard, springy pads on the lower side of the feet which give support and impetus for leaping. The flexible, curved vertebral column and powerful muscles also encourage rapid movement and help to soften the impact of the feet against the ground.

Stamina also depends on the animal not being overweight. The carnivores are mostly slender, streamlined animals which have done away with what might be termed heavy weapons, relying exclusively on the strength of their jaws and the sharpness of their teeth. And since their comparative lack of weight puts them at a disadvantage in single combat, they normally hunt in groups. Superior numbers usually prove more than a match for the strongest single individual, and several pairs of jaws are more deadly than one. Such methods demand a high degree of intelligence and the acceptance of a rigid social hierarchy or system of leadership within the group. Discipline and co-operation yield the desired results. A pack of African hunting dogs can attack and destroy animals far more powerful than themselves, including lions, and a small band of wolves is capable of outflanking and bringing to the ground a creature as massive as a moose.

The Felidae, who comprise the other main group of carnivores, adopt a contrasting method. With their large, heavy muscles, economical and precise movements and superbly co-ordinated reflexes, they elect for direct, individual aggression. An attack is therefore carefully planned and perfectly timed in order to take the intended victim unawares. Their weapons are powerful fangs and curved retractile claws. These claws, attached to special tendons, can be sheathed when not in use so as to avoid wear and tear, and are fully projected when the animal hurls itself upon its prey.

The predatory wild cats, concentrating all their energy on a sudden, short attack from ambush, must be able to conceal their identity and whereabouts from a potential victim. Padded paws enable them to stalk their prey in silence while the colour and pattern of their coats provide effective natural camouflage. The extraordinary versatility and variety of the Felidae—ranging from the diminutive domestic cat to the huge lion and tiger—testifies to the successful manner in which they have evolved through the ages.

The terrestrial carnivores are divided by some authorities into two superfamilies—Canoidea and Feloidea. The classification is based on the comparative structure of one part of the anatomy—the ear section known as the auditory bulla. The Canoidea have a simple undivided tympanic (ear drum) cavity, and include the dog family (Canidae), bears (Ursidae), raccoons (Procyonidae) and weasels (Mustelidae). The Feloidea have a partitioned tympanic cavity and comprise the cat family (Felidae), civets (Viverridae) and hyenas (Hyaenidae). This arrangement into two major groups, one allied to dogs, the other to cats, is convenient, but unfortunately the various familes within each group do not always resemble one another, either in appearance or habit. Thus the Ursidae, to which all bears belong, have certain affinities with dogs and related species; but instead of supporting themselves on the fleshy tips of their toes in the manner of the digitigrade Canidae, they walk on the soles of their feet, plantigrade fashion. Furthermore, they are largely vegetarian. The Procyonidae, including such clever, attractive animals as the raccoon (which is an omnivore, eating all types of food), also has in its ranks one very rare species, the giant panda. This shy, retiring creature lives on a vegetable diet of bamboo shoots. The Mustelidae, with such typical representatives as the weasel and the stoat, are powerful little animals with long bodies, short limbs and semi-retractile claws. Few people would relate them to the dog family.

There is just as much diversity among the carnivores who are allied to the cats. The tigers and leopards, with their typical cat-like appearance and characteristics, present no problem; and the graceful genet with its black-ringed tail—one of the Viverridae—clearly has much in common with the wild cats. The spotted hyena, however, looks like a rather ungainly dog.

The Canoidea and Feloidea belong to the suborder of carnivores called Fissipeda—animals with paw-like feet and divided toes. The other suborder is the Pinnipeda—animals whose feet are formed like fins or flippers and function as paddles. Such creatures are of course better suited to an aquatic way of life. They include the walrus, feeding mainly on molluscs and crustaceans, and the seals and sea-lions, whose diet consists chiefly of fishes, cephalopods (squids and octopuses) and small invertebrates.

Fissipeda and Pinnipeda alike are probably derived from the primitive suborder of creodonts—carnivores of the Paleocene, Eocene and Miocene periods. Some were as large as bears, but with hooves rather than claws, and all had small brains and large canine teeth. Different families were found in North America, Europe and Asia.

Although zoologists apply the term 'carnivore' to all these predators, both large and small, it is clear that the label is in some ways misleading. Not all of them are exclusively meat eaters nor is there any single, readily identifiable characteristic that all, without exception, have in common.

Although these young cheetahs still show the distinctive markings of immaturity, one of them has intuitively adopted the wary pose of adulthood. Long narrow chest, muscular limbs and short pricked ears give these handsome animals some of the characteristics of dogs as well as of cats.

CHAPTER 3

The wandering nomads of the Serengeti

The animal communities of East Africa's great national parks and game reserves have the good fortune to live in protected regions. The only humans privileged to be there are wardens, and zoologists working in institutes and laboratories specifically designed for the study of local fauna. It is on these animal sanctuaries that we too intend to focus our survey of the wild-life of the continent.

Serengeti is reached by the road which crosses Ngorongoro Crater and the immense, semi-arid plain where, during the rainy season, some half million large herbivores gather like an enormous army. The track, of firmly packed volcanic soil, continues straight ahead. On our right we have passed the famous Olduvai Gorge where in 1959 the anthropologist Dr L. S. B. Leakey discovered the skull of *Zinjanthropus*, a primitive ape-like man claimed to be 1,700,000 years old. Before us extends a boundless plain, empty of trees and covered with short, grey-green grass, where dense groups of animals graze, losing shape and definition as they recede to the horizon. It has been estimated that this gigantic army is made up of some 350,000 gnus, 180,000 zebras and 500,000 antelopes. It is hard to imagine animals in such numbers as these, but they are not exaggerated. There are more ungulates packed into this great area of grassland than in any other comparable region on the earth's surface. Yet if we were to visit this same area five months later, the landscape would be transformed, no longer fertile but bleak and inhospitable, dotted with clumps of dry thorn and bush, empty of wildlife apart from isolated groups of gazelles or ostriches.

It is an extraordinary phenomenon, this double face of the Serengeti, where the annual cycles of rain and drought turn

Facing page : One of the most breathtaking sights for visitors to the East African savannah is the seasonal trek of the gnu and zebra herds in search of fresh pastures. Huge armies of nomadic animals move endlessly across the dry plains of Serengeti, furrowing the soil and its now-sparse vegetation with tens of thousands of trampling hooves. The animals seem to head instinctively towards the distant hills where massing clouds hold out promise of rain and broad acres of fresh, succulent grass.

lush green pastures into dusty wastes, then back again into nutritious grazing. And it brings in its train an amazing spectacle – the massed migration of the herds. With the approach of the dry season a colossal horde of herbivores abandons the high plain for the more humid regions of parkland. Five months later, when the rains begin, the animals trek back to their former pastures.

Continuing in a northerly direction, we cut across Tanzania's great parkland region, shaped roughly like a tomahawk, its handle the corridor of Lake Victoria. We pass through all the characteristic biotopes of the country, each with its own distinctive fauna. The grassy plains are inhabited by hartebeeste and gazelles, while the largest concentration of lions is found in the wooded savannah terrain of the Seronera river valley. Here too are the much-photographed leopards in their acacias. Farther north, in the hilly bush country, many of the migrating animals gather in the spring with buffaloes, large antelopes and elephants; and in the neighbourhood of the rocky *kopjes* are small antelopes, hyraxes and the black Verreaux eagles.

Our detailed study of the different animal communities of these parts will begin with the gnus and zebras, together with their lion predators, some 200 of which follow the herds back to the high plains when the rains return. Although gnus are artiodactyls (cloven-hoofed) and zebras perissodactyls (single-hoofed), both are herbivores, grazing on the same land and sharing the same migration routes and patterns. So it makes good sense to consider them together, starting with the gnus which, because they are more numerous, exert a greater influence on the movements of the community as a whole. And although there are some similarities in the social behaviour of both species there are many important distinguishing features demanding separate attention.

The gnus' spring fever

From the beginning of March until the middle of May – the period which in East Africa approximates to our spring – the clouds gather ever more thickly above the Serengeti, bringing heavy rain to stimulate the new growth of grass, including those species particularly favoured by herbivores – *Digitaria macroblephara, Sporobolus marginatus* and *Themeda triandra.*

The rain water forms pools in the shallow depressions on the plain where for the past six months the animals have gathered to quench their thirst. Now, in mid-May, the dense masses of cloud give way to wispy trails of vapour, promising only an occasional sharp shower. Twelve hours of scorching sunshine soon absorbs the moisture in the soil and dries out the vegetation. But for two or three weeks, while the plain is brilliant with green grass and bathed in golden sunlight, we are rewarded with an astonishing and wonderful sight. During the rainy season the gnus have remained in fairly small bands of five to ten or in larger concentrations of several hundred individuals. They have wandered from one piece of pasture to another, then back to spots affording them shade or water. Now,

Facing page : Although in outline and build it looks rather like a young buffalo, the brindled gnu – also known as the blue wildebeeste – is a genuine member of the antelope family. Indeed, with its long, slender legs, it is one of the fastest moving herbivores of the African savannah. At the slightest hint of danger the animal starts to drum its hooves against the ground and lash the air with its long tail.

Male gnus challenge one another to fierce contests to win and hold territory. These fights rarely culminate in death, for the weaker animal usually concedes victory to his adversary. Once he has staked out his territory, the triumphant male will defend it bravely—even against predators.

all of a sudden, they seem to be activated by new energy. A feeling of nervous tension pervades the herds as spring comes to the savannah.

Six months of feeding on lush grass has made the gnus fat and sleek. The adult males, heads held high, regard the world with a bold, challenging eye. Dark, shining horns, shaped like crescent moons, point to the blue sky, flowing beards are gently tossed by the breeze. Their manes fall in handsome brown locks on either side of neck and shoulders. They stand motionless and then abruptly, without warning, begin to emit a series of snorts, drumming their hooves and lashing the air violently with their horse-like tails. The next moment they are in full flight, the rest of the herd following suit.

It is quite nerve-wracking to find oneself surrounded by a herd of these huge beasts, each weighing close on a quarter of a ton. Fortunately they soon make it clear, by galloping off, that they would rather remove themselves from danger than face it. Yet they are far from cowardly, for some males can fight off a hyena or put a lion to flight.

The horns of the female gnu are less heavy than those of the males. The calves are rather lighter in colour than the slate-grey adults and their short horns are more like those of cattle, though upward-pointing. Indeed, a herd of peacefully grazing gnus bears a closer resemblance to domestic cattle than to antelopes, so much so that the Boers of South Africa call them wildebeeste—Afrikaans for 'wild cattle'.

The gnus' springtime restlessness gives a superficial impression of utter confusion, but closer observation reveals a consistent and strange pattern of behaviour. Here and there, some thirty yards distant from the others, a particularly massive, proud-looking individual stands foursquare on a piece of ground, usually where the grass has been heavily trampled. Although it has no visible boundaries, this is his own section of territory which he may have fought to possess and is ready to defend. Other wandering gnus will be encouraged to join him as if he intended to surround himself with a miniature herd of his own.

These so-called territorial males make up not more than fifteen per cent of the herd but exercise a disproportionate influence in the community at large, unlike the solitary gnus which have never succeeded in forming groups or which have become separated from them in the course of their wanderings. The latter spend the greater part of their time challenging one another to spectacular ritual combats, such fights usually being staged within the bounds of their jealously guarded domains. Fights to the death are uncommon, for this could pose a grave threat to the future of the species.

Harems of the savannah

Territorial rights, frontiers, native land—concepts which we humans accept as basic and which cause all our petty disputes and major wars—are also features of most animal societies. The cheerful singing of garden birds is their method of asserting

ownership rights in a small portion of territory where court-ship, mating and rearing of young take place – an area into which no other adult member of their own species is permitted to intrude. Carnivores too, as we shall see, defend their own hunting precincts, staking out the boundaries by leaving scent signals. Such areas may be reserved either for an individual – as with the leopard – or, in the case of lions or wolves, for a clan.

Ethologists – scientists concerned with the study of animal behaviour – have recently concluded that among certain ante-lopes, notably the kobs and topis, the males do not fight for possession of the females but in order to win a small section of savannah where they can establish themselves and exclude invaders of the same sex and species. Once securely installed in his own territory, but only then, the ardent male will attempt to attract the rutting females. When these are finally enticed into his domain, mating will take place, but always within the area reserved for this specific purpose. Young, though sexually mature, animals and adults which have failed to claim a corner of territory often become sexually inhibited, so that even when they have the opportunity of coupling with a female they will not do so. This conquest and defence of 'nuptial' territory is of fundamental importance in the breeding ritual of gnus. Ousted males, leading a solitary life, play no part in the reproductive activity of the species. Moreover, those which have successfully conquered a section of ground seem to be endowed with un-usual energy, so that they almost always emerge victorious from contests with other would-be suitors and intruders. This too makes it difficult for many males to enter the ranks of potential reproducers and at the same time facilitates the operation of natural selection mechanism within the species, so that it is always the strongest and most dominant which breed.

Having fought for and won his small area of savannah, the adult male gnu has no trouble in attracting a number of rutting females. Only then, in the security of what is virtually 'nuptial' territory, does mating take place.

Following page: White-bearded gnu and calf. Most of the females give birth when they return each year to the high plains and new pastures around the beginning of January. The calves born at this time of year are better protected than others against prowling carnivores such as hyenas.

Sentinels and decoys

Unlike the kobs, which are sedentary antelopes, the gnus' mating season usually coincides with the beginning of their migration to summer quarters. The adult males, who have taken such pains to acquire their territory are therefore compelled to keep on shifting ground according to the movements of the herd. Thus they are engaged in a series of personal duels with other bulls, having to assert their superiority over potential rivals and less fortunate neighbours with an eye on their hallowed piece of territory, and also to ensure that members of their harem do not stray off to join other family groups. The frequent fights are highly interesting to watch. Two bulls may, for example, come face to face and kneel down in one of some twenty ritual positions adopted in such contests, and then proceed to butt each other so violently with their horns that the ominous crashing and cracking sounds can be heard for miles. Astonishingly, these spectacular fights usually end without either animal suffering the slightest scratch. Sometimes both are losers, for while they have been locked in battle a third contender may calmly have taken possession both of territory and of fickle, complaisant females.

Amid all this noise and confusion mating eventually takes place. Almost all the male's energy has been expended in ostentatious fighting, keeping a watchful eye on the females and rounding up stray members of the herd, sheepdog fashion. The final act of copulation is in fact brief and unremarkable, far less noteworthy than the preceding courtship rites.

Yet this elaborate mating pattern has an important bearing on the welfare and continuity of the species. In the first place, immature males and adults who do not own territory keep well away from the zones reserved for courting and mating—invariably the best pastures—which they leave to the really active members of the community—the virile males, the females and the young. Secondly, since the dominant males are continually shifting their territories, and refrain from fighting one another once these are established, the entire savannah—apart from sections occupied by solitary males—consists of individual nuptial zones, through which the cows wander at will, thus increasing mating and breeding prospects.

The aggressive temperament of the dominant males, kept at fever pitch by the constant need to fight for territory, is formidably displayed when an intruder threatens to disturb the peace of the family unit—whether it be a predator or a member of the same species. Hyenas, which are especially partial to newborn calves, can expect no mercy from the infuriated bulls; and one angry male was even seen launching a whirlwind attack on a young, inexperienced lion, putting it to ignominious flight.

Des Bartlett, in his marvellous book of photographs *Nature's Paradise*, describes a similar incident, though on this occasion with a less determined enemy. One morning, on the sun-drenched plain of Amboseli, a herd of gnus headed towards a watering place. A cheetah, hidden in the tall grass, had its eye on a calf following closely on the heels of its mother. Suddenly it pounced, covering some 300 yards at lightning speed. Before

As soon as the young male brindled gnu has reached adulthood he will engage in fierce contests to win himself a portion of savannah. He will try to attract into his territory individuals that happen to pass near it and more particularly rutting females. Once he has gained possession of this territory he will deny access to other mature males of the species.

This winding furrow—worn deep into the
bone-dry soil of the plain—has been
formed by the thousands of trampling
hooves of migrating gnus as they advance
in orderly procession towards their new
feeding grounds at the start of the rainy
season.

the calf knew what was happening, a cuff from the cheetah's paw had bowled it over and sharp fangs were poised at its throat. The rest of the herd promptly took to its heels, disappearing in a cloud of dust—apart from two bulls, who unhesitatingly charged the cheetah. The astounded animal beat a hasty retreat and the tiny gnu, quite unharmed, clambered to its feet and trotted after its mother.

This tendency of the dominant males to intervene whenever a member of their group is threatened is so marked that should a female sense the presence of danger, she will at once summon her calves and gallop from one block of territory to another, until a bull comes to the rescue.

The gnus' close attachment to their territories does, however, lay them open to attacks by their special enemies—lions and hunting dogs. They are at a disadvantage on three counts. Firstly, their capricious movements automatically attract the attention of carnivores; secondly, because reluctant to quit their territory they lose precious seconds before deciding to take flight; and finally, because they use up so much vital energy in their daily activities they have little strength left to repel an attack—especially by a pack of hunting dogs who will chase a victim for miles until it sinks exhausted to the ground.

The males responsible for reproducing the species thus have additional duties. They must act as sentinels, ready to repel attacks on the herd by the less powerful predators, and sometimes as living decoys, bearing the brunt of aggressive forays by the larger carnivores. This triple responsibility of reproduction, defence and self-sacrifice is found among animals in no way related to the antelopes. Both the male robin and the partridge, for example, equally territory-conscious during the mating season, will quite deliberately sacrifice their lives defending females and young—the real guarantors of the community's future.

Annual migration

It is not often that we have the opportunity of watching the wonderful sight of the migrating herds. Although it usually occurs towards the end of May, when storms are frequent, rains may either delay or advance the date. A week before departure the small, self-contained groups gather together in one part of the high plateau, forming a single immense herd comprising adult males, females and calves over the age of six months. A ripple of nervous energy courses through the throng of animals who set up the characteristic *gnuu . . . gnuu . . .* bellowing that gives the species its name. The sound echoes endlessly from thousands of throats to create a strange, majestic chorus, interspersed with the shriller whinnying of small groups of zebras mingling with the solid brown mass of great antelopes.

Soon the animals are on the move, and as we watch from the roof of our Landrover the long, dark ribbon curving its way towards the horizon, we can still smell the acrid stench of the herds and hear the faint echo of their concerted bellowing. For several hours our vehicle follows in the wake of the huge army

which holds its course steadily to the north-west. Heads lowered in the face of the stiflingly hot wind blowing off the plain, the animals move forward in an unchanging rhythm, as if driven by an irresistible force, the very symbol of certainty and determination. Thick clouds of dust swirl up from the parched soil and the air still reverberates with the distant murmur of their voices and the drumming of their hooves. There is nothing chaotic about this mass exodus, nothing reminiscent of the wild stampeding of American bison – as so often depicted on the cinema screen. The mighty herd marches forward in long, disciplined lines, its passage marked by the deep, undulating trenches furrowed out by the animals' hooves. Here and there, between the parallel lines of furrows, a few stray gnus trot along, seeking an opportunity to fall in with the advancing columns.

The spectacle of large collections of animals on the move is awe-inspiring and at the same time faintly disturbing. Probably no other zoological phenomenon has so intrigued man throughout the ages as the regular occurrence of these huge population

These drawings illustrate three varieties of gnu. *Above, left :* adult black-bearded brindled gnu (*Connochaetes taurinus taurinus*). *Above, right :* adult white-tailed gnu (*Connochaetes gnou*). *Below, left to right :* white-bearded brindled gnu (*Connochaetes taurinus albojubatus*) – adult; year-old calf; calf a few days after birth.

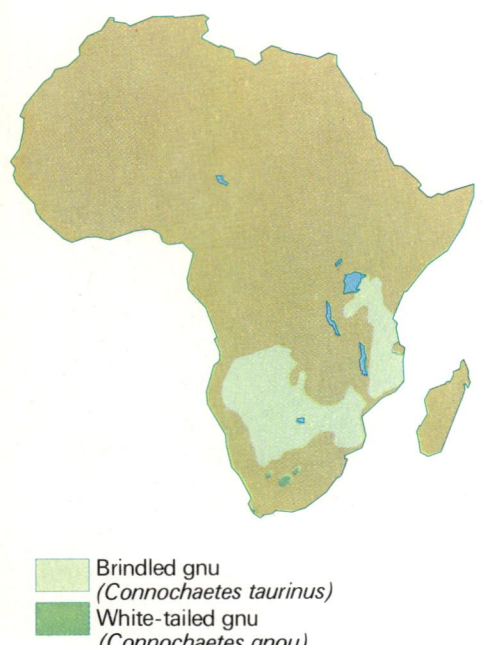

Brindled gnu
(Connochaetes taurinus)
White-tailed gnu
(Connochaetes gnou)

Geographical distribution of gnus.

BRINDLED GNU
(Cannochaetes taurinus)

Class: Mammalia
Order: Artiodactyla
Family: Bovidae
Length of head and body: 72-96 inches
  (175-240 cm)
Height to shoulder: 50-57 inches (125-142 cm)
Weight: 360-580 lb (165-265 kg)
Food: grass
Gestation: 8-8½ months
Number of young: one
Longevity: 16-20 years

Adults
Withers higher than rump. Colour slate-grey
with vertical black bands on neck and shoulders.
Hooves slender; tail black, extending to fetlocks.
Horns, with a maximum span of 32 inches, are
shaped like those of domestic cattle.

Young
Colour sandy-brown with dark dorsal line. No
horns. Soon grows abundant mane and other
hairy appendages.

Subspecies
The beard, consisting of long hairs, is the
distinctive feature of two subspecies. The
black-bearded gnu *(C. taurinus taurinus)* is
darker and more mottled; the white-bearded gnu
(C. taurinus albojubatus) paler and more uniformly coloured, the feet being lighter than
the rest of the body. Both have flowing manes.

---

shifts among different species. The migration of Holarctic birds, the geometrical flight formations of cranes and wild geese, the autumn departure of swallows and their punctual return as harbingers of spring—these events are a constant source of wonder and scientific interest.

A quest for food

What in fact are the reasons dictating the resolute march of the herds of gnu towards their summer feeding grounds? This displacement of an entire community is not a typical example of what scientists call 'gametic' migration or breeding—stimulated by reproductive urges—as in the case of salmon returning from the ocean to the rivers where they were born, but an instance of 'trophic' or 'alimental' migration, commonly called the feeding migration, dictated in the main by the quest for food.

These mass dispersals of wild ungulates take place at more or less the same time each year, depending on the pattern of rainfall, most of the animals abandoning their feeding grounds when the vegetation is exhausted and moving on to fresh pastures. It is a simple matter of survival for if the herds failed to move large numbers of animals would surely die of starvation as they quickly consumed all available vegetation.

The herds of the Serengeti advance steadily towards their summer quarters, pausing if need be until a storm has passed (several days may be spent resting in the savannah of Seronera) and splitting up into two separate groups. The first heads north, following the tracks of innumerable former generations in the direction of the shrub-covered hills and valleys north of Benagui and Kilimafeza. They cross swampy areas infested by tsetse flies and eventually reach the banks of the Mara river. The second group turns westward, following the beds of the various rivers that flow into Lake Victoria. Here, in the corridor region of the great lake, there is an abundance of fresh grass to be found in the hollows that remain flooded throughout the rainy season. During the five and a half months of summer the herds and smaller family units reorganise. The calves grow to maturity alongside their mothers and most of the adult females become pregnant once more, ready to give birth when they return to the Serengeti.

Thus the migration of the gnus, though essentially of the trophic type and undertaken for the primary purpose of food, is at the same time closely bound up with the breeding cycle, culminating in the birth of the young in the traditional home-lands when the rains return.

The gregarious gnus

The migrating animals who leave the high plain of the Serengeti around the end of May do not all head automatically to the north and north-west. Another group moves southward, assembling at Oldangua preparatory to climbing the gentle slopes of Ngoron-goro. Eventually they wind their way down to the floor of the enormous volcanic crater, spending the entire dry season feeding on the vegetation growing in the crater's swampy zones and mingling with other gnus that form Ngorongoro's sedentary population.

This latter group, which we shall now examine, is confined to an area–easily supervised–measuring a mere ten miles in diameter. The social structure of these herds–on which the following description is based–has been exhaustively studied by the American zoologist Richard Estes.

Most of the permanently based gnus of Ngorongoro organise themselves for the whole year round into bands varying from fifty to a little over a thousand head. Some twenty or thirty per cent of the animals, however, remain in even smaller groups scattered at random over the crater floor. These groups consist of about ten females with young of various ages and are generally situated inside a network of plots occupied by territorial males. Certain unusual variations of behaviour lead to the conclusion that the small groups and the large herds have adapted themselves in different ways to their environment.

As happens in most gregarious ungulates, the gnu community of Ngorongoro is divided into two main classes. The most numerous and tightly-knit groups are made up of females with calves of less than a year old. Males of one year and upward join older bulls in 'bachelor' groups which are far less stable than the other type. As in Serengeti, there is also a small group of territorial males, sometimes distinguishable from other bulls according to surroundings and season. The main groups appear to be semi-exclusive and of long duration as a result of the females' instinctive tendency to remain together with their offspring. An animal attempting to join the group from outside is normally barred entry or expelled.

Head and neck of a white-tailed gnu.

Two different life patterns

During the rainy season, from November to June, each small group functions independently, confining itself to a limited area of less than a square mile. But in the dry season, although they retain fixed bases, the various groups meet during the daytime to graze on the best available pastures, dispersing in the evening to their respective territories.

The behaviour pattern of the larger groups of gnu in Ngorongoro is essentially the same as that of a migratory community while the smaller groups display the behaviour typical of a sedentary population. The co-existence of these two life patterns among animals of the same species is explained by the region's special ecological structure.

The highly fertile volcanic soil of the crater, together with its comparatively heavy incidence of rainfall, provides favourable conditions for herbivores, but the relatively restricted area sets a limit to the animal population. Nevertheless about 25,000 large herbivores are able to live in the region permanently, thanks to the swamps which never dry up and the rain water oozing from the crater walls. In the long summer season about eighty per cent of Ngorongoro's animals graze together on the rich swampy pastures.

WHITE-TAILED GNU
(Connochaetes gnou)

Class: Mammalia
Order: Artiodactyla
Family: Bovidae
Length of head and body: 68-88 inches (170-220 cm)
Height to shoulder: 36-48 inches (90-120 cm)
Weight: 400 lb (180 kg)
Food: Grass
Gestation: 8-8½ months
Number of young: one or two
Longevity: 16-20 years

Adults
Long flattened head with tufts of thick hairs on muzzle, throat and chest, forming a beard. Small eyes, ringed with white hairs; flexible, pointed ears. Horns grow forward and downward, turning upward at tips, maximum span 30 inches. Coat dark brown, almost black. Long, flowing white tail.

Young
No horns; colour light brown; long, delicate limbs.

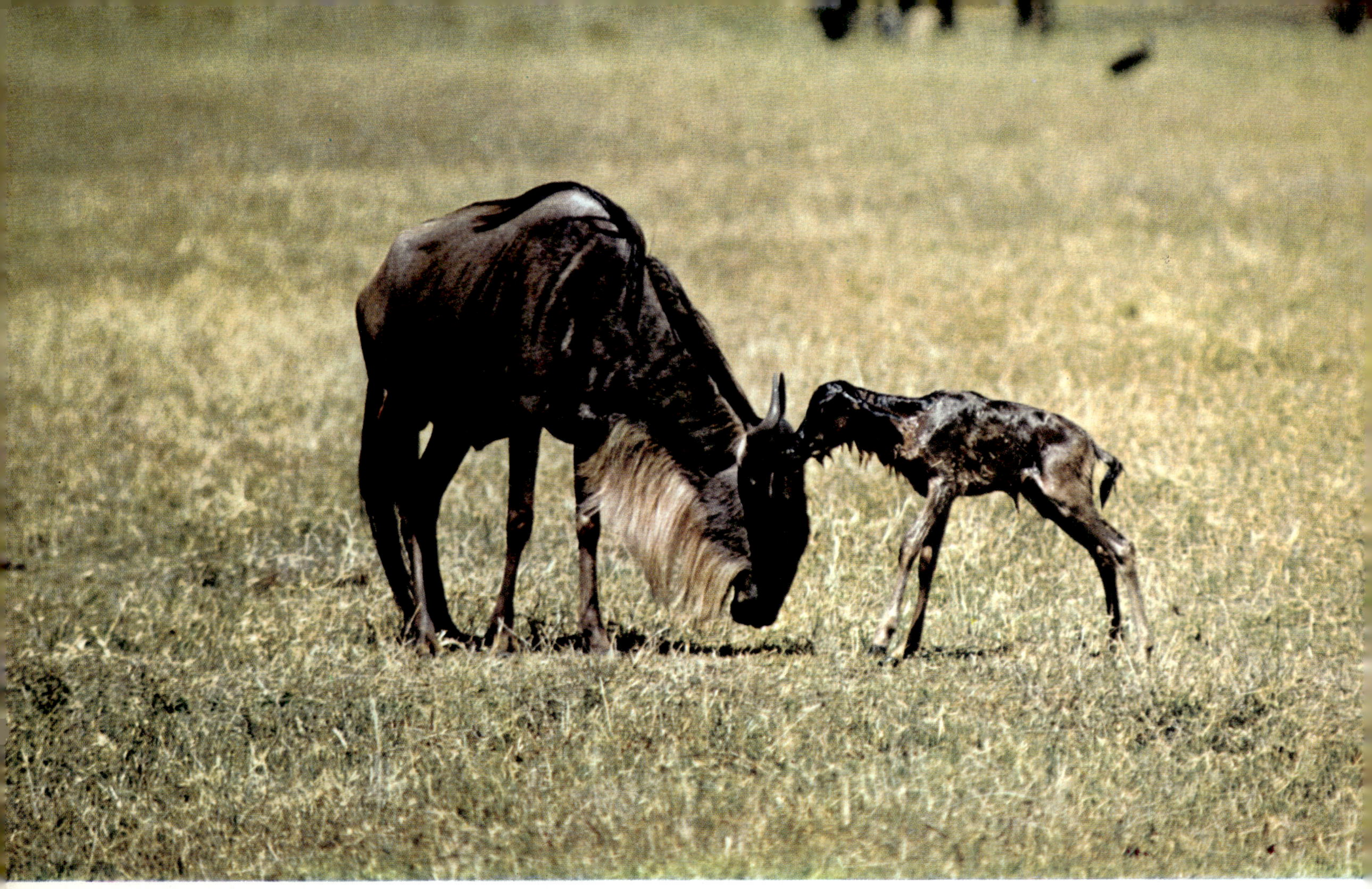

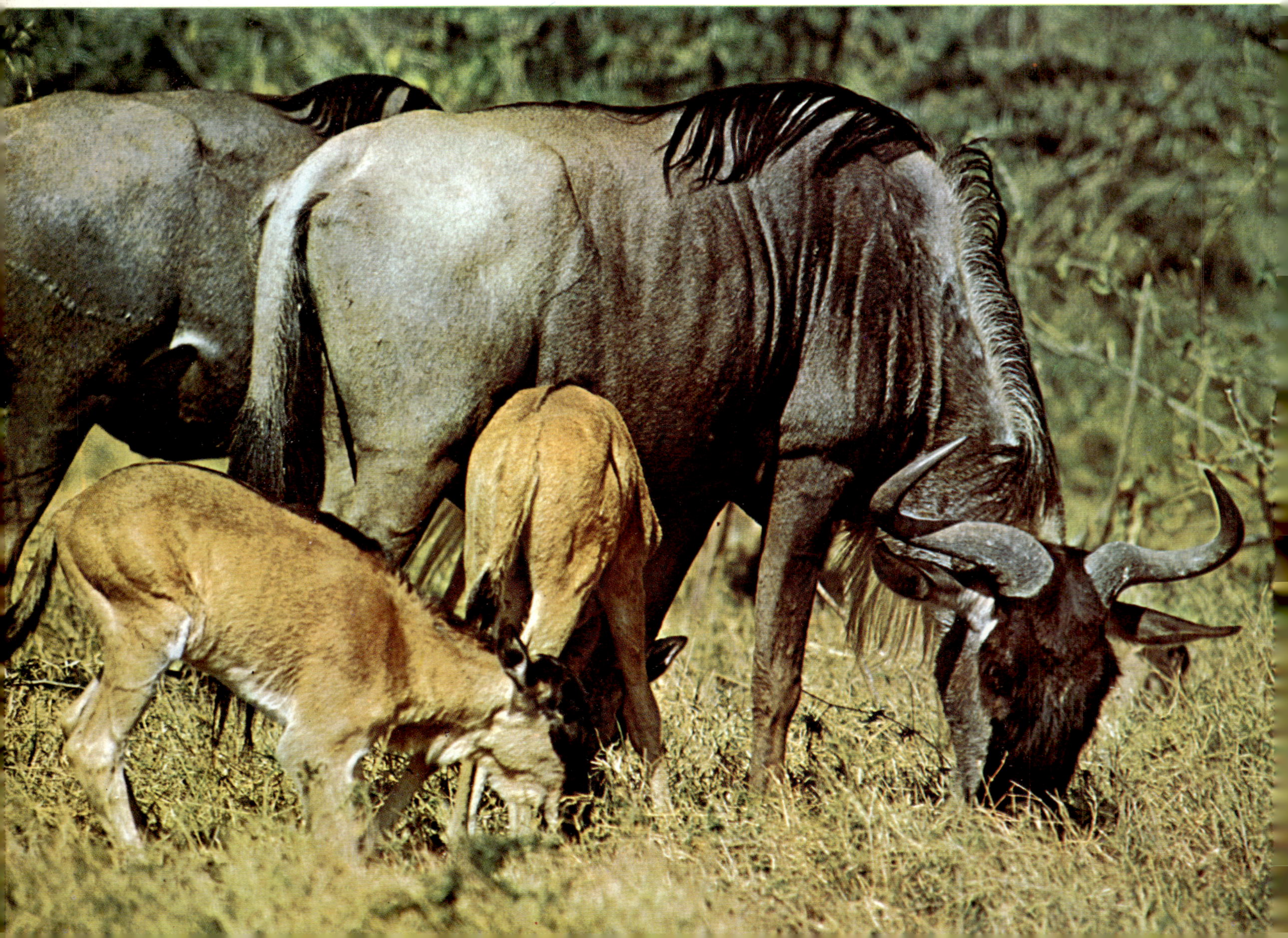

So we find a small community of gnus leading the type of sedentary existence common to the animal populations of all regions where there is a permanent supply of food and water, alongside a much larger community of migratory gnus from the Serengeti plains. But even among the sedentary gnus there are local variations of social behaviour. About twenty per cent of them leave the region of the crater in November, journeying some seven or eight miles to the moist valley which separates the plain of the Serengeti from the uplands of their home territory. At the end of the rainy season they make their way back to Ngorongoro.

Life returns to the Serengeti

At the beginning of October the plains of the Serengeti crack open like the bed of a vast, dried-up lake. Even the antelopes have vanished, to regather in the parched savannahs of Seronera in search of new shoots of vegetation. Under the merciless sun the only living creatures in view are flocks of sand-grouse, circling the sky twice daily in their massed flights to distant water-holes. The empty spaces that once echoed to the bellowing chorus of migrating antelopes are now rent by the harsh, guttural cries of the soaring birds.

All this soon changes as the light rains of November and December signal the return from summer quarters of the great army of gnus. As the huge animals gather once more on the plains carnivores and scavengers alike greedily await one predictable seasonal event—the simultaneous births of thousands of baby gnus.

The females' gestation period of about 240 days normally ends around the beginning of January, when the herds return to the plains—now carpeted with fresh green grass and a multitude of white and yellow flowers. Back in the region where, at the end of the previous May, mating and conception took place, the cows now prepare to give birth.

The gnus, however, have not returned alone. Spotted hyenas, jackals and wandering lions have tracked the antelopes at every stage of their travels and they too are back in the Serengeti. The hunting dogs too, who for five and a half dry months have had to live on gazelles, can now resume the hunt for their most favoured prey—gnus and zebras. The five species of vulture common to the region take wing again, describing large circles in the sky in their quest for freshly-killed game; and the lions of the Seronera valley journey to the plains where once more the ungulates graze.

Life and death have returned together to the savannahs of the Serengeti. As the predators and carrion-eaters converge it seems only too probable that they will take a heavy toll of the newly born animals. But nature appears to have foreseen the contingency. The females, according to their stage of gestation, instinctively group together. Those on the point of giving birth separate themselves from their companions to form one large herd in an empty part of the plain where the grass is shortest—denying lions and hyenas the opportunity of a surprise attack.

A photo-safari in Ngorongoro Crater. Zebras and gnus mingle peacefully on the high plain, apparently undisturbed by the presence of interested observers.

Facing page (above): The tiny gnu is able to stand on its feet only a few minutes after birth but needs its mother's care for several months. *(Below)* Chances of the young gnus' survival depend to a large extent on their remaining in the midst of the herd, where they are better protected from predators.

This young gnu, lacking horns and still with the light brown coat typical of infancy, has wandered a little distance away from the adults. Though not yet weaned it already feeds on grass.

Most of the births take place at dawn and are quickly over; but provided the head has not appeared, the expulsion of the fetus can be interrupted or delayed at will for any desired length of time. This marvellous natural ability allows the mother to hold back the moment of birth until conditions are sufficiently peaceful and undisturbed. The newborn calves are pale sandy-brown and within seven minutes of birth they are able to lift themselves to their feet and take their first tentative steps, instinctively keeping close to the cows, although they probably do not recognise their mothers until they are suckled for the first time. The placenta is expelled about three hours later, usually being devoured by vultures and jackals.

Most births take place during the first two or three weeks of January, then tail off gradually, with about twenty per cent occurring over the following four or five months. Everything points to the fact that although the gnus' breeding cycle is closely related to the climatic pattern—and may be modified by seasonal variations—climate alone does not determine the timetable. The other influencing factor is the survival instinct and the need to escape the clutches of predators, above all the spotted hyena.

Hyenas are especially fond of newborn gnus, the greatest period of danger being the first few hours of life. After a couple of days the calves are already able to run fairly fast and, provided they do not stray too far from the herd, are difficult to catch. Those born in the midst of the groups formed by the pregnant females are pretty well protected, for predators find it almost impossible to infiltrate the herd. The one calf in five born at other times of the year, lacking group protection, is far more exposed to marauding hyenas and hunting dogs and the resultant death toll is far heavier.

A gruesome spectacle

If we were to venture out at dawn on to the Serengeti plain during one of these periods we might catch a glimpse of the furtive shadow of a spotted hyena darting in and out of the herd. Its intended prey may be a newborn calf or even a half-expelled fetus, ripped from the mother's belly. For an animal not renowned for speed, the attack is carried out with astonishing pace and precision. The carnage completed, it drags the victim heavily away in its jaws.

The human spectator, powerless to intervene, may well feel sickened by such a cruel and unpleasant spectacle. Yet once again we have to remind ourselves that this is the price to be paid for maintaining the natural equilibrium of the species. The onslaughts of hyenas stimulate the gnus' gregarious instincts and help to regularise their breeding cycle. By banding together they ensure survival. Isolation, coupled with births spread over the entire year, would long ago have doomed the species.

Once the first dangerous days are past, the calves frolic in carefree exuberance. They grow fast. They weigh 150 lb a five months, 220 lb at ten. By now they have much of the adults' speed and stamina and are a match for most predators. By the

time they are three years old the males are capable of staking out their own piece of territory.

For some sixteen years or more they will take part in the seasonal migration of the herds, journeying from the high plain of the Serengeti to the banks of Lake Victoria. They must be alert enough to guard against the group hunting of lions and the solitary attack of the leopard, with sufficient endurance to throw off the relentless pursuit of packs of hunting dogs. And should misfortune or disease deliver them to the claws and fangs of their natural enemies before they have completed their normal lifespan, they will only be fulfilling nature's purpose – providing life for lions, hyenas, jackals, vultures and all the other carnivores and carrion-eaters of the savannah.

The white-tailed gnu is threatened with extinction. Today there are only a few left, all of them protected in South African reserves, especially in the Orange Free State.

Following pages : A sequence of pictures – from left to right and from top to bottom – showing successive stages in the birth of a Grant's zebra. After a gestation period of about a year the female drops a single foal. After emerging from the embryonic sac the foal is licked clean by its mother, moving forward on shaky feet to suckle.

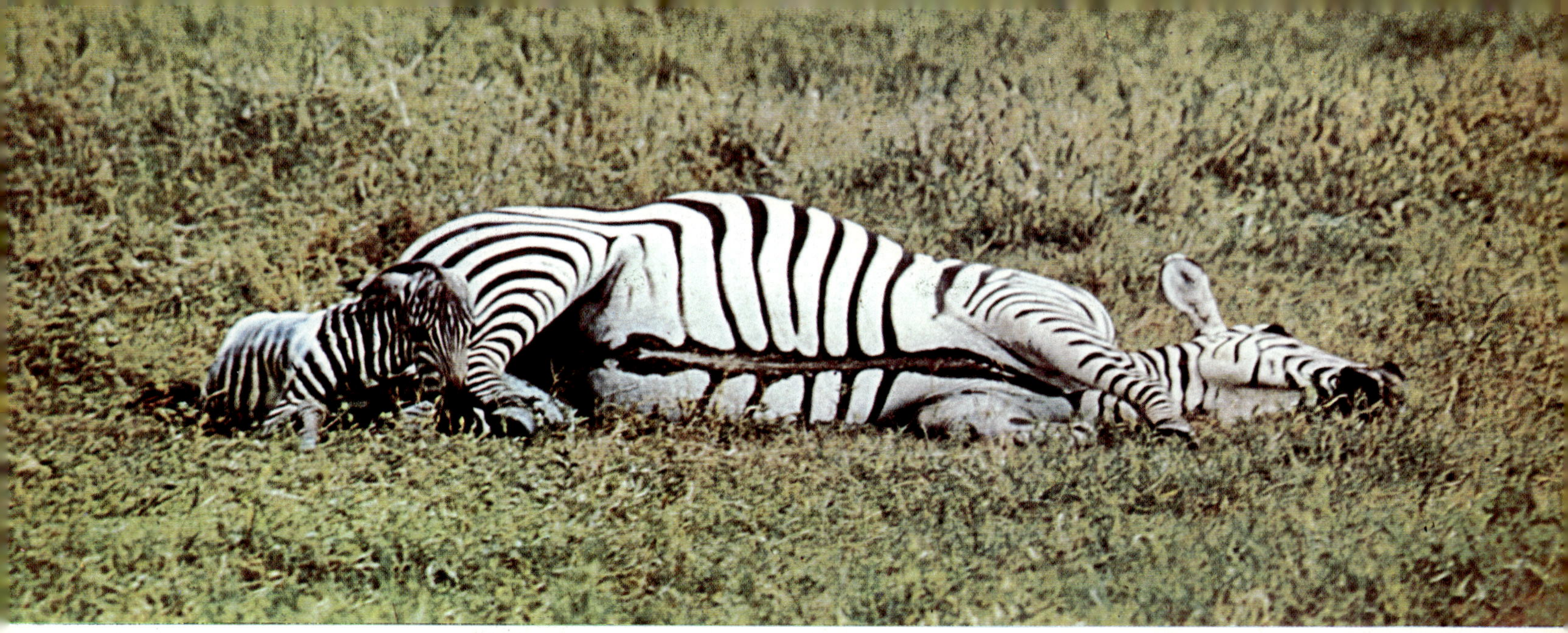

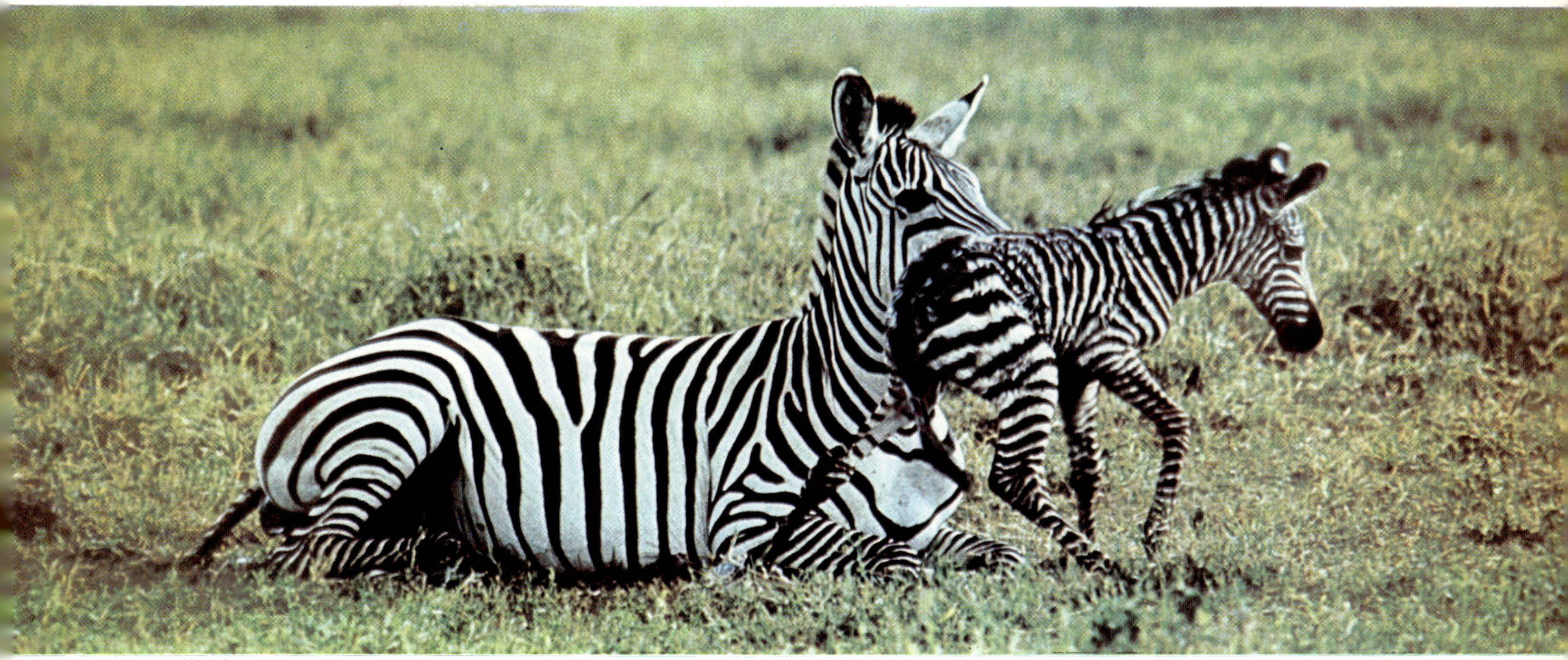

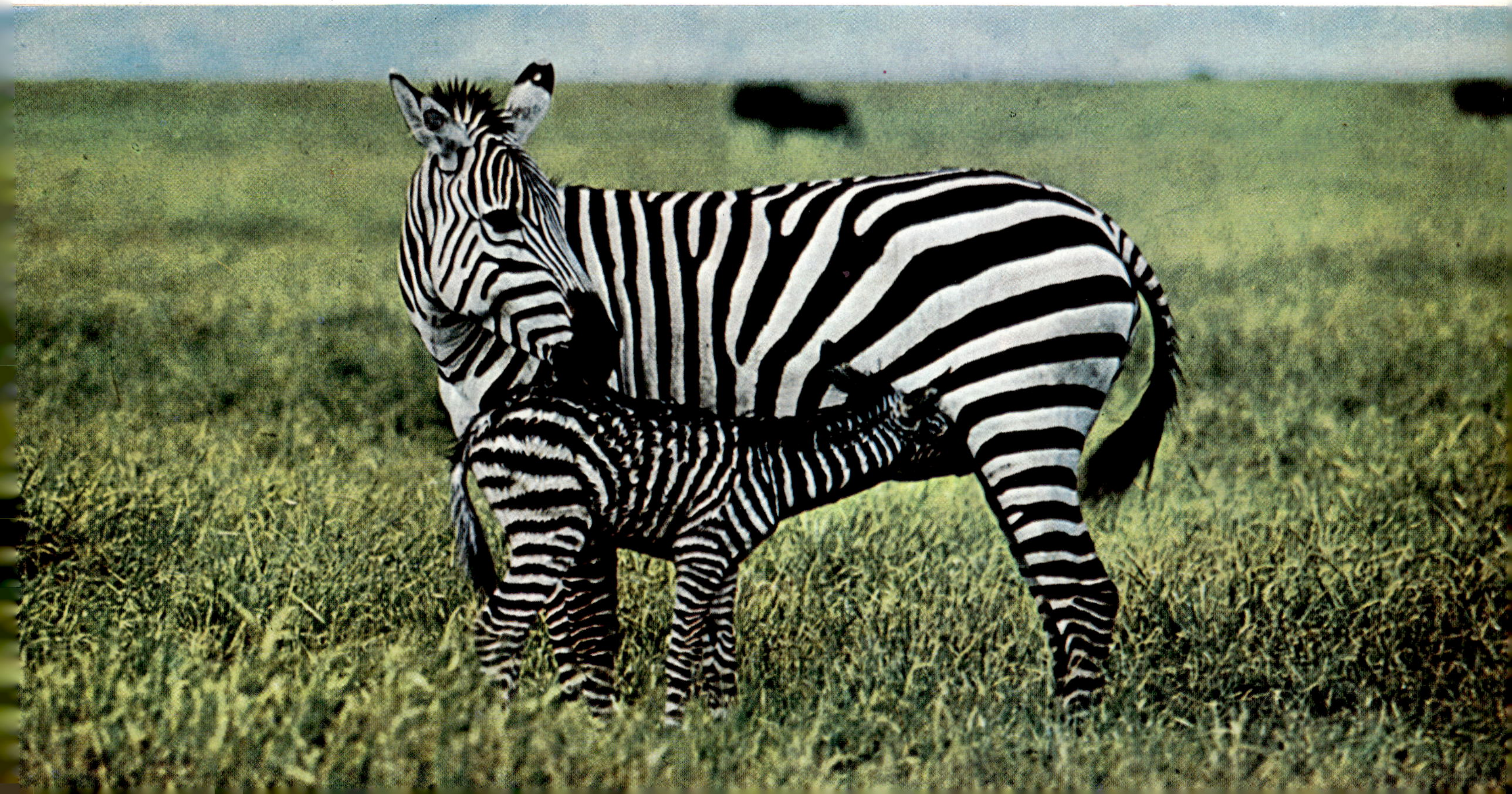

Zebras

Photographed in the shade of acacias with Mt Kilimanjaro in the background, filmed from every angle in the most characteristic and picturesque surroundings that Africa can offer, zebras have appeared on the covers of books and magazines and on television screens all over the world. These attractive animals are understandably popular but never so impressive as when seen in their natural environment—mingling, for example, with herds of gnus in the Serengeti.

Here, as throughout the East African steppes and savannahs, the two herbivores, so dissimilar in appearance, share the same feeding grounds. To a European observer there is something strange and remote about the unwieldy gnu, whereas the zebra, with its horse-like build and cheerfully striped coat, seems friendly and familiar. Yet man, who by means of selective breeding has so profoundly influenced the development of the domestic horse—producing animals as far removed from the primitive prototype as the massive Percheron and the graceful English Thoroughbred—has in no way affected the evolution of the wild zebra. Nature alone has been at work here and the zebra no more resembles the domesticated horse than the wolf does the household dog. And like the wolf, its perfectly proportioned body presents an image of balanced strength, agility, stamina and speed.

The zebra population, like other zoological groups spread over a wide geographical area, is relatively homogeneous in appearance and behaviour. Yet environment and comparatively secluded conditions have combined to produce local variations. Thus in the most northerly zone of population distribution—northern Kenya and the dry scrub-covered steppes of Somaliland—the large Grévy's zebra (*Equus grevyi*), less horse-like in appearance than other species, has adapted itself to almost desert-type conditions and is able to withstand extremes of heat and thirst. Its long hairy ears are reminiscent of a donkey's and its stripes are relatively narrow.

The herds of zebra inhabiting the East African and South African savannahs are smaller than the Grévy species and inter-related. Experts disagree on certain details of classification so that some confusion still arises over the identification and naming of the various subspecies. The common zebra (*Equus quagga*) ranges over the entire region and is sometimes called Burchell's zebra—the original Burchell's zebra (*Equus quagga burchelli*), once a resident of South Africa, now being extinct. The coat colour and pattern of the common zebra varies according to locality and the subspecies are not easily distinguished from one another. They include Grant's zebra (*E. quagga granti*), Boehm's zebra (*E. quagga boehmi*) and Chapman's zebra (*E. quagga chapmani*), but even the stripe patterns of these show variations within the subspecies itself. Some authorities in fact consider Grant's zebra and Boehm's zebra to be one and the same animal. Those who separate them give northern Kenya as the habitat of the former, with Boehm's zebra being found in the Serengeti, Ngorongoro Crater and other

Facing page : Zebras and gnus quench their thirst in the river that flows through Ngorongoro Crater in Tanzania. Mixed herds of these herbivores are seen all over the savannah.

The clearly defined black and white pattern of Grant's zebra would seem to make it a natural target for predators. In fact it provides an excellent form of camouflage, the stripes blurring the animal's outlines when in motion.

reserves in southern Kenya and Tanzania. The coat of Grant's zebra is described as pure black and white whereas Boehm's zebra shows a lightish brown shadow stripe as well. More distinctive is Chapman's zebra from south of the Zambezi, with defined shadow stripes between the larger black stripes.

Finally, in South-west Africa we find two subspecies of small zebra descended from the herds living there prior to the arrival of European colonists—the Mountain zebra (*E. zebra zebra*) and Hartmann's zebra (*E. zebra hartmannae*).

The visitor to the Serengeti, confronted by the uniformly coloured, ungainly gnus and the neatly striped, handsome zebras, will probably point his camera first at the little animals that remind him so forcibly of miniature racehorses. If he happens to find them huddled together alone in a tightly bunched group it will probably be the result of real or imagined danger on their part. At most other times they will be seen grazing amicably with the gnus, attracted by the same kind of vegetation.

Study of the hunting methods of hyenas and lions suggests, however, that the carnivores make a fine distinction between the two species, preying on gnus in preference to zebras. If this theory is correct it may explain why the latter find it safer to stick closely to their gnu companions instead of risking exposure in isolated groups. While the predators are preoccupied with chasing gnus the zebras can take advantage of the breathing space offered to remove themselves to comparative safety.

It is probably for similar reasons that gnus, zebras and hartebeeste welcome the company of ostriches on their grazing grounds. Thanks to their height and remarkable vision, these earthbound birds of the savannah are the first to spot a potential predator and to give hissing warning of an impending attack. And in those parts of savannah and steppe where thick grass limits the ostriches' field of vision, the herbivores reciprocate the service – the hartebeeste and gnus with their finely developed sense of smell, the zebras with their exceptionally acute hearing.

The colour of zebras

Some animals are especially well equipped by nature to attract the attention of others of their own kind as well as of other species. Millions of years of evolution – in the course of which mutation and natural selection have been influencing factors – have endowed these creatures with distinctive colours, spots and stripes which function as signal systems. Such hues and patterns are to be found throughout the animal kingdom. The linnet, for example, sometimes builds its nest in hedges on the perimeter of cornfields. The gaudy crimson breast of the male bird is just as conspicuous as the bright red wild poppy that grows nearby. The flower attracts insects and in the autumn, after pollination is concluded, its petals fade. The bright plumage of the bird warns others of his kind that this is his chosen territory. By autumn the brood has been reared and the red feathers moult. Neither flower nor bird has further need of recognition colours.

Many animals prefer to conceal rather than expose themselves, blending with their surroundings so that predators are unable to discover and attack them. The plumage of the nightjar and the brown coat of the hare are two of many such examples. Other animals possess mimetic coloration which enables them to take on the guise of species better armed than themselves against potential enemies.

We are possibly so accustomed to seeing zebras in photographs and in zoos that we do not consciously realise how very conspicuous they are with their striking black and white coats, in comparison with the predominantly dark, uniformly coloured asses, donkeys and horses. It is almost as if nature had selected them from the ranks of other savannah dwellers as special targets. Yet most zoologists who have studied the question have concluded that the zebra's stripes do not make the animal stand out with startling clarity and that the unusual pattern that distinguishes them from other equids serves for concealment rather than recognition. In fact the stripes tend to blur the outline of the animal when it moves. (The combatants of the second world war took a leaf out of nature's book by painting their warships with broad stripes of contrasting colour for precisely the same purpose!) On the African plains atmospheric disturbances caused by overheating help to blur and fuse shapes and colours still further so that naturalists taking a wildlife count often have difficulty in picking out the zebras from the less gaudy antelopes and other species.

Obviously the type of protective coloration that helps an animal such as the hare to evade pursuers by flattening itself

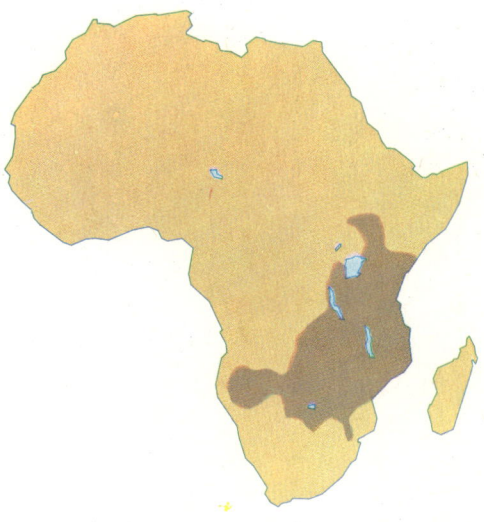

Geographical distribution of the common zebra.

COMMON OR BURCHELL'S ZEBRA
(Equus quagga)

Class: Mammalia
Order: Perissodactyla
Family: Equidae
Length of head and body: 80-92 inches (200-230 cm)
Height to shoulder: 50-52 inches (125-130 cm)
Weight: 495-660 lb (225-300 kg)
Food: grass
Gestation: 330-375 days
Number of young: one
Longevity: 25 years

Adults
The many local races of common zebra are of medium height, horse-like in appearance, with small heads and flexible ears. The colour and pattern of coat varies according to location, becoming lighter with progress from north to south.

Young
Newborn foals are light chestnut-brown with some darker stripes. The coat is thicker than that of adults.

Subspecies
Grant's zebra *(E. quagga granti)* is the most northerly, the large black stripes being clearly defined, extending to feet. Boehm's zebra *(E. quagga boehmi)* is very similar in appearance. The zebras south of the Zambezi are of variegated colour and pattern, some of them, such as Chapman's zebra *(E. quagga chapmani)* having a subsidiary brown stripe through the white areas.

Geographical distribution of Grévy's zebra.

GREVY'S ZEBRA
(Equus grevyi)

Class: Mammalia
Order: Perissodactyla
Family: Equidae
Length of head and body: about 100 inches (250 cm)
Height to shoulder: about 60 inches (150 cm)
Weight 550-715 lb (250-325 kg)
Food: grass
Gestation: about 390 days
Number of young: one
Longevity: 25-30 years

Adults
Narrow black stripes on very pale or white ground. Stripes larger on elongated head, muzzle greyish or almost black. The upper edge of the large, rounded ears is white. Stripes on neck are particularly broad—up to three inches compared with about one inch on rest of body—and form closed rings. The thick, stiff mane reaches from ears to withers, continuing along the dark brown dorsal line which, outlined in white, extends to root of tail. Belly white; feet completely ringed. The black stripes on rump take on semi-circular formation around upper part of tail, which ends in a tuft.

Young
Newborn foals have long limbs in proportion to body size. Ground colour of coat is light brown, with dark brown stripes visible on head, neck and limbs. Several weeks elapse before distinctive features of the species appear.

---

against the ground is of little benefit to the lesser kudu, motionless in a clump of bushes, or the zebra, creature of the open plains, relying on speed to escape from danger. In fact the zebra's unique coat pattern serves a double purpose, concealing it from the watchful gaze of predators and making it more clearly visible to its companions.

The communal life of zebras

Ngorongoro Crater is an ideal place to study the behaviour of the East African ungulates, for the animal population of this great natural bowl is for the most part sedentary, or failing that, migratory only within certain bounds. The herds of Boehm's zebras that live here are most attractive beasts, best studied at dawn when they gather to drink, together with gnus, along the banks of a stream.

The casual observer will immediately note the presence of a number of zebras which are both larger and livelier than the rest of the herd. These animals emit strange cries—sharp barks rather than whinnies—and frequently kick out at one another with their hooves. These are the stallions, whose size and behaviour easily distinguishes them from the young males and the females.

About 5,000 zebras live in the crater and their social structure is surprisingly compact and organised. They form themselves into self-sufficient family groups of some fifteen individuals, each of them comprising a stallion, with up to six mares and their foals. These family units are fairly stable although a mare will sometimes leave it for another group. Should the stallion abandon it for any reason he will probably be replaced by another and the family bond remain undisturbed. But the structure of the group undergoes change as soon as the young zebras are old enough to fend for themselves. Females reach puberty at about two years of age and promptly depart to find a stallion of their own; and the immature males separate from their parents at any time between the age of one to three years, or perhaps later, joining other males until they are ready to mate. Some, however, may lead a solitary life or group together with fifteen or so other single males.

Sex and reproduction

All the subspecies of the common zebra display more or less the same kind of behaviour during the mating season. Rutting females try to attract future partners by repeatedly opening and closing their mouths. All of them are sexually mature at two years; Grant's zebras mature two or three months earlier but do not foal for the first time before the others. Males, which are mature at two and a half years, are generally compelled to fight older stallions for territory before being allowed to take a mate.

The gestation period of the mares varies according to species from eleven to thirteen months; the process of giving birth takes seven or eight minutes. The newborn foal, weighing 65–75 lb, has a short body with long, delicate limbs. Clear

Equus grevyi

Equus zebra zebra

Equus zebra hartmannae

Equus quagga granti

Equus quagga boehmi

Equus quagga chapmani

Equus quagga burchelli (extinct)

Equus quagga quagga (extinct)

This little Grant's zebra takes its ease while the adults graze. It has the long hair and brown stripes typical of infancy. The latter will become darker as it grows older.

chestnut-brown stripes are visible on the forehead, nape and fore- and hind-quarters, becoming darker on the neck and legs. The average length of the coat is a little over an inch but the thicker hair of the back and haunches measures over three inches. The foal grows rapidly and at a month old weighs more than 100 lb. Meanwhile the brown stripes become gradually darker until the coat takes on the overall black and white pattern of the adults. Though able to crop grass after only a few days and capable of grazing without difficulty before they are two weeks old, they continue to be suckled by their mothers until about their seventh month, by which time the mares may be well advanced into another pregnancy.

In Ngorongoro, as in other parts of East Africa, the baby zebras are born at any time of year, the majority of births occurring between October and March, with a peak figure in January. Scientists studying one group observed female zebras in heat only a few days after dropping their foals, the shortest recorded period between successive births being between 365 and 378 days. Statistics also indicate that among the Boehm's zebras of Ngorongoro the numbers of newborn males and females are about even. There is no reason to suppose that the other subspecies show a substantial variation.

Survey of a mixed herd

The layman may find it difficult to understand why zebras should not mingle more freely with one another and be able to trace a line of common descent. The reason is that in spite of close links different species have for thousands of years been prevented by insuperable geographical barriers from making easy contact with one another. Nevertheless in certain regions zebras of different species have come together to form mixed herds, though not to the point of producing hybrid offspring. This is a strange phenomenon for zebras are homogeneous creatures and there would seem to be no reason why the kind of inter-breeding that exists among other members of the horse family – both in the wild and experimentally in captivity – should not naturally occur.

In the fertile river valley of Usao Nyiro in northern Kenya, where there are particularly favourable climatic conditions, Grévy's and Grant's zebras live together for the greater part of the year. Close study of their respective behaviour patterns has led experts to conclude that although they mingle quite happily there is no natural inclination on the part of either species to inter-breed.

When zoologists began their survey of the mixed zebra population of this region there were, apart from a few small, virtually autonomous groups of both species, two large herds comprising 160 and 250 animals respectively, of which about three-quarters were Grévy's zebras. The herds were seen to spend the morning a mile or so from the river, would rest in the shade of trees during the mid-day hours and would then troop out to drink late in the afternoon. At sunset they would make their leisurely way back from the river, pausing at intervals along the track to graze.

During the entire course of the survey the smaller of the two herds contained on average 124 Grévy's zebras and 36 Grant's zebras, the latter remaining together and breaking away from the main body of the herd to form two or three small groups of about a dozen animals. They grazed some distance away from the others but came together quite readily when trekking to and from the river or when forced to flee from a predator. Contrary to the normal custom of a mixed herd of gnus and zebras, the two species did not go their own separate ways at such times of crisis; in fact the less numerous Grant's zebras were usually to be seen in the centre of the composite herd. So although the two species remained entirely independent at most times – and especially so during the mating season – they showed an unmistakable sense of community at others. This was clearly the result of adaptation and by no means accidental or dictated by circumstance.

Stallions, whether in mixed or single herds, are often seen to engage in fights, rearing up on their hind legs, lashing out with their hooves and trying to bite each other. These battles, fought

Mother and foal are inseparable up to the time of weaning – around the age of seven months. After that the young zebra is left to its own devices.

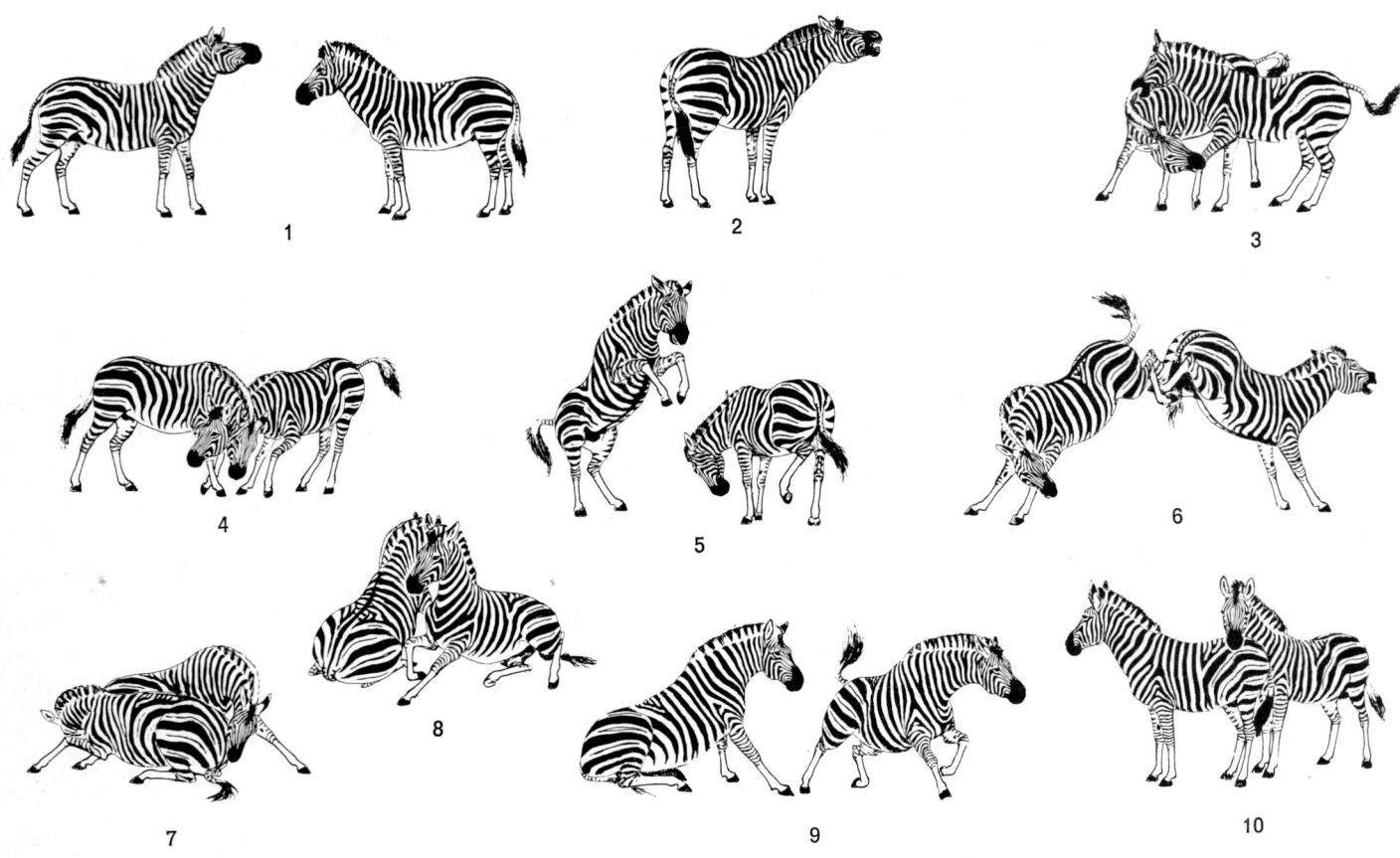

These drawings of two Chapman's zebras in characteristic combat attitudes were based on studies made in South Africa's Kruger National Park. 1. The stallion on the left defies his rival by raising his head. 2. Stretching his neck and baring his teeth, he challenges his opponent. 3. The combatants try to bite each other on neck and legs. 4. The stallion on the left avoids his rival's teeth by forcing him to lower his head. 5. The stallion on the right bends his head, drumming a hind hoof against the ground. 6–7. The two zebras attempt to castrate each other with kicks and bites. 8. Exhausted, both animals take a short rest. 9. One of them quits the fight. 10. As a sign of victory, the winner leans his head against the loser's rump.

for dominatory or territorial reasons, follow an established ritual, as in most animal societies, and are no mere random encounters, decided on the basis of brute strength. Formalised attitudes and movements clearly signify each contestant's intentions and the outcome is invariably the unconditional surrender of the vanquished animal.

These ritual combats occur most frequently among Grévy's zebras and often take place when they are disturbed by humans. Indeed the presence of a stranger is often enough to provoke behaviour which temporarily shatters the calm of the family unit. The stallions make strenuous efforts to prevent the mares drifting off to other groups and generally display considerable nervous tension. Any form of outside intrusion is enough to trigger off a show of aggression that may culminate in fighting, the main purpose of which is to impress the potential enemy with a demonstration of strength.

Despite such moments of tension and crisis, Grévy's and Grant's zebras have never been known to attack each other, indicating that in all internal matters affecting the herd – domination, territory and sex – each species confines its quarrels to its own kind. When in the afternoon the two species were seen to mingle on the river bank there were some tentative approaches towards each other but no observed instance of a Grévy's stallion actually coupling with a Grant's mare, or vice-versa.

The conclusion is that in a mixed herd the stallions of one species display little if any interest in the rutting females of another species. Although they have common anatomical links, there would appear to be specific differences that preclude cross-breeding in the wild state.

The extermination of the South African zebras

Africa once boasted a rich and varied fauna, distributed over the entire continent. Today only a small remnant of this natural splendour remains, confined to the parks and reserves of East Africa. Reports by the first white settlers as well as the later field work done by naturalists confirm that South Africa, for example, was a veritable paradise of wild animals. When the Boers landed on the continent and struck inland from the coasts they found immense herds of antelopes, giraffes, elephants, white rhinoceroses and zebras. The variety and healthy equilibrium of these animal communities had been unaffected by the activities of the primitive tribes of Bushmen who had hunted them for thousands of years. But the arrival of the Europeans violently altered the situation. Soon South Africa became the arena for one of the most savage and senseless massacres of wild animals ever known. Within a few decades many herds had been decimated beyond recovery.

One animal in particular was singled out for speedy destruction – a species of zebra not very different in appearance from the horses which the settlers had brought with them from Europe. It had a reddish-brown coat, paling almost to white on the legs, with brown stripes on head, neck and fore-quarters. The local Hottentots called it 'quagga'. Vast herds grazed on the plains which the Boers decided to convert into farmland. The quaggas were doomed to extinction. They were deprived of their traditional pastures and were ruthlessly killed to provide food for the colonists' hired Hottentot labour. According to a report dated 1758 the animals were shot in such numbers that hunters were instructed to recover the bullets from their corpses to make good the ammunition shortage. Exactly 100 years later the last wild quagga was shot at Aberdeen in Cape Province. Only a handful of animals, especially reared for commercial purposes, remained, and by 1879 they too had disappeared.

A few quaggas had been despatched to zoological gardens in Europe but their fate was no happier. The only specimen in the Paris zoo died in 1793 while those in London, Berlin and Amsterdam survived until 1872, 1875 and 1883 respectively. Thus, within an unbelievably short period – barely 200 years – human greed, callousness and stupidity had succeeded in exterminating an animal species that had for thousands of years lived in perfect harmony with its environment.

Only a few years later Burchell's zebra (*Equus quagga burchelli*) had suffered a similar fate. This common zebra from Bechuanaland and the Orange Free State had formerly been plentiful yet the last representative of the species in captivity died in the London zoo at the beginning of the present century.

The two subspecies of mountain zebra – so named because they inhabit the mountainous regions of southern Africa and are not creatures of the plains like the common zebra and Grévy's zebra – are likewise threatened with extinction. *Equus zebra zebra* was apparently never very numerous. When the British settled in Cape Colony in 1806 they introduced laws to

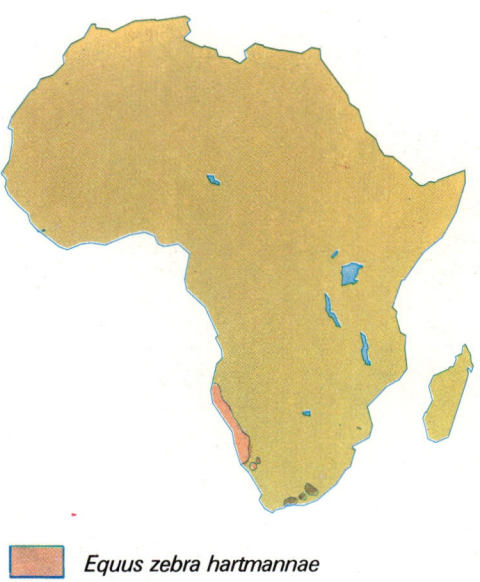

■ *Equus zebra hartmannae*

■ *Equus zebra zebra*

Geographical distribution of Hartmann's and mountain zebras.

MOUNTAIN ZEBRA
(Equus zebra)

Class: Mammalia
Order: Perissodactyla
Family: Equidae
Length of head and body: about 88 inches (220 cm)
Height to shoulder: 50-56 inches (125-140 cm)
Weight: 460-505 lb (210-230 kg)
Diet: grass
Gestation: 340-360 days
Number of young: one
Longevity: 25 years

Adults
Smallest of all zebras, recognisable by the way the black stripes are distributed on the rump— forming a kind of triangular gridiron pattern ending at the tail.

Young
Assuming characteristic adult features shortly after birth.

Subspecies
There are two subspecies. The typical Mountain zebra *(Equus zebra zebra)* is nowadays almost extinct. Hartmann's zebra *(E. zebra hartmannae)* is much like and a little larger than an ass. Both subspecies have large ears and a short, thick neck. In the former the black stripes are more numerous, especially on the hind legs, while in the latter white and black are evenly divided.

Above, left : The zebra's rallying call is more of a bark than a whinny.
Above, right : The slender, elongated head of Boehm's zebra is much like that of a horse. *Below :* Grévy's zebras frequent both mountain and steppe.

restrict hunting, which resulted in a new wave of massacre and the almost total disappearance of the dwindling herds of Mountain zebra. A few individuals kept by Boer farmers were spared but confinement did not encourage them to breed freely. In 1937 a small national park was opened in Cape Province to protect this species. Early results were not encouraging but success was more marked after the park had been enlarged and dwindling numbers reinforced by new blood. Today more than 50 zebras live and breed there in ideal surroundings.

Prospects are somewhat brighter for the second subspecies, Hartmann's zebra (*Equus zebra hartmannae*), less ass-like and slightly larger than its relative. It is estimated that about 7,000 of them still live in the steep mountain areas east of the Namib Desert in South-west Africa; but their numbers too must be steadily on the wane as a result both of poaching and the encroachment of domestic cattle on their grazing grounds.

Although the zebras of East Africa have also vanished from many regions now given over to agriculture and livestock raising they still survive in fairly large numbers, thanks to the protection afforded by the great national parks and reserves. Their population is relatively stabilised, the comparatively low incidence of deaths caused by predators more or less balanced by the birth rate. But the overall picture reflects little credit on man. Prior to his colonising ventures in Africa the great zebra herds had coursed freely over steppe and savannah, their numbers relatively stable despite the continuous assaults of lions, hyenas and hunting dogs. It was man alone who was responsible for upsetting the delicate natural equilibrium, coming perilously close to destroying the entire zebra population.

Can the zebra be tamed ?

Man has always tried to make animals serve him either for his profit or pleasure. Even the wild zebra was not exempted. The quagga, when the first European settlers arrived in South Africa, was evidently used as a beast of burden. But its extermination was so rapid that nobody can say whether the animal was truly domesticated.

Attempts to tame other species of zebra have met with rebuffs. One settler in Cape Province who tried to harness a couple of zebras to his cart was severely mauled. Determined use of teeth and hooves makes the animal difficult and dangerous to catch; nor is the front part of its body really strong enough to support the weight of a rider. So the zebra has gained a reputation for being untameable. In some countries zebras have been trained to appear in circuses, performing the kind of simple acts usually expected of horses. This indicates only that in exceptional circumstances they may be persuaded to respond instinctively to certain stimuli.

It might be possible to make some progress with animals born and reared in captivity, but final conclusions could not be reached unless efforts to domesticate them were continued over several generations. Judgment on the question must therefore remain suspended.

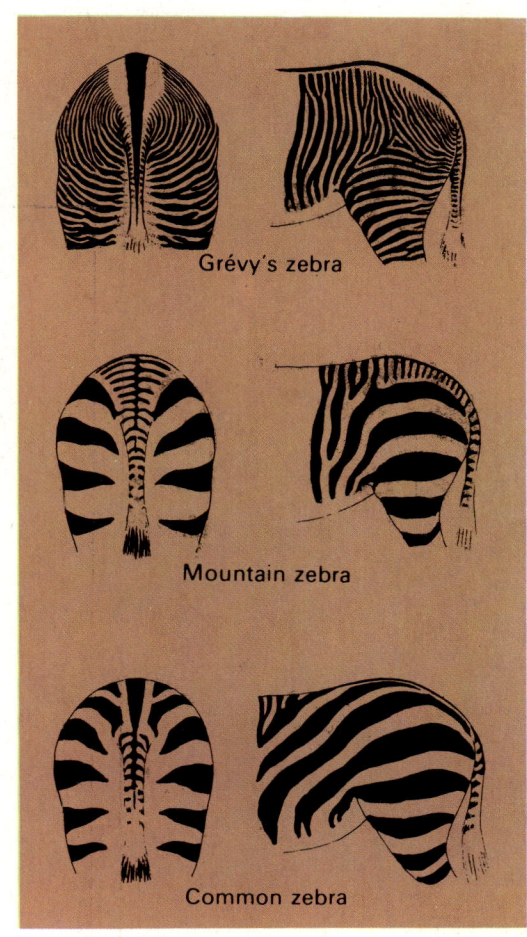

Grévy's zebra

Mountain zebra

Common zebra

The three main species of zebra may be distinguished by the pattern of stripes on their rumps. In Grévy's zebra these stripes are narrow and numerous. In the mountain zebra they are wider, while smaller stripes are visible on the back and at the base of the tail. The common zebra is also noteworthy for its broad stripes, the horizontal lines adorning the upper part of the legs extending down the whole length of the flanks; in the mountain zebra the lines change direction as they move down the body.

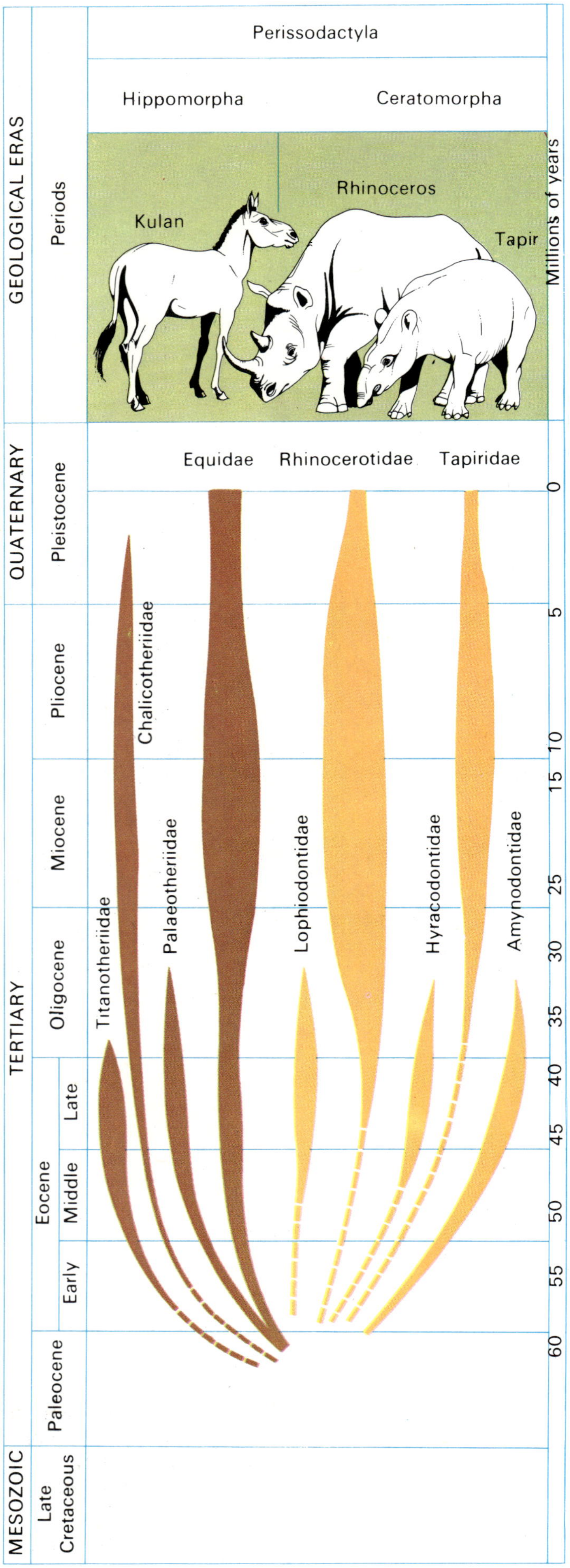

GEOLOGICAL ERAS		Periods		

Perissodactyla

Hippomorpha   Ceratomorpha

Kulan   Rhinoceros   Tapir

Millions of years

Equidae   Rhinocerotidae   Tapiridae

Chalicotheriidae

Palaeotheriidae

Titanotheriidae

Lophiodontidae

Hyracodontidae

Amynodontidae

MESOZOIC — Late Cretaceous

TERTIARY — Paleocene, Eocene (Early, Middle, Late), Oligocene, Miocene, Pliocene

QUATERNARY — Pleistocene

0   5   10   15   25   30   35   40   45   50   55   60

ORDER: Perissodactyla

The perissodactyls, or odd-toed ungulates, are mesaxonic mammals – the axis of the foot passing through the middle toe. This toe is well developed whereas the others have gradually degenerated or disappeared in the course of time.

The fossil remains of perissodactyls indicate that this has always been an important group of mammals, and it is thanks to the work of paleontologists that we know so much about their history and evolution.

During the Early Eocene – approximately 70 million years ago – there lived in North America an animal called *Eohippus*. A similar creature named *Hyracotherium* was found simultaneously in Europe. These small animals, probably little larger than modern foxes, were grouped together in the same genus by the American paleontologist G. G. Simpson. Some authorities claim that they were the ancestors of all the perissodactyls; others consider them to have been the forerunners only of the Equidae – modern horses and their relatives.

Between 40 and 60 million years ago – precise dating is impossible – there were large numbers and innumerable species of these mesaxonic mammals. A favourable climate, with regular dry and rainy seasons, as well as abundant vegetation stimulated the development of several new families of mammals – the Lophiodontidae, Hyracodontidae and Amynodontidae – all of which disappeared during the Oligocene. We also possess fossil records from the Eocene of the Titanotheres – gigantic creatures that walked on padded feet like modern elephants. They too had become extinct by the Early Oligocene, about 38 million years ago, but during that time they underwent rapid and spectacular evolutionary development. Four other families lasted much longer. The Chalicotheres, small animals that browsed on leaves and branches, standing on their hind legs and resting their fore limbs against the tree trunks, died out as recently as the Pleistocene – about a million years ago. The others – Tapiridae, Rhinocerotidae and Equidae – overcame successive climatic and geological upheavals to survive alongside man. All of them have living representatives today.

Generally speaking, members of these three families are heavy, fairly tall animals. Their thick skin may be either hairless or nearly so – as with the rhinoceroses – but is more frequently covered with abundant and variously coloured hair. The coat may either be of a uniform colour (as with many horses) or striped (as in the case of zebras). Their vegetarian diet has resulted in a dentition in which the canine teeth have been reduced in size and the premolars have become almost indistinguishable from the molars. It has equally affected the structure of their digestive system which is characterised by a comparatively simple stomach and voluminous caecum (part of the large intestine).

Of the three extant families of Perissodactyla, the Equidae – made up of six main species – include animals whose anatomy is particularly well adapted for running, such as horses, wild asses and zebras. The Tapiridae or tapirs are the smallest of living perissodactyls and also among the most primitive. The head of the tapir is long, with a short, flexible proboscis formed by the prolongation of the upper lip and nose. This animal, once widespread in distribution but now confined to the American continent and a part of Asia, has four toes on its fore limbs, three on the hind limbs.

The Rhinocerotidae, by contrast, are the largest living perissodactyls. Their huge, heavy bodies are enveloped in thick hide and the short, sturdy limbs are all three-toed. Today there are five species of rhinoceroses – the Indian and Javan (*Rhinoceros*), the Sumatran (*Dicerorhinus*), the black (*Diceros*) and the white (*Ceratotherium*). None are found in large numbers and they are in urgent need of protection.

CLASSIFICATION OF PERISSODACTYLA	
Suborder	Family
Hippomorpha	Equidae
Ceratomorpha	Tapiridae Rhinocerotidae

Facing page (left): Phylogenetic tree of the Perissodactyla. The greatest number of families, genera and species flourished during the Eocene, but only four families – the Chalicotheriidae, Equidae, Tapiridae and Rhinocerotidae – survived through the Oligocene and into the Pleistocene. The first of these became extinct before historic times. (*Right, top to bottom*) typical representatives of the three surviving families of perissodactyls – the kulan, the tapir and the rhinoceros.

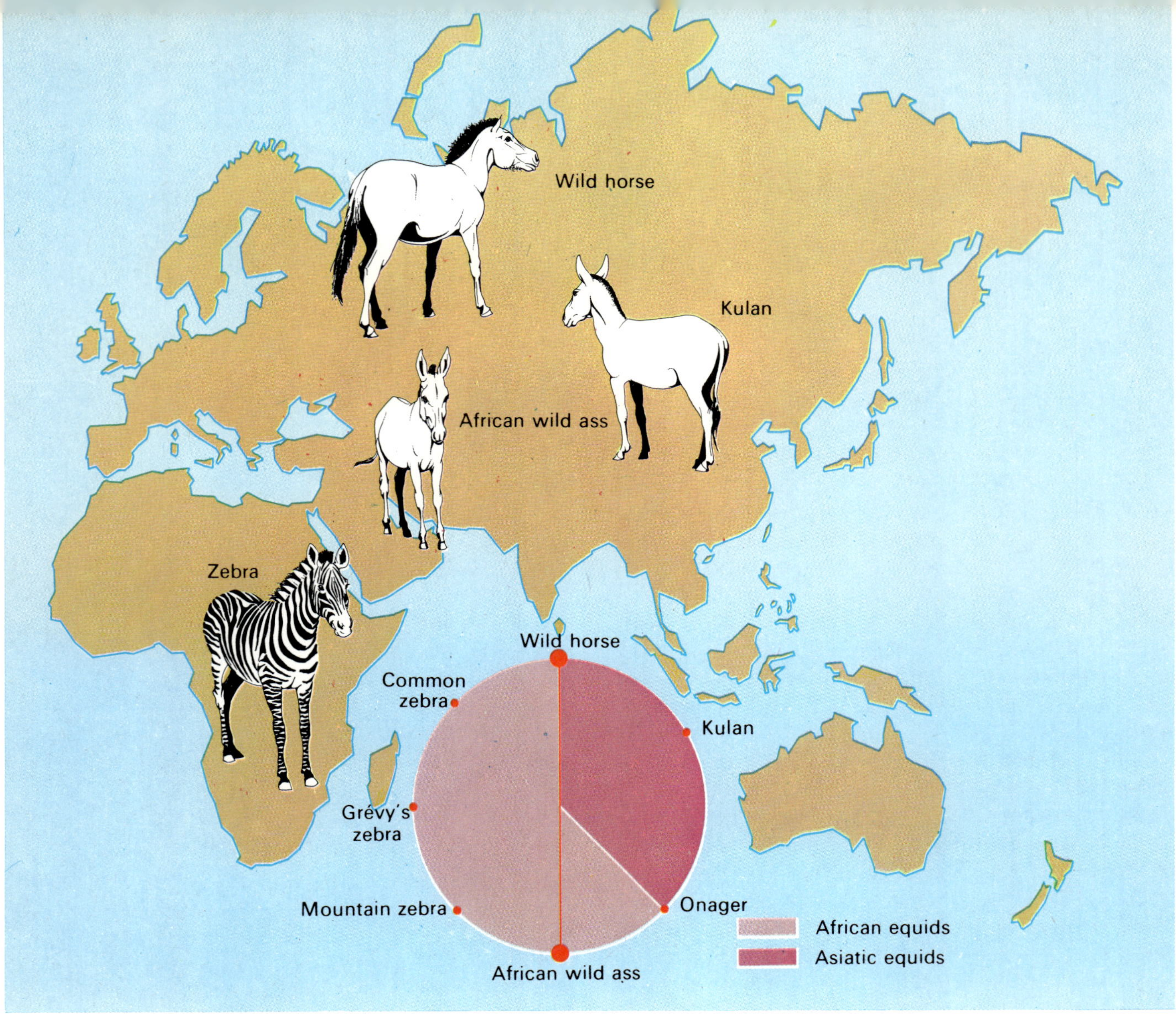

Wild horse

Kulan

African wild ass

Zebra

Wild horse

Common zebra

Kulan

Grévy's zebra

Mountain zebra

Onager

African wild ass

African equids
Asiatic equids

FAMILY: Equidae

Paleontologists, with the aid of fossil remains, have laboriously reconstructed the histories of long-extinct animals. Thanks to their amazingly precise work we are able to trace, almost without interruption, the extraordinary processes of evolution which led to the development of the modern horses, zebras and related species—creatures with marvellous powers of speed and endurance.

We have to go back almost 70 million years to see how it all began. At that remote period the 'dawn horse' *Eohippus* and the very similar *Hyracotherium* were browsing in the forests of North America and Europe respectively. These medium-sized mammals were the distant ancestors of modern horses and zebras. The crowns of their molars and premolars were low and rounded, unsuitable for cropping grass. They had four toes on the fore feet and three on the hind feet, all terminating in small hooves.

Hyracotherium, perfectly adapted to a forest existence, flourished and spread to many parts of Asia, although its numbers remained far higher in Europe, where the vegetation cover was ideal. There it gave rise to the family Palaeotheriidae, the best known genera of which were *Palaeotherium* and *Lophiotherium.* Both these creatures vanished from Eurasia about 36 million years ago, leaving only fossil remains.

The North American *Eohippus* was responsible for a more direct and continuous line of descent. About 30 million years after its original appearance during the Early Eocene, there emerged another animal known as *Mesohippus.* This creature of the Oligocene was not much larger than a lamb but its molars were provided with ridges of enamel. This made it possible to include grass in its diet although the low-crowned teeth were not yet ideal for cropping this form of vegetation. *Mesohippus* had lost two toes from its fore feet and of the remaining three, though all touched the ground, the middle toe was already evolving as the main supporting digit.

During the Miocene—about 25 million years ago—*Merychippus* made its appearance. This was a creature much better adapted to the life of the open plains—the type of terrain found on the African savannahs—and its size was approximately that of a modern pony. By now the outer toes had atrophied, leaving only the middle toe of each foot touching the ground. The crowns of the molar teeth, with their folded enamel ridges, were now higher, permitting the efficient cutting of grass into small pieces. The teeth were in fact similar to those of modern horses.

Cumulative rises and falls in the level of the earth's crust, causing shifts of continental masses, made it possible for groups of primitive American equids to cross the Bering Straits—by a convenient land link—and to set foot in Eurasia. A number of them, notably *Anchitherium* and *Hypohippus* during the Miocene and *Hipparion* during the Pliocene, migrated to the Eurasian steppes and prairies, only to become extinct. On the other hand *Pliohippus* continued to flourish in the New World and was found in large numbers on the North American plains about 10 million years ago.

In later representatives of the Equidae the two outer toes of the hind feet had dwindled into mere bones under the skin—as in modern horses—and it was from the sturdy *Pliohippus* that the genus *Equus* evolved. This genus is indisputably the direct ancestor of all modern horses, asses and zebras, with all their variations of shape, size, colour and habit. But curiously, although members of the genus adapted well to new conditions and spread all over America from north to south, they eventually disappeared entirely from the continent.

The fossil traces of primitive equids found in Tertiary layers in North America provided the paleontologists with many missing clues, enabling them to piece together the history of the Equidae in both the Old and New Worlds. Much emphasis was placed on the vital role played by the Bering

Facing page (above) : The distribution of wild equids (zebras, horses, asses, kulans, etc) is concentrated in Africa and Asia. Differences between the species of the two continents are slight. This chart, based on an original by Professor Bourdelle, shows how the equids may be classified according to whether they bear a closer anatomical resemblance to horses—as do the common zebra and the kulan—or to asses—as do the mountain zebra and the onager. *(Below)* herds of semi-wild horses are still found on the North American prairies, descendants of the mustangs, feral horses that were numerous in the 19th century. Although the Equidae evolved to a certain stage throughout the American continent they had disappeared when the first Europeans arrived. The Spaniards introduced domestic horses from the Old World, some of which wandered wild and became the ancestors of the mustangs.

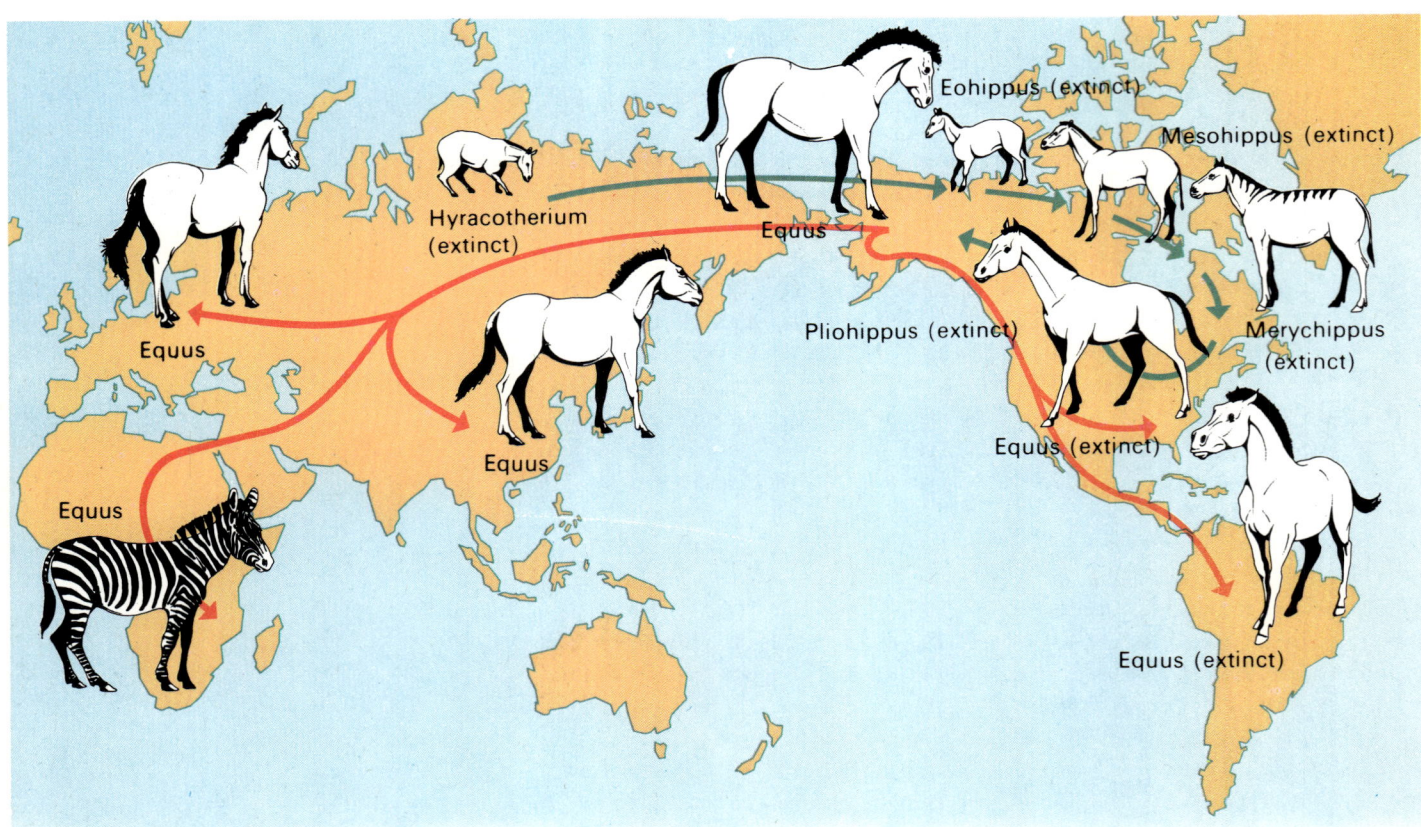

The Equidae, after becoming extinct in Eurasia, populated the American continent. From there they crossed via the Bering isthmus back into Asia, Africa and Europe, vanishing from North and South America. *(Right)* a chart illustrating features of equid evolution – skull formation, foot structure, dentition and diet. The dates are very approximate, indicating only when each period started.

			High crown		
PLEISTO-CENE 1,000,000 years	Equus	One toe	With cement		Grass
PLIOCENE 12,000,000 years	Pliohippus	One toe	High crown With cement		Grass
MIOCENE 29,000,000 years	Merychippus	Three toes	High crown With cement		Grass & leaves
OLIGOCENE 40,000,000 years	Mesohippus	Three toes	Low crown Without cement		Leaves & grass
EOCENE 70,000,000 years	Eohippus	Four toes	Low crown Without cement		Leaves

isthmus – whose level varied at different times – in determining their subsequent evolution and distribution.

The various representatives of the Equidae were always singled out as the special prey of carnivores and it was partly because of the activities of predators that certain weaker species failed to survive. Others, however, were better able to cope with the situation as progressively streamlined bodies and longer limbs made them faster moving. Whereas their primitive ancestors had possessed four toes and three toes respectively on fore and hind feet they gradually dispensed with superfluous digits and this adaptation too made for greater speed.

Modern equids are virtually one-toed animals. The first and fifth toes are completely missing and the second and fourth have atrophied to the point where they survive nowadays only as splint bones. The whole weight of the body is carried on the remaining middle toe and the end of the foot is encased in a protective sheath or hoof. The hooves are very strong yet comparatively supple – perfect for running over hard terrain or for use as a weapon when the animal lashes out in self-defence. Even a domestic species, such as a horse or a mule, is capable of causing serious injury if it kicks out in anger or alarm.

Endurance and speed are also guaranteed by a powerful combination of muscles and tendons – especially in the legs. Additionally all equids possess extremely acute senses of smell and hearing, essential for detecting the presence or approach of enemies under cover, zebras in the wild being particularly reliant on such perceptions.

Their dentition is perfectly adapted to a diet which is now largely herbivorous. Adults have between 36 and 42 teeth. Six incisors in each jaw form a kind of pincers, ideal for cropping grass, and are separated by a wide space or bar from the cheek teeth. Canines are, as a general rule, found only in males. The premolars and molars – 12 in the lower and 14 in the upper jaw – are similar to each other, high-crowned and furnished with deep ridges, admirably suited to chewing fibrous vegetable matter.

All the wild members of the Equidae live on grassy plains and need no further supplement to their diet, though most domestic breeds of horse have gradually become accustomed to a granivorous diet. It is grass that has conditioned their digestive structure. The stomach is notable for the very long caecum – its liquid capacity being more than 17 pints – and it is here that the cellulose content of the ingested vegetable matter is decomposed by the action of the microscopically small protozoa and bacteria which are present in the intestines.

The Equidae are all gregarious animals living in herds and the various species of horses, asses, kulans and zebras are grouped together in the single genus *Equus*.

Horses and foals galloping in a meadow. The horse is both friend and helper of man, who has been instrumental in developing many different breeds.

CHAPTER 4

Buffalo, topi and kongoni

As the plains of the Serengeti begin to lose their green carpet of grass we follow in the footsteps of the migrating herds. Continuing northwards we find the vegetation pattern changing. As we come across greater numbers of umbrella-like acacias — those familiar trees of the African landscape known by the scientific name *Acacia tortilis* — the grass becomes progressively taller. Among the stalks, golden as summer corn, there are glimpses of strange, rather conspicuous animals, whose handsome, satiny coats take on yellow, red and even bluish tints — changing with the angle of the sunlight. Their withers are higher than the rump and since their feet are hidden they give the illusion of standing on pedestals. Zoologists have given this interesting antelope the somewhat unflattering name of bastard hartebeest, but the local appellation of topi is both more convenient and more attractive. Not far away we spot another group of grazing antelopes, light sandy-brown in colour. These are Coke's hartebeeste, otherwise known as kongonis, and their fore- and hind-quarters seem even more disproportioned than those of the topis.

We shall not stop for the time being to look more closely at these strange herbivores for our destination is the northern part of the reserve, on the banks of the Mara river. It is here that the African or Cape buffalo — reputed to be the most aggressive of herbivores — is found. But we are grouping together these dissimilar kinds of animal — topi, kongoni and buffalo — because all of them feed exclusively on grass and inhabit the same type of savannah terrain. Furthermore they have a common enemy — the lion.

The journey of almost 200 miles across the Serengeti has now brought us to the most northerly fringes of the park, where the

Facing page : A line of African porters carrying buffalo skulls, their massive curving horns still intact. The African buffalo was the tragic victim of the white man's greed. In the mid-19th century hunters, dealers and settlers combined to decimate the herds and cattle disease took a further toll. Laws to restrict hunting, as well as the natural prolificity of the species, averted total disaster and today there are herds of over a thousand animals in protected areas.

An African buffalo emerges from a muddy pool. Wallowing in the mud is not merely a pleasant way of keeping cool but also an effective method of getting rid of the hordes of irritating parasites that lodge in the skin. The mud, as it dries, forms a crust in which the ticks are imprisoned and they fall off when the buffalo rubs itself clean against a tree.

scenery is again different from that previously encountered. The grass in the hollows is even taller here and the plain is dotted with miniature acacias whose branches are swollen by rounded, dark, acorn-like growths. Each of these protuberances is armed with three or four hard, pointed spines. The peculiar whistling of the wind in the neighbourhood of these trees is caused by the vibrations of these spines and the air rushing into the openings of the peculiar growths. It is unwise to try to break off a branch for not only are the thorns unpleasantly sharp but they are liable to be swarming with colonies of ants of the *Crematogaster* species, whose bites are extremely painful. This is an interesting example of a plant-insect relationship, formed for mutual convenience and defence. The acacia's thorns help to protect the ants' larvae in their nests and the ants in their turn ward off any voracious insects who might otherwise destroy the tree's foliage. Moreover, the combination of thorns and ants is a partial deterrent to leaf eaters such as the rhinoceros and giraffe, though the latter animals still succeed in inflicting some damage to the tree when they feed.

Along the banks of the Mara other enormous trees—fig, mahogany and yellow wood—form as dense and impenetrable a natural barrier as the acacias. It is in the shade of this gallery forest—wallowing in the muddy pools and puddles formed during the rainy season—that we find the largest concentrations of buffalo in the whole of East Africa.

A myth exploded

A large herd of these buffaloes is an unforgettable sight. The huge black cattle stand motionless, as if carved from stone. When they lift their flat muzzles and fix their unflinching gaze on an intruder they make even the fiercest domestic bulls look weaklings by comparison. They are especially dangerous when they attack for they charge with head raised, so as to retain both the sight and smell of an enemy.

Some African animals are noteworthy for their size, such as elephants, some—like lions—for their courage, and others—such as the gazelles—for their incomparable grace and beauty. But for those who have made an intensive study of the African fauna or whose experience may be confined to a safari, there is one animal which has assumed almost legendary status as a result of its allegedly aggressive nature—the massive buffalo. No other herbivore has been credited with inflicting so many casualties on its foes—lion and human alike. It has a fearsome reputation.

Our attitude towards this animal has probably been influenced by hunting anecdotes, leading us to suppose that we are dealing with a truly ferocious beast, predisposed to charge blindly at any creature crossing its path and as dangerous to its own kind as to its enemies. Yet sober and objective surveys by naturalists—undertaken in tranquil surroundings and conditions—have revealed a very different picture. This powerful bovine turns out to be a surprisingly placid creature, admittedly well endowed by nature to deal with any enemy on two or four legs, but—when left to itself—sociably and peacefully inclined.

African buffaloes display extreme variations in colour of coat and span of horns. Their size too depends on habitat. The enormous Cape buffalo *(Syncerus caffer caffer)* is a creature of the southern part of the continent while the dwarf or forest buffalo *(Syncerus caffer nanus)*, small by comparison, is an inhabitant of equatorial Africa.

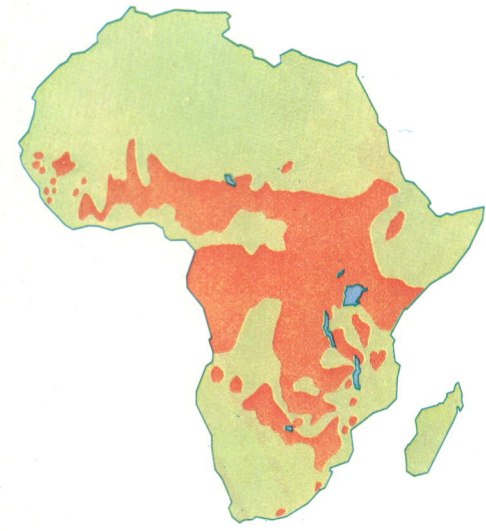

Geographical distribution of the African buffalo.

The African buffalo

No African mammal seems to have adapted itself better to its natural surroundings than the buffalo. It may be encountered in a variety of habitats—forest, swamp, wooded mountain country, pasture land, scrub-covered steppe or open savannah. The only prerequisite is that there should be enough food and water and that man should intrude only on rare occasions. Because of the buffalo's ability to live comfortably in so many contrasting types of terrain, its distribution ranges over nearly the entire continent south of latitude 15°N (the Sahara Desert). And as a result of this wide distribution it shows considerable diversity in size, colour and formation of horns. The little reddish-brown dwarf or forest buffalo (*Syncerus caffer nanus*) is little more than 3 feet in height, weighs about 450 lb and has vertical horns measuring some 12 inches long. By contrast, the enormous black Cape buffalo (*Syncerus caffer caffer*) stands over 5 feet tall at the shoulder, weighs anything between three-quarters of a ton and a ton, and has immense curving horns that measure up to 58 inches from tip to tip.

It was formerly believed that there were a number of different species of African buffalo but it is now recognised that there are merely two subspecies of *Syncerus caffer*. Comparative studies have shown that, as in breeds of domestic cattle, the reddish coat colour corresponds to that of the buffaloes' primitive forebears, whereas the black is a specialised departure from the original.

The solitary bulls

It is now time to take a closer look at a typical herd of African buffaloes—from a discreetly safe distance of, say, 50 yards. The bulk of the herd consists of females with their young, together with males that have just reached maturity. There may be as few as 150 animals but in Serengeti it is not uncommon to find as many as a thousand buffaloes herding together. Smaller groups of 20 or 30 animals are also frequently seen, but their community structure is less cohesive.

Among the animals, which stand and gaze at us with more curiosity than animosity, it is usually possible to pick out an impressive-looking male grazing among the tall acacias some distance away from the main herd. This is the solitary bull—something of a living legend. Should we be fortunate enough to observe him at close quarters, perhaps newly emerged from a mud bath, he would look like a massive black statue, hewn from a block of basalt, against the brilliant green of the savannah background. This magnificent old bull is power personified, strong but not top-heavy, built on such a colossal scale that he can defend himself where he stands against predators, without having to run away from them.

The larger a herbivore happens to be, the more likely it is to attract the attention of carnivores as a valuable source of protein. But if the animal in question is fleet-footed enough to outrun its pursuer or strong enough to challenge it to a fight, the

predator will use up a large amount of extra energy and its inclination and capacity for food will be significantly reduced. The buffalo of course comes into the second category. With its powerful muscles, solid build and horns which, as with most antelopes, are not mere ornaments but also effective weapons, it possesses all the defensive apparatus necessary for the survival of the species. Obviously its behaviour is conditioned by its physical structure and it will not shy from direct attack. This is an instinctive reaction in the face of danger, particularly when cornered or wounded by an enemy closing in for the kill. But this does not mean to say that it will charge automatically and without provocation. As often as not it will take the prudent course and flee. Only when this proves impossible or when every attempt to reach shelter and safety has failed will it turn to the attack.

The relationship between these solitary males and the rest of the herd is absolutely normal and harmonious. When a male buffalo has reached his maximum physical development (by which time he may stand 5 feet high and weigh almost a ton) he is a truly imposing creature but too large to follow the other animals in the herd as they go about their daily activities. So he adopts his own special life pattern, confining himself to a piece of territory where there is enough grass for nocturnal grazing, sufficient shade for sheltering from the cruel mid-day sun and, if possible, provided with a watering place and muddy pool where he can lie for hours and get rid of irritating parasites. Only contingencies such as drought or imminent danger will force him to move away from his territory. So it seems that the chief reasons for his deciding to lead a solitary existence are his size and weight, which are not conducive to the kind of restless

Two adult male buffaloes clash in single combat. These ferocious fights, spectacular but rarely fatal, may have a female or a piece of territory at stake but also serve as outlets for aggressive feelings and release of nervous tension.

activity preferred by the more agile, highly-strung cows and calves. It is highly improbable that it should be dictated by any anti-social feelings.

The solitary bull will have been born and reared in the heart of the herd, protected from daily dangers by his mother, by the other cows and even by adult males who, when conveniently close by, take on the responsibility of defending the community at large. When old enough to cope for himself he will have sought out the company of males of the same age, dispensing with the protective female attentions and gradually becoming more and more sedentary in habit. The final stage comes when he attains the full power and majesty of adulthood, cutting himself off from the other animals to take up solitary residence in his own corner of savannah.

The solitary bull always chooses a portion of ground as near as possible to the track used every day by his former herd as they go to and fro between their pastures and communal water sources. Should a group of roving young males cross his territory he will show no signs of aggressiveness but will graze quietly with them. If he runs short of water he will journey twice a day to the drinking places patronised by the rest of the herd, accompanied by the owners of adjoining pieces of territory. So it is not uncommon to find groups of three or four such males tramping the plain or bathing together in a muddy pool.

Follow the leader

It used to be taken for granted that buffaloes automatically obeyed a single leader but apparently this is not the case. They tend to follow the individual who happens to be best acquainted with a particular locality. It may be a male, but more often it is an adult female, simply because the cows remain with the herd whereas the bulls leave it at a certain age. When the herd, in their search for a water-hole, cross the boundaries of a solitary bull's territory, the latter will advance to meet them, placing himself at their head and guiding them over his home ground. He will subsequently hand them over to a neighbour who in his turn takes over the role of guide in his terrain. Finally the herd will be passed on to the safe keeping of the bull whose territory lies closest to the watering place, he too joining their ranks.

Looking at these huge solitary bulls, it is hard to imagine them being motivated by any feelings of sociability or co-operation. Yet their sense of communal responsibility is astonishingly highly developed. These enormous beasts, with their lazy, measured movements, are alert to the least sign of danger and not slow to face it. If the wind is in the right direction they can detect the scent of an enemy as much as 500 yards away. In such cases the guiding bull will suddenly 'freeze', this serving as a signal to the others. The herd will then show obvious signs of disquiet, turning themselves face to the wind, raising their heads to sniff the distant scent, taking a few hesitant steps forward and then stopping to scan the horizon. The females remain some distance behind. Should they decide to take to their heels, the

Facing page (above) : The summit of Mt Kilimanjaro, rising above the clouds, forms a picturesque background to this shot of three male buffaloes in the acacia-studded savannah of Kenya's Masai Amboseli game reserve. *(Below)* a group of buffaloes raise their heads in the presence of a stranger. They will probably resume their peaceful grazing when the stranger has gone. Alternatively they may take flight or, if danger is imminent, charge to the attack.

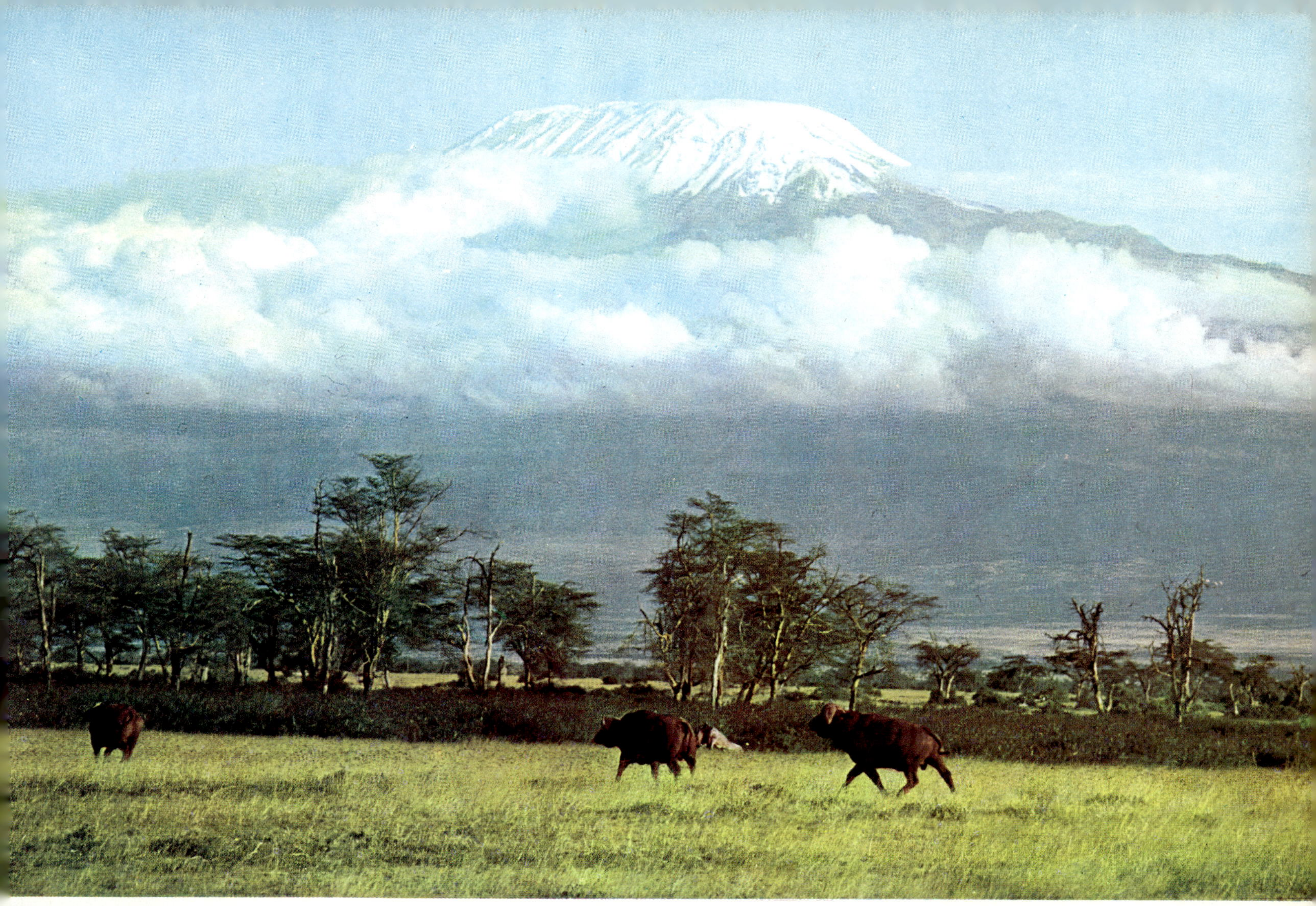

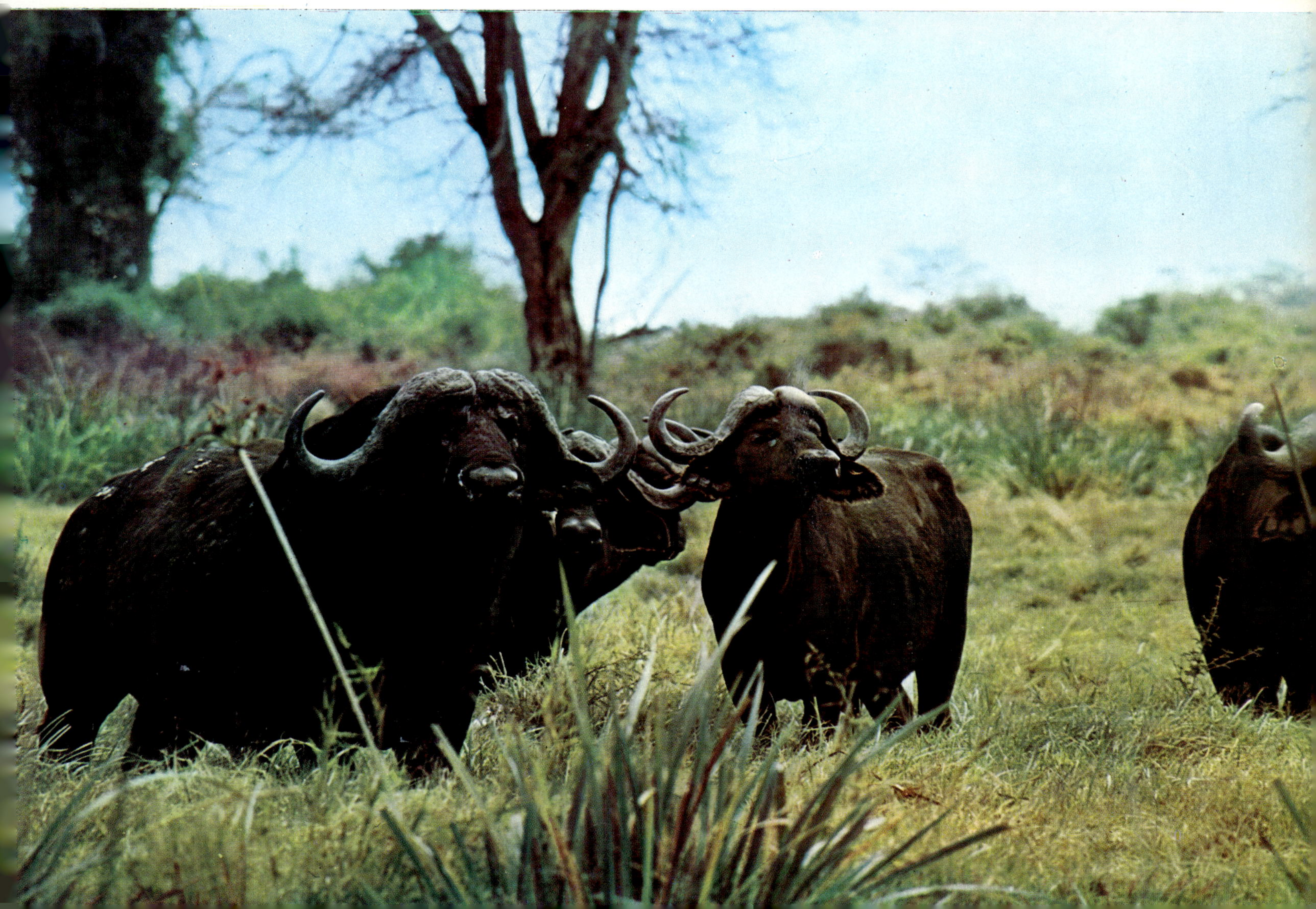

bulls will be the last to flee—instinctively safeguarding the cows and calves.

The mere presence of an enemy or stranger does not necessarily produce a general panic and massed withdrawal. Sometimes the females will retreat a few yards while the males, tense and edgy from their efforts to identify and pinpoint the danger, will start fighting furiously among themselves. The main purpose of this unexpected activity is to relieve nervous tension and give expression to pent-up feelings of aggression. It is a form of defence mechanism common to many living creatures, and indeed a familiar aspect of human behaviour as well. This kind of 'substitute' action may be seen when a person who is speechless with anger hammers his fists against a table top, stamps on the floor or utters a violent swearword. These are simply indirect methods of resolving an internal conflict, avoiding direct confrontation with an adversary—probably undesired—which might result in serious hurt to either party. Animals, which are naturally incapable of reacting in this way, have no alternative but to engage in actual combat.

Battle of the bulls

One such furious contest between two bull buffaloes was witnessed and described by a member of a film crew. The spectacular fight was doubtless sparked off by the presence of the cameramen, although they had taken normal precautions to conceal their whereabouts.

'We were half-hidden behind a thick clump of bushes, our cameras at the ready. The nearer the herd approached, the more nervous they seemed to become. There was hardly a breath of wind in this part of the deep Rift Valley yet, thanks to their remarkable powers of smell, the buffaloes had apparently scented us at a distance of about 250 yards.

'Displaying understandable caution, the enormous beasts stopped dead. As soon as their leader raised his muzzle in warning the animals began herding together. We sat there, motionless, holding our breath. All of a sudden, the females and young, who had been grouped in the centre of the herd, beat a retreat, while the males began to show renewed signs of tension and alarm. Then two of the bulls, without any warning and for no evident reason, closed with each other in a spectacular duel.

'Standing face to face at a distance of no more than ten yards, they began pounding the ground with their fore feet, breathing noisily and swinging their heavy heads in menacing fashion. The enormous, curving horns glinted in the sun. Then they both hurled themselves forward, clashing their heads so violently that they seemed momentarily stunned. They retreated several yards and then charged once more, repeating the process over and over again. From our hiding place we could clearly hear the sounds of battle shattering the morning peace of the empty plain.

'In the course of the long and furious contest one of the bulls was clearly getting the worse of it. He was steadily giving ground and his opponent was driving him back towards a barrier of

A buffalo may wallow for hours in its mud bath, only its head showing above the surface.

Facing page : These dwarf or forest buffaloes, seen in their natural habitat in an equatorial forest, are smaller than their Cape buffalo relatives. Reddish-brown rather than black, they have shorter horns which are not joined on the forehead as are those of the huge buffaloes from the south.

The lion, although the buffalo's most dangerous predator, will only tackle the massive herbivore as a last resort. Fights between lions and buffaloes are often of unparalleled ferocity.

bushes. In the meantime the herd appeared to be quite hypnotised by the sight and had completely lost interest in us. Eventually the weaker animal, shaken by another unbelievably violent charge, became absolutely still. His head dropped, his limbs were trembling and he was utterly at the mercy of his rival. We were unable to decide whether these were standard attitudes signifying surrender—similar to those employed by many animals who engage in such ritual combat—but from that moment the winner did not aim another blow at his conquered opponent, merely thudding his hooves against the ground and proudly tossing his head, horns glittering like the helmet of some medieval knight. When the two rivals had trotted off to join the rest of the herd we saw that the ground had been churned up as if by a bulldozer.'

Not every fight between adult male buffaloes ends so peacefully. Sometimes—and especially at a time when the females happen to be on heat—one of the combatants may be severely wounded or even killed outright. In most cases, however, the fights end with the loser beating a prudent retreat and leaving the winner in sole possession of the battlefield.

The mating season

Although buffaloes customarily mate at any season of the year, they very often choose a period towards the end of the rainy season. The process appears to be in some way connected with the protein-rich grass then growing so plentifully.

We have seen how the breeding cycle of gnus is closely linked with the activities of their predators—particularly spotted hyenas—which attack the calves. Buffaloes, however, have little to fear from such enemies and their breeding pattern seems to be directly related to the nutritive content of the grass on which they feed. The female's appetite increases throughout the gestation period until about two months prior to the birth. The quantity of ingested food then shows a marked decline, doubtless due to the growing fetus.

Gestation lasts about eleven months and as with all ungulates, whose pelvic cavity is small in comparison with the size of the fetus, the birth is slow and painful—unlike that of carnivores, who find the process easier. The fore limbs are the first to appear, followed by the head, and the expulsion of the fetus may take 30 minutes or more. The newborn calf, if male, weighs 120–130 lb, females being somewhat lighter. The calves are able to stand about ten minutes after birth. During the time she is suckling them the female will eat a larger quantity of food in order to ensure an adequate supply of milk.

Because of the size and weight of newborn buffalo calves, they are virtually immune to the attacks of predators such as hyenas, though lions may pose a threat.

It seems probable that the process of natural selection has tended to concentrate the majority of buffalo births in the latter part of the rainy season—a period when the grass is not only more plentiful but also possesses a higher than average nutritive content.

The buffalo's enemies

Apart from man himself, who hunts it for sport or profit, the buffalo has to contend with only one serious enemy – the lion. But even the redoubtable 'king of beasts' will consider carefully – making certain that all the odds are in its favour – before tackling this most formidable of herbivores. If there are plenty of gnus, zebras and antelopes around, the lion will leave buffaloes alone, knowing from experience that they cannot be relied upon to run away but are just as likely to band together in face of an attack. The bulls are more than a match for the hyenas and leopards which would otherwise cause havoc among the calves, and a cornered buffalo will not yield ground to any carnivore, however powerful.

It was once thought that buffaloes, having scented danger, would come together to form a kind of defensive 'square', with the cows and calves in the centre and the bulls strategically placed around the perimeter. This is not quite true, but it is well established that buffaloes will co-operate and come to one another's aid when imperilled. Surveys carried out in Tsavo National Park – as described by Mervyn Cowie – have illustrated this point.

In 1960 a group of rangers encamped at Sala, on the banks of the Galana river, had an unusual experience. Four male buffaloes happened to live in the particular area patrolled daily by the rangers and these animals were quite accustomed to the human presence. One night a band of nine or ten lions invaded the district and killed one of the buffaloes. Next morning the rangers came across the lions devouring their victim. The second buffalo had prudently abandoned the area while the remaining two bulls – a young and an old one – had resumed their everyday activities as if nothing had occurred.

About a week later the rangers saw a young lion rashly attacking the younger buffalo, succeeding in flinging it to the ground.

Although buffaloes have a reputation for being aggressive they are as likely to flee as to attack when danger threatens. At any time, however, these massive animals give an impression of tremendous power and are not to be trifled with.

Before it could sink its fangs into the victim's throat, however, the larger buffalo had charged and, with a single blow from its massive head and horns, had sent the lion hurtling through the air, ripping its side open against a tree stump. In the meantime the felled buffalo, still bellowing with fright and rage, had found its feet and now the two avenging bulls joined in pursuit of the lion which was trying to limp painfully away. Wounded and cornered, the frustrated carnivore crouched in the long grass, snarling defiance at the two buffaloes as they closed in for the kill. Then, with a desperate leap, the lion shot past them and found temporary refuge behind a clump of bushes. The buffaloes wheeled and followed, ripping away the branches with hooves and horns, forcing the lion from cover and back on to open ground. Every time it attempted to make another sudden dart for shelter the implacable bulls used similar tactics, returning to the charge again and again. The lion's situation seemed hopeless.

At that point the lion caught a glimpse of two of the rangers who had been watching the progress of the battle from the high branches of a nearby tree. It let out a terrible roar of alarm and the buffaloes, momentarily distracted, lifted their heads. The brief lapse of concentration was enough to save the lion's life. In a flash it had bounded off towards the river and was soon safely across to the opposite bank. The buffaloes, refusing to admit defeat, charged away in pursuit but the culprit had vanished, leaving no traces. Instead, the enraged bulls stumbled across the other lions taking their ease in the shade of some nearby acacias. One lion being much like another, the indomitable pair hurled themselves at the peaceful group, scattering them in all directions. The rangers too judged it a convenient moment to make themselves scarce, not waiting to discover how the story eventually ended.

This eye-witness account proves that buffaloes do indeed come to one another's assistance when attacked or otherwise endangered. It also testifies to the cunning and tenacious methods they will employ to track down and punish an enemy. Hunters and zoologists are well aware of this tendency but take comfort from the fact that they will normally only attack in self-defence. Courageous the animal certainly is, but it is an over-simplification of its complex behaviour to label it as cruel or ferocious.

Disease and parasites

The most persistent and dangerous enemy of the buffalo is in fact disease. It is especially prone to the cattle plague known as murrain, probably introduced into Africa by domestic cattle. The epidemic that swept the continent at the end of the 19th century almost wiped out the wild buffalo herds. That they survived was due to their natural adaptability and wide distribution range, a high breeding rate, preventive vaccination measures and legislation to restrict hunting. As a result there are today good-sized herds of more than a thousand animals in a number of national parks.

An oxpecker perched on the head of a buffalo. These small birds are invaluable allies of the huge bovines, ridding them of parasites and giving warning of danger.

Less dangerous, but a constant source of irritation, are the hordes of ticks, fleas, flies and mosquitoes that embed themselves in the skin or plague the huge beasts remorselessly wherever they go. Measures to rid themselves of such insect pests can at best be makeshift and temporary.

The celebrated Treetops Hotel in Kenya, so named because it is literally built among the branches of an enormous tree, is a wonderful vantage point for observers and photographers of African wildlife. Elephants, rhinoceroses and antelopes troop down to drink in the floodlighted pool barely 30 yards from the hotel terrace. Here, at dusk, an enormous buffalo will sometimes wander down to take a dip before the other animals arrive. It will plunge its massive body repeatedly into the mud—a leisurely process that may take more than two hours—until it is completely covered with a coating of red clay. In the rays of the setting sun it looks like a great bronze statue. Other buffaloes meanwhile will tentatively put their feet into the water, though none will venture in too deeply or for too long.

There is a practical purpose in this solitary bull's bathing ritual—to get rid of the ticks and fleas lodged on the skin and to keep off noxious flying insects. The drying crust of mud entraps the parasites which are then removed as the buffalo rubs itself against a tree trunk.

A more effective and less tiring method of ridding the great ungulates of their skin pests is undertaken by certain birds, particularly the oxpecker *(Buphagus africanus)*. These birds, about the size of starlings, perch on the buffaloes' backs, hang below their bellies or delve into their ears, as agile and sure-footed as woodpeckers on a tree. Not only do they pick off the ticks, but they also cleanse, as efficiently as any surgeon, the skin sores and abscesses caused by the parasites.

Another bird performing a similar service is the buff-backed heron or cattle egret *(Bubulcus ibis)*. In addition to pecking parasites from the skin of buffaloes and other ungulates, and feeding

The buff-backed heron or cattle egret is another valued companion of the buffalo, quick to give the alarm when predators are in the vicinity and warding off flies and other troublesome insect pests.

The bastard hartebeest, closely related to the hartebeest, is strangely shaped, with a steeply sloping back. The two genera differ in coat colour and horn formation.

Facing page : Hartebeeste and bastard hartebeeste are keenly aware of danger and speed is their best means of defence. The hartebeeste may be distinguished by their high, more noticeably hooked horns.

on the insects dislodged by their trampling hooves, these birds act as sentinels, warning the herds of imminent danger. When they take to the air in their hundreds, flapping their large white wings, the herbivores are well advised to take heed of the alarm signal.

In many African villages, particularly those in the neighbourhood of reserves and parks, buffaloes may sometimes be seen lying near the huts while chickens peck round them and even clamber on to their backs, just as the oxpeckers and buff-backed herons do on the open savannah. In fact experiments are currently going on to see whether this animal can be domesticated, in the belief that its nature is basically peaceful and friendly. Should these experiments succeed it will be of inestimable benefit to the local and national economy for it has already been demonstrated that buffaloes make far more efficient use of grassland than domestic cattle, that they are more resistant to disease and that the quality of their meat is superior to the beef of ordinary cattle.

The bastard hartebeeste

Several races of antelope found in savannah country all over Africa are grouped together under the generic name *Damaliscus*. Some of these so-called bastard hartebeeste are migratory animals, others sedentary. Like the hartebeeste proper (genus *Alcelaphus*) they have long faces, prominently ringed horns and withers that are higher than their rump. With their reddish-brown coats—which in certain lights give off a bluish or purplish sheen—they are very conspicuous in the green grass, and the pronounced slope of their back (even more marked than that of the hartebeeste) makes them easy to identify.

This unusual body shape, which is hardly likely to have come about accidentally but rather as a natural adaptation to the environment during the course of evolution, stands the animal in good stead against predators. The high angle of the head and relatively high, lateral positioning of the eyes provides them with a wide field of vision—much greater, for example, than that of a grazing gnu. Although the quality of their vision is not remarkable—their sense of hearing is far more highly developed—they are much quicker than the larger herbivores to spot a lion lurking in the long grass and that much prompter to make their escape. Since they are also fast runners, with considerable staying powers, they are less vulnerable to sudden attack than gnus and zebras. The horns are useful supplementary weapons in an emergency. These are not as markedly curved as those of hartebeeste, rising more or less vertically and separately from the head, whereas those of *Alcelaphus* are more closely joined and grow from a bony pedicle at the top of the skull.

'Operation Hunter'

One of the rarest of East African antelopes, Hunter's hartebeest (*Damaliscus hunteri*) lives in a narrow ribbon of territory, about 60 square miles in area, bounded on the south by the Tana river

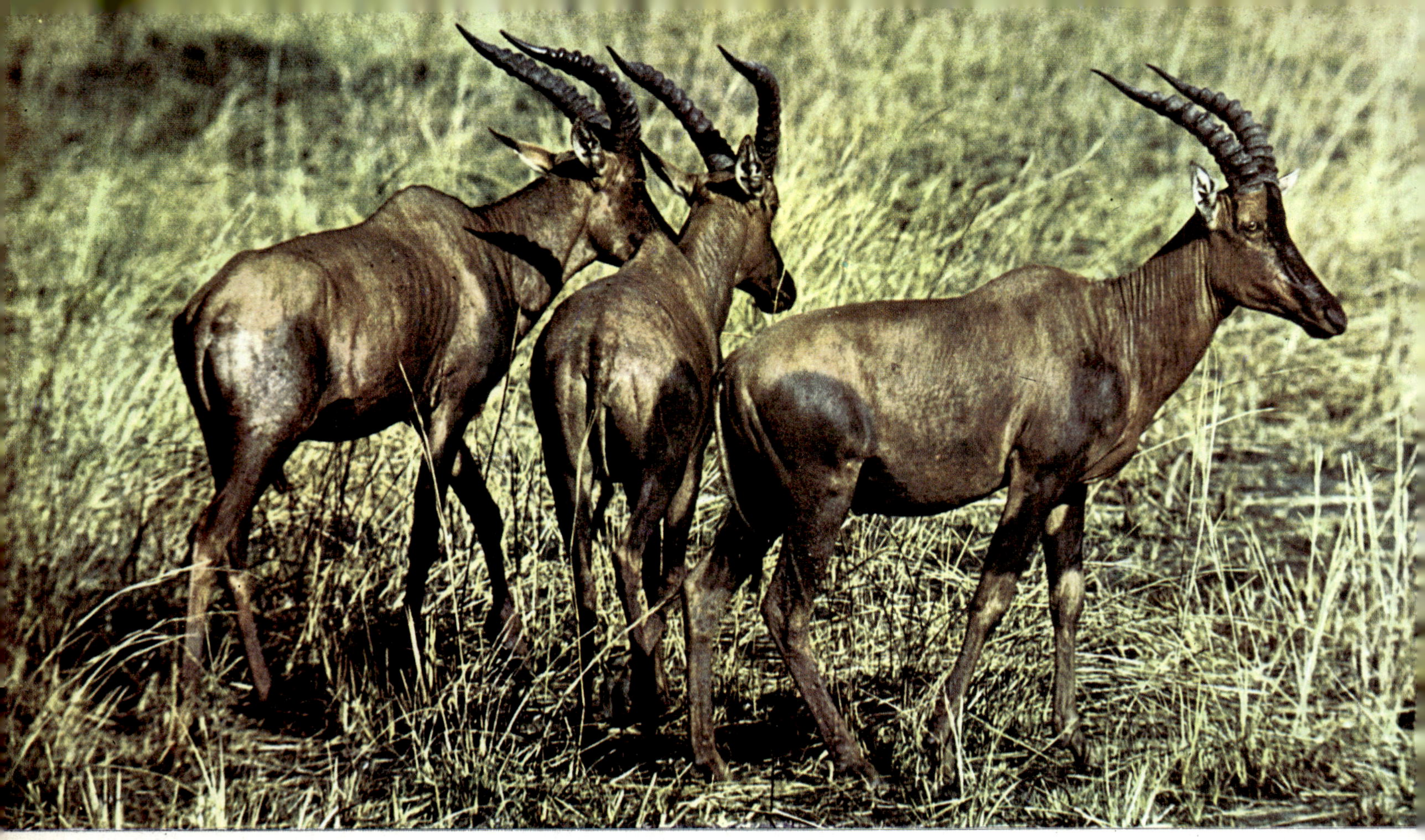
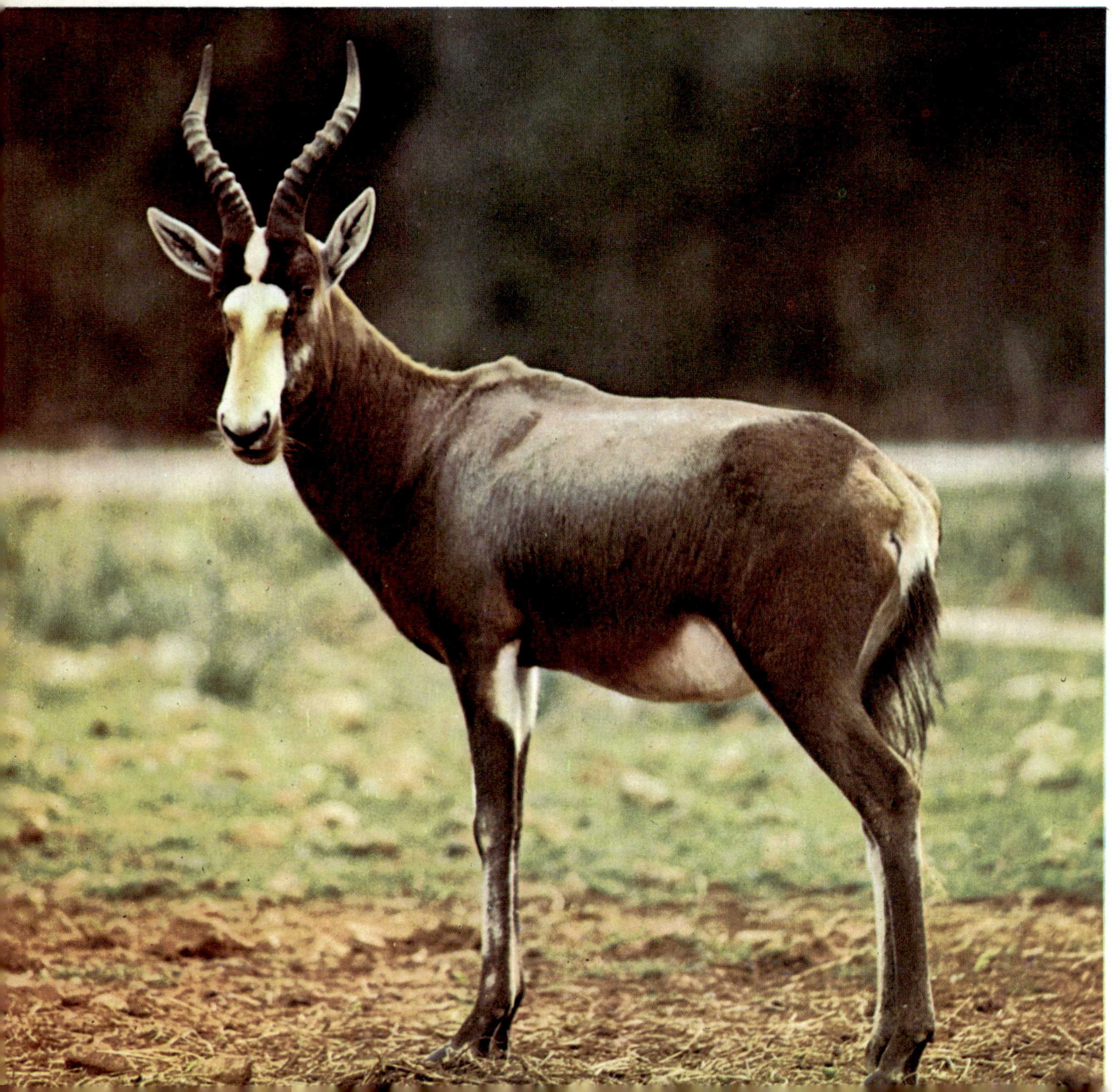

in Kenya and on the north by the Juba river in Somalia. The terrain is a cross between desert and savannah.

Because of its restricted habitat—as well as for political reasons—life has always been rather precarious for this antelope with the white chevron pattern on its forehead. Few are now left in Somalia and about 1,500 animals on the Kenya side of the frontier—the last remaining examples of the species—were in 1962 threatened with extinction. The crisis was the result of a decision by the United Nations' Special Fund for Economic Development to send a survey team to Kenya to examine the feasibility of an irrigation scheme in the Tana river valley. This would entail flooding the area inhabited by the Hunter's hartebeest and naturalists were quick to express their concern.

Work on the irrigation project was not due to commence for three years, giving ample time—so it seemed—for experts to make emergency plans for a rescue operation. The first step was to despatch a team from the Wildlife Society to study every aspect of the animals' life and behaviour. It was particularly important to know exactly what type of grass they fed on and the minimum acreage of grassland necessary to support the existing population. At the same time plans were initiated to transport as many animals as possible to another game reserve or national park in Africa in order to ensure the survival of the threatened species.

Internal political and economic problems in the newly independent republic of Kenya added to the difficulties, with the ecologists unable to move about freely in the sensitive frontier area. But the conservationists were much encouraged by the offer of an anonymous American benefactor, transmitted through the World Wildlife Fund, to underwrite the expense of transporting the animals to their new home.

The site chosen was a plain in Kenya's Tsavo National Park, well provided with the type of vegetation favoured by the antelopes. It was decided to transfer a small nucleus of animals for breeding purposes as soon as possible, placing them for a few days in an enclosure in order to get them acclimatised, then setting them loose in their new grazing grounds. It was simply a matter of rounding the animals up and transporting them to their destination.

This proved a more complicated business than had been imagined. The beginning of August was selected as the most opportune time to carry out Operation Hunter so that the animals could be released at the start of the rainy season. Unfortunately the Kenyan authorities had forbidden any camping in the sector of the Wallu plains, north of the Tana river, where the antelopes were being herded together. The distance from here to the projected reception enclosure was 50 miles as the crow flies, but considerably farther by road because of the absence of bridges over the river. The antelopes could not be expected to survive such a journey.

The British army came up with what looked like the solution, building a pontoon of empty petrol drums over the river, solid enough to take the weight of a small truck. The vehicle to be used had been specially constructed for a similar wildlife rescue programme in Arabia known as Operation Oryx. It was simple enough

The blesboks and bonteboks of the species *Damaliscus dorcas* are South African animals, with white facial markings. This individual is a bontebok, *Damaliscus dorcas dorcas*.

Facing page (above): Three bastard hartebeeste of *Damaliscus lunatus*, identifiable by their brown coats with darker patches. *(Below)*: an interesting variant, or a true hybrid? This animal has a bontebok face, but the body colouring of a blesbok.

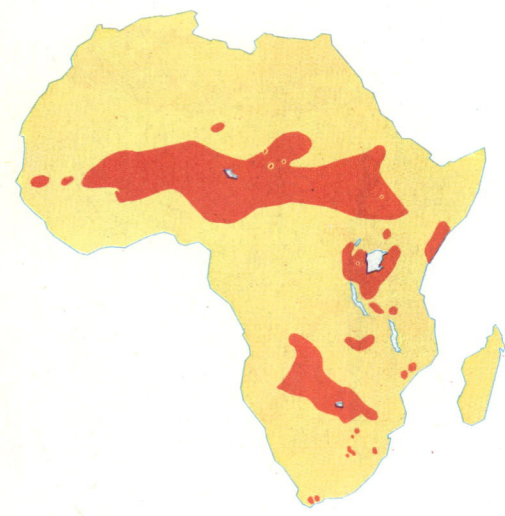

Geographical distribution of bastard hartebeeste.

TOPI
(Damaliscus lunatus topi)

Class: Mammalia
Order: Artiodactyla
Family: Bovidae
Length of head and body: 60-82 inches (150-205 cm)
Height to shoulder: 48-50 inches (120-125 cm)
Weight: 310-330 lb (140-150 kg)
Diet: Grass
Gestation: 225-300 days
Number of young: one, rarely two

Adults
Head long, with characteristic blackish mark between eyes extending down to mouth. The horns, prominently ringed and roughly lyre-shaped, are short (maximum recorded length 20 inches) but solid, curving gently backwards along their entire length, the tips pointing forwards. The line of the back slopes from withers to rump. The narrow ears are pointed. Coat reddish with blue or purple sheen, and dark blue bands on haunches and upper part of hind legs, becoming yellowish-white towards ankles. Tail slender, terminating in tufts of black hairs.

Young
No horns. Sandy-brown coat, like that of young deer.

---

to catch the two adult male antelopes selected to initiate the experiment, but both animals proved too delicate to survive the 150-mile journey and died shortly after arrival. Following consignments were made up of younger animals, but although these seemed to settle down quite happily in the midst of their new surroundings, there were still unacceptably heavy casualties. Antelopes that appeared to be perfectly healthy suddenly collapsed and died for no apparent reason. Post mortems revealed the cause of death to be muscular dystrophy. Clearly new and safer methods had to be devised, with delays cut to a minimum and the whole operation carried out in a single stage. Six more animals were loaded into a cage for the southward journey. But it still proved impossible to cover the difficult route in less than 15 hours and this was too much. Two of the six antelopes died on the way. Although later consignments seemed to be more successful, with all the animals arriving safely at their destination, many of these also died shortly afterwards. Operation Hunter seemed doomed to failure. The antelopes were apparently too delicate to survive the stresses and strains of their enforced move.

Then, by chance, a solution was found. The British aircraft carrier *Ark Royal* had just berthed at the port of Mombasa. Having read about the rescue operation, Captain Pollock placed both his ship and his men at the disposal of the wildlife authorities. A supply and maintenance team was despatched to the area where the antelopes were concentrated and a temporary base set up. Some hours later three helicopters took off from the carrier to begin airlifting the antelopes one by one. Twenty animals were flown to Tsavo and deposited, in perfect condition, in the allotted reception area. A few days later the rains arrived and the Hunter's hartebeeste were set loose to graze on the newly-grown grass.

Operation Hunter had at last been crowned with success. Today the rare species of *Damaliscus hunteri* is one of the many antelopes roaming freely in Tsavo National Park, though few visitors have any idea of the difficulties involved in getting them there. Yet this was not the final act of this strange drama. The United Nations—mainly as a result of political difficulties—shortly afterwards announced the abandonment of all plans to launch an irrigation scheme in the Tana river region. So it had all been a false alarm and Hunter's hartebeest was again free to remain in its traditional habitat. Yet for once human resourcefulness had taken precedence over human stupidity and the decision marked the ironic but happy ending to a remarkable adventure.

Topis, blesboks and bonteboks

The genus *Damaliscus* is widely distributed throughout Africa and represented by several species and subspecies. *Damaliscus lunatus*, for example, is divided into two main groups, broadly speaking western and eastern. The former inhabits the dry regions of Central and West Africa. Because of the hard climate and scant vegetable cover the antelopes have to migrate in the dry season in search of food and water. Among these animals, distinguished from the eastern subspecies of bastard hartebeeste by the concave shape of the skull, long narrow muzzle and flatter, more curving horns, are the tiang *(Damaliscus lunatus tiang)*, the korrigum *(D. lunatus korrigum)* and the subspecies *D. lunatus purpurescens.* When summer comes to the arid subdesert savannah region known as the Sahel, groups of tiang – consisting of 30 or 40 animals – set out on their travels, forming themselves into larger herds, 100 strong, along the banks of the lakes and rivers of Oubangui-Chari and Chad, ready to turn north as soon as the first rains arrive.

The second group of bastard hartebeeste inhabits the much more fertile regions east of the Sudanese savannah, where a plentiful supply of water makes it unnecessary to migrate. Like almost all the herbivores, however, they lead a nomadic type of existence. The most typical East African subspecies is the topi *(D. lunatus topi)*, found in large numbers on the plains of the Serengeti and particularly in the areas of short grass known as 'mbugas'.

One of the best regions for studying the behaviour of the topi herds is the Rukwa valley – a broad plain, on the floor of which is one of the many lakes that stud the great Rift depression. Steep mountains hem the plain in from east to west and the highlands extending from Lake Nyasa to Lake Tanganyika enclose it on the south. Another natural barrier is formed in the north by forests. There are some 500 square miles of excellent grazing in this region, which supports more topis than any other ungulate. Since man has hardly penetrated this remote district the equilibrium of its animal community has remained unaltered for thousands of years. The antelopes wander at will on the valley floor during the summer and graze along the banks of the lake. Although they drink plentifully when water is available they can do without it provided there is a sufficient quantity of green grass. As the dry season continues, small groups wander away over the plains, while the males either go off on their own or seek the company of other species. With the arrival of the rainy season the level of the water in the lake gradually rises, transforming the lowest parts of the valley into swamps. The topis then leave the soft wet ground for the drier outlying plains which are covered with short grass. In April, when the rains cease, the herds move back to their summer pastures.

Most of the births take place in September. The calves develop amazingly fast and at three months already show adult characteristics. By the end of the year they are sexually mature and together with the older males – returning from their summer isolation – young males join the females in the annual trek to winter quarters.

Two other subspecies are found in South Africa – the blesbok *(Damaliscus dorcas phillipsi)* and the bontebok *(D. dorcas dorcas).* The former is hardly seen any more in the wild, its status now reduced to that of a domestic animal, raised for its venison. The bontebok, which, like the blesbok, was once found in large numbers in its wild state, is also comparatively rare. At the beginning of the 19th century the government was already

Profile of a bontebok, showing the characteristic formation of the horns.

Following pages : A herd of hartebeeste come down to a pool to drink. Several of them are posted as sentinels, ready to warn their companions of hidden predators.

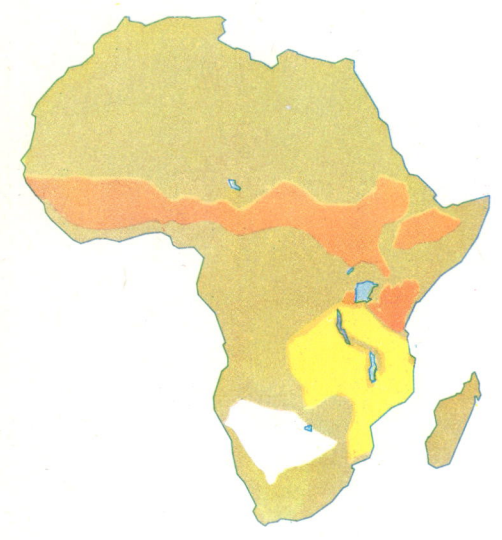

■ Other subspecies of *Alcelaphus buselaphus*

■ *Alcelaphus buselaphus lichstensteini*

□ *Alcelaphus buselaphus caama*

Geographical distribution of the different subspecies of *Alcelaphus buselaphus*.

KONGONI OR COKE'S HARTEBEEST
(Alcelaphus buselaphus cokei)

Class: Mammalia
Order: Artiodactyla
Family: Bovidae
Length of head and body: 70-98 inches
   (175-245 cm)
Height to shoulder: about 48 inches (120 cm)
Weight: about 320 lb (145 kg)
Diet: Grass
Gestation: 8 months
Number of young: one

Adults
Head exceptionally long and narrow, with relatively short (maximum 20 inches), ringed horns, rising from skin-covered bony pedicle. High, laterally positioned eyes; long pointed ears. Withers higher than rump. Coat reddish-brown, becoming paler towards hindquarters which, together with the hind legs, are almost white. Slender limbs, short tail, terminating in tuft of dark hairs.

Young
Calves have vertically directed horns, curving outwards then up and back at tips, as animal grows.

---

making plans to protect the dwindling herds, imposing heavy fines on anyone guilty of killing them without permission. But no real progress was made until 1864 when Alexander der Byl collected 300 of these wild animals together in a vast enclosed area. In 1937 Bontebok National Park was opened in Cape Province, but even here there is not adequate space to support more than 500 animals. That is why from time to time groups of these antelopes are distributed to local farms specialising in livestock rearing.

The differences between the typical bontebok and the typical blesbok are easy to state. The two animals are closely related, being at best only subspecies of the same species, and are very similar. In the blaze of the blesbok the two white patches are usually separated by a brown bar across the face, just below the eyes. In the bontebok this bar is incomplete so that the two white patches are continuous, as in the picture on page 107. In addition, the bontebok has a conspicuous white rump patch. In the blesbok the rump patch is pale brown with white around the base of the tail only. The bontebok has white lower legs whereas in the blesbok there is a brown stripe down the outside of the lower leg.

These are, however, the typical characteristics and, as is usual, in all animals, there are individual variations, and in these two antelopes they tend to blur the distinctions between the two subspecies. Added to this there are reports of hybrids between the two. These reports are usually treated with a certain degree of scepticism simply because the remnants of the two subspecies are located 200 miles apart and have been so for some years past. That they could hybridize is highly likely because the two subspecies are so closely related, and that they have actually done so is also claimed by some writers. It is, however, less easy to say whether any individual animal, such as the one in the photograph on page 106, is a hybrid between a bontebok and a blesbok or a bontebok that through individual variation in the pattern of its coat looks more like a blesbok. It has the typical blaze of a bontebok but the rest of it, taking into account the absence of the distinctive white rump looks like a typical blesbok.

The bontebok, although most are on private farms, is not 'domesticated' because it has not been tamed. Therefore, except in a zoo, where a male of one subspecies has been quartered with a female of the other, hybridization would be unlikely.

The vanishing hartebeest

The hartebeeste of the genus *Alcelaphus* are closely related to the topis, blesboks and bonteboks and like the members of the genus *Damaliscus* are widely distributed. They too have adapted to many different environments, ranging from Morocco to the Cape of Good Hope and they too have been tragically affected by man's colonising activities.

In the 18th century large herds of these antelopes still roamed the plains of southern Algeria and many parts of the Atlas

mountain chain. These North African ruminants were typical of a varied fauna—including lions, wild boars, wild sheep and deer—living in a transitional zoogeographical zone, part Palearctic and part Ethiopian, extending through the northern belt of the continent to the southern limits of the Sahara.

The North African hartebeest was described and given the name 'bubal' by the Roman colonists of the Barbary Coast. Scientists later classified it as *Alcelaphus buselaphus bubal* and the creature roamed the open plains with the predatory black-maned Barbary lion. But the development of agriculture, the invention of more destructive firearms and the increasing numbers of domestic animals were among the main factors combining to push them ever farther northward in the direction of the Atlas Mountains, a region far less suited to their ecological needs. The lions were forced to follow them, but in this inhospitable environment neither herbivore nor carnivore was able to survive. By 1925 there were only a few bubals left in Morocco and Algeria, and in a few years even they had vanished, together with their

A mother kongoni and her young calf. After a few weeks the latter will form a group with companions of the same age, but will not stray far from its mother.

The lyre-shaped horns of hartebeeste are formidable weapons. Despite the sideways-facing eyes, giving all-round vision, the quality of eyesight is not exceptionally keen; but hearing is especially acute.

Facing page : A group of hartebeeste, an adult and three calves a few weeks old. The horns of the newborn calves are at first imperceptible, but they grow rapidly.

lion predators. The extinction of these two species was an irreparable loss in an already dwindling animal population.

The Abyssinian bubal suffered a similar fate, though there is a hope of it being saved from extinction. Its name appears in the pages of the red book listing endangered species, published by the International Union for Conservation of Nature and Natural Resources (IUCN). Swayne's hartebeest (*A. bucelaphus swaynei*), which formerly lived in Somalia, has also disappeared from that country and only a few individuals of the subspecies remain in Ethiopia.

The decimation of the hartebeest herds was accelerated by the outbreak of cattle plague at the end of the 19th century. Unlike the buffaloes they were not sufficiently prolific to recover from this catastrophe nor were any laws introduced to restrict their hunting. The tora *(A. buselaphus tora)*, once plentiful in Ethiopia, the valley of the Blue Nile and the region lying between the Nile and the Atbara river in Sudan, is nowadays rarely seen. Isolated groups that fled for refuge to an almost inaccessible region along Sudan's eastern frontier are the last representatives of the species.

The present-day distribution of the surviving hartebeeste is conditioned by their ecological preferences. All of them, before man and nature combined to reduce their numbers so drastically, were creatures of semi-arid regions. The kongoni or Coke's hartebeest *(A. buselaphus cokei)* favours short-grass savannah with steppe-like features, whereas Lichstenstein's hartebeest *(A. buselaphus lichstensteini)* prefers wooded 'miombo' terrain. The Cape hartebeest *(A. buselaphus caama)* roamed the South African veld before being decimated by the Europeans while in the west *A. buselaphus major* and other subspecies are found in a zone extending from the semi-arid Sahel savannah to the more lush grasslands of Guinea.

The kongoni

The kongoni or Coke's hartebeest has proved hardier and also more fortunate than many of its more northerly relatives. Herds of these antelopes are to be found in the Serengeti and provide excellent opportunities for the detailed study of the behaviour of the species.

The kongonis are sociable, adaptable creatures grazing either in tall or short grass, or even among the denser vegetation of the Mara river region in the northern part of the Serengeti. Although they form themselves into herds of anything from ten to a hundred individuals such groups are far less compact or permanent than those of gnus or zebras. Yet there is a semblance of communal organisation, with adult females, newborn calves and immature animals tending to remain together.

There are also groups of 'bachelors' and territorial males that indulge from time to time in ritual duels every bit as spectacular as those of gnus. Their chief weapons are the curved horns, capable of inflicting damaging wounds. A favourite method of fighting is to force the adversary to the ground by bearing down with the whole weight of neck and body, then, from a kneeling

All hartebeeste have long, narrow heads, with horns rising from a bony, skin-covered pedicle attached to the skull. The structure of the horns varies in width and shape according to subspecies.

position—much in the manner of gnus—to deliver a series of violent head butts in order to enforce submission.

The calves may be born at any time of the year, though as with other antelope species, births reach a peak figure at the beginning of the rainy season. Clear nut-brown at birth, the calves grow rapidly and are gambolling joyfully and noisily together—though never far from their mothers' sides—when they are a few weeks old.

During the rainy season the kongonis leave the tall grasslands and districts of thick vegetation—ideal hiding places for predators—for the 'mbugas' or shorter grass of the Serengeti plains. There they graze peaceably with zebras, gnus and ostriches. Among the other herbivores the sand-coloured coat stands out against the green grass while even more conspicuous is the large white mark on the rump. The shape too is unmistakable—long face, high hump-like withers and very wide, lyre-shaped horns. These, as with all hartebeeste, rise from a cylindrical bony pedicle on the head and curve gently backwards. The disproportion between the fore- and hindquarters is very evident, and aids rapid identification.

The kongoni is not a true migratory animal but does tend to move about according to season, both in Serengeti and elsewhere. In Nairobi National Park, for example, it has been recorded that the population of kongonis is more than a third higher in the dry season than in January, after the rains.

Jackson's hartebeest (*A. buselaphus jacksoni*) has an even more elongated head than other subspecies and also a fiery reddish coat, rather like that of a fox. In the Murchison Falls National Park herds of from 40 to 100 animals are commonly seen. In this region, with its thick cover of vegetation, one may sometimes see an elderly female mounting guard on top of an anthill while her companions graze nearby, lifting their heads from time to time to scan the horizon. These antelopes were once found in large numbers on the rolling grasslands on the Mau plateau in Kenya, which is at a height of between 5,000 and 6,000 feet above sea-level.

Lichstenstein's hartebeest

This subspecies was discovered on the lower Zambezi by a German naturalist who named it after Dr Lichstenstein, an ardent student of the African fauna and later director of the Natural History Museum in Berlin.

It is a shy animal, pale in colour, with short horns. Because of its protective coloration and the fact that it does not remain long in any one place it has proved difficult to make an accurate count of it. It is often found in wooded 'miombo' country, especially in open stretches covered with short grass. It is certain that it has vanished from regions where it was once prevalent though it is often seen from October to July in the savannahs of south-eastern Congo and the savannahs and parks of Katanga.

Farther south one may find small numbers of Cape hartebeeste mingling with the sassaby *(Damaliscus lunatus lunatus)* but both have disappeared from parts where they were once common.

Hartebeeste and their predators

The size and variety of the African hartebeeste and bastard hartebeeste make them particularly tempting and appetising sources of food for the hungry lion. Yet given the choice, the lion will undoubtedly go first for the gnus and zebras. Not only are the latter herbivores even larger and heavier than the hump-backed antelopes but they are far less alert and vigilant. Despite their somewhat lopsided appearance, the antelopes are sur-prisingly fast runners, displaying great powers of acceleration and capable of maintaining a fairly high speed for several miles on end.

According to a survey conducted in Nairobi National Park, in the course of which an accurate count was taken of the animals habitually preyed on by lions, it was found that the percentage of kongonis killed was very low compared with the gnu and zebra victims. Speed would appear to be the decisive factor. One eye-witness described an attempted attack on a kongoni by a lioness. The carnivore was hidden behind the trunk of a dead tree, gaze fixed on her intended victim calmly drinking from a nearby pool. The distance between the two animals was not more than 50 yards. Suddenly the lioness sprang, but the antelope, showing lightning reactions, was too quick for her. With a couple of prodigious leaps it was off. The lioness gave chase but, falling farther behind with every stride, soon gave up the pursuit as hopeless.

Both the hartebeeste and bastard hartebeeste, when at full gallop, keep the head stretched rigidly forward. If need be, they are capable of changing direction without losing speed. Yet mere pace is not always enough to save them from those most implac-able and savage of predators—the hunting dogs. The latter employ the same tactics as they do with all other herbivores—relentless pursuit of the victim to the point of eventual exhaus-tion and collapse. They usually select an adult male, but the outcome is not inevitable. The fierce dogs have been observed to call off the pursuit after a few miles when the victim shows exceptional determination.

Leopards will also attack these antelopes should the opportu-nity arise, showing a marked preference for newborn or young animals, which are easier to hoist into the branches of a tree, there to be devoured at leisure. Solitary cheetahs will rarely bother to attack a hartebeest but may occasionally band together for the purpose. As for spotted hyenas, they normally confine themselves to killing newly-born antelopes—as they will any herbivore inadequately protected either by the parents or the rest of the herd.

Despite all this activity, the fact that many species of harte-beest have become extinct or been driven to the point of extinc-tion is in no way attributable to their natural predators. Herbi-vores and carnivores have evolved along parallel paths wherever they have shared the same habitat, and had it not been for outside human interference the size of the antelope populations would have remained as steady as it was before the settlers came to Africa.

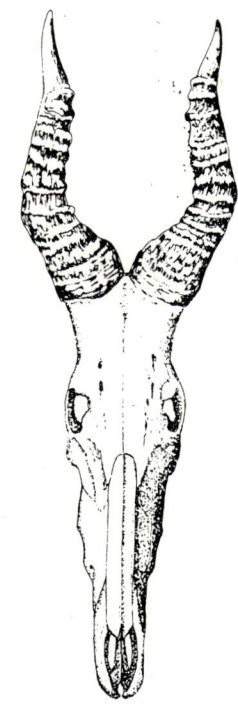

Skull of Jackson's hartebeest, showing how the horns are supported by a pedicle formed by prolongation of the frontal bones.

A charming picture of a female kongoni or Coke's hartebeest suckling her young.

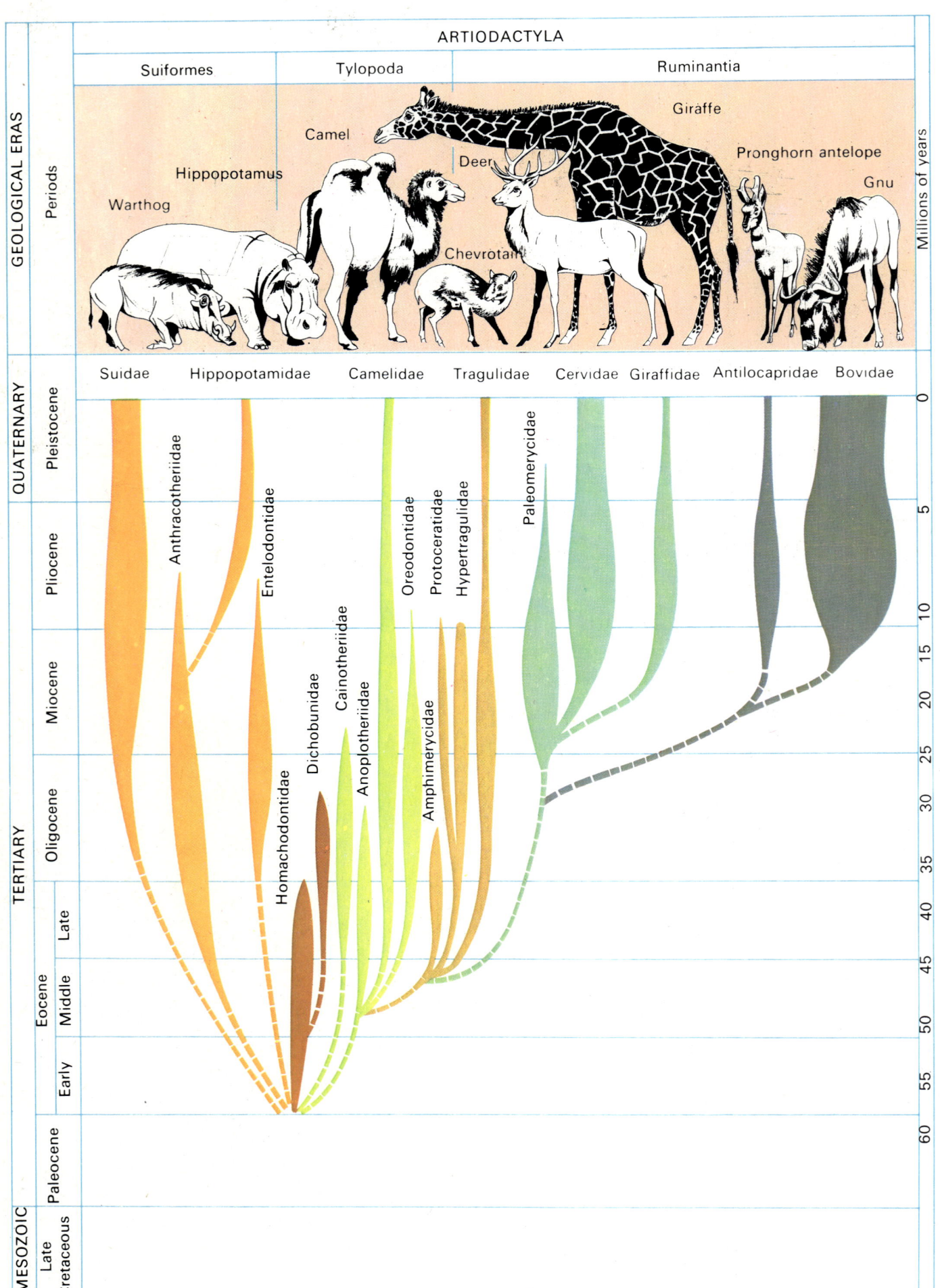

ARTIODACTYLA

Suiformes | Tylopoda | Ruminantia

GEOLOGICAL ERAS

Periods

Millions of years

Warthog Hippopotamus Camel Deer Chevrotain Giraffe Pronghorn antelope Gnu

Suidae Hippopotamidae Camelidae Tragulidae Cervidae Giraffidae Antilocapridae Bovidae

QUATERNARY

Pleistocene

Pliocene

Anthracotheriidae

Enterdontidae

Paleomerycidae

TERTIARY

Miocene

Oligocene

Oreodontidae

Protoceratidae

Hypertragulidae

Dichobunidae

Cainotheriidae

Anoplotheriidae

Amphimerycidae

Homachodontidae

Eocene — Late — Middle — Early

Paleocene

MESOZOIC

Late Cretaceous

0
5
10
15
20
25
30
35
40
45
50
55
60

ORDER: Artiodactyla

The Artiodactyla or even-toed ungulates are paraxonic mammals—the main axis of the foot passing through the third and fourth toes. It is these toes that rest on the ground, the others having either become reduced in size or completely disappeared. As for dentition, the premolars are smaller than the molars and the canine teeth may be missing. Today they are divided into three suborders—Suiformes, Tylopoda and Ruminantia.

The ancestors of the artiodactyls, like those of the perissodactyls, made their appearance in Europe and North America during the Early Eocene, about 65 million years ago. The Paleodontidae were small pig-like creatures, with undifferentiated teeth, and they became extinct during the Oligocene. Another family, the Entelodontidae, flourished during the Oligocene and Miocene; they also were pig-like but somewhat larger than the paleodonts, with strong canine teeth and two toes on each foot. They disappeared during the Pliocene. The Anthracotheriidae were another pig-like family of the Oligocene, large creatures adapted to an amphibian existence, originally with five toes, later reduced to four, and possibly the ancestors of the hippopotamus. Other families of Suiformes, probably dating from the Eocene, flourished for shorter periods—a mere ten or twenty million years. They included the Homachodontidae, the Dichobunidae, the rabbit-like Cainotheriidae (all inhabiting both Europe and North America), the Anoplotheriidae (from Europe only) and the Oreodontidae of North America.

The Tylopoda are also of primitive stock, the Camelidae (the only surviving family) dating from the Late Eocene in North America. They are thus probably older than the Ruminantia, of which the long-extinct Amphymericidae (from Europe) and the Hypertragidae (from North America) are the most primitive. The chevrotains of the Tragulidae, with living representatives, date from the Oligocene, and the Cervidae, Giraffidae, Antilocapridae and Bovidae from the Miocene. All have survived until modern times but the hornless Paleomerycidae vanished during the Pleistocene.

The surviving artiodactyls are distributed unevenly all over the globe, with the exception of Australasia, and the three suborders are broken down into nine families.

The Suiformes contain the families Suidae, Tayassuidae and Hippopotamidae. The Suidae are medium-sized animals with a long head terminating in a snout. In some species the canines are prolonged into tusks. Genera include the domestic pig *(Sus)*, the river hog *(Potamochoerus)*, the warthog *(Phacochoerus)*, the forest hog *(Hylochoerus)* and the babirusa *(Babirusa)*. The Tayassuidae are represented by two species of American peccary *(Tayassu)*. The Hippopotamidae are massive creatures with strong limbs adapted to a semi-aquatic mode of life. The canines are well developed but have remained inside the mouth. The two living genera are both African— the hippopotamus *(Hippopotamus)* and the pygmy hippopotamus *(Choeropsis)*.

The Tylopoda comprise only one family, the Camelidae. Unlike other artiodactyls, which have hooves, they are digitigrades with padded toes. Their stomach is divided into three chambers so that although they chew the cud they are not considered true ruminants. The two genera are the camels and dromedaries *(Camelus)* and the llama, vicuna, etc *(Lama)*.

The Ruminantia comprise five families—Tragulidae, Cervidae, Giraffidae, Antilocapridae and Bovidae. The Tragulidae are the chevrotains or mouse deer of the genera *Tragulus* and *Hyemoschus*, the most primitive of ruminants, with a three-chambered stomach instead of the typical four. The upper canines of the males are formed into tusks.

The Cervidae are the deer, with 14 genera and numerous species. Most of them have branching antlers (shed annually) and representatives include the

Facing page (above) : All the artiodactyls or even-toed ungulates are herbivores and all, apart from the camels, have hooves. Despite such common features they vary enormously in size and appearance, ranging from the tiny dik-diks and mouse deer to the tall giraffe and massive Cape buffalo. The contrasts can be readily appreciated by comparing typical representatives of the existing families. *(Below)* phylogenetic tree (based on the work of Dr Romer) of the Artiodactyla. The different families, many of which became extinct after several million years, are shown extending over a period of some 60 million years, from the Early Eocene until modern times. The bands are narrow or broad in proportion to the number and variety of species. Thus of all artiodactyls, past or present, the Bovidae are by far the most prolific and varied.

The gemsbok *(Oryx gazella)* is one of many handsome members of the family Bovidae.

roe deer, fallow deer, reindeer, caribou and moose.

The Giraffidae contain two genera, the giraffes *(Giraffa)* and the okapi *(Okapia)*, the former characterised by a long neck, prehensile tongue and small horns.

The Antilocapridae are represented only by the American pronghorn antelope *(Antilocapra)*, with horns that are shed and renewed annually.

The most numerous and diversified of the ruminants are the Bovidae, with hollow horns (never shed), four-chambered stomach and two toes on each foot. There are 46 genera, among which are many familiar animals of hill and grassland—including domestic cattle, sheep and goats as well as antelopes.

FAMILY: Bovidae

Millions of years of evolution have brought the Bovidae to a high level of self-reliance and efficiency. They have the means of escaping from the clutches of their enemies, of defending themselves with confidence and determination, and of deriving the maximum advantage from their exclusively herbivorous diet. Long and supple limbs, each equipped with two hoof-encased toes, enable them to run rapidly over hard ground; and the situation of the udders and teats in the females' groin region affords minimum hindrance of movement.

The teeth of the Bovidae—varying from 28 to 32—are not intended to be defensive weapons, the canines and upper incisors generally being absent. This lack is compensated by the presence of horns—usually in both sexes but always in the male. These are unbranched and hollow, consisting of a horny sheath covering a bony core which is attached to the skull.

The cutaneous (skin) glands—situated in different places according to the subfamily concerned—play an important role in the community life of the Bovidae. The scent of their secretions is a key factor in the demarcation of territory, mutual recognition, sexual attraction and the maintenance of order within the herd. As in all other grass-eating animals the sense of smell is extremely well developed. So too—and no less valuable—is the sense of hearing.

A four-chambered stomach enables the Bovidae to store large quantities of food within a short period in order to digest at leisure.

The Bovidae are customarily divided into ten subfamilies.

The Cephalophinae are small or medium-sized animals with a long narrow head, naked muzzle and short conical horns—usually common to both sexes. They possess both preorbital (in front of the eye-socket) and interdigital (between the toes) glands. The hooves are well developed. They live in forest galleries bordering rivers in equatorial and southern Africa. Typical representatives are the duikers of the genera *Sylvicapra* and *Cephalophus*.

The Boselaphini are a tribe belonging to the subfamily Bovinae. There are only two representatives—both Indian—the nilgai or blue bull (*Boselaphus tragocamelus*), largest of all Asiatic antelopes, and the much smaller four-horned antelope (*Tetracerus quadricornis*).

The Neotragini are also African animals, including such dainty creatures as the royal antelope (*Neotragus*) and the dik-dik (*Madoqua*). The latter is a typical member of the tribe—very small, with a long head and enlarged muzzle which is prolonged into a snout. The eyes are large and the limbs very slender, with rudimentary lateral hooves. The fawn coat is mottled with white and chestnut-brown markings. The females (unusually) are somewhat larger than the males and the latter have small, ringed, backward-pointing horns.

The Tragelaphinae include larger animals with long spiral-shaped horns. They are all forest dwellers and the two genera *Tragelaphus* and *Taurotragus*, though of common parentage, show distinct anatomical differences. Among the former are the harnessed antelope (*Tragelaphus scriptus*), the nyala (*T. angasi*) and the lesser kudu (*T. imberbis*). The latter genus includes the largest of the African antelopes, the eland (*Taurotragus oryx*).

The animals belonging to the subfamily Bovinae are large with naked, well-developed muzzles. Strong cervical vertebrae support a powerfully muscled neck, a dewlap hanging from the throat of certain species. The shape and size of skull as well as the curvature and direction of the horns display many variations. The long rounded tail ends in a tuft of hair. Only a few of the Bovinae possess lachrymal glands but all of them have interdigital glands. Among the many members of the subfamily are the Cape

Two typical members of the family Bovidae. At the top is the eland (*Taurotragus oryx*), largest of all African antelopes. A male may stand 6 feet at the shoulder with horns 3 feet long. *Below* is the nilgai or blue bull (*Boselaphus tragocamelus*), largest of Indian antelopes, standing up to 4½ feet tall but with very short horns.

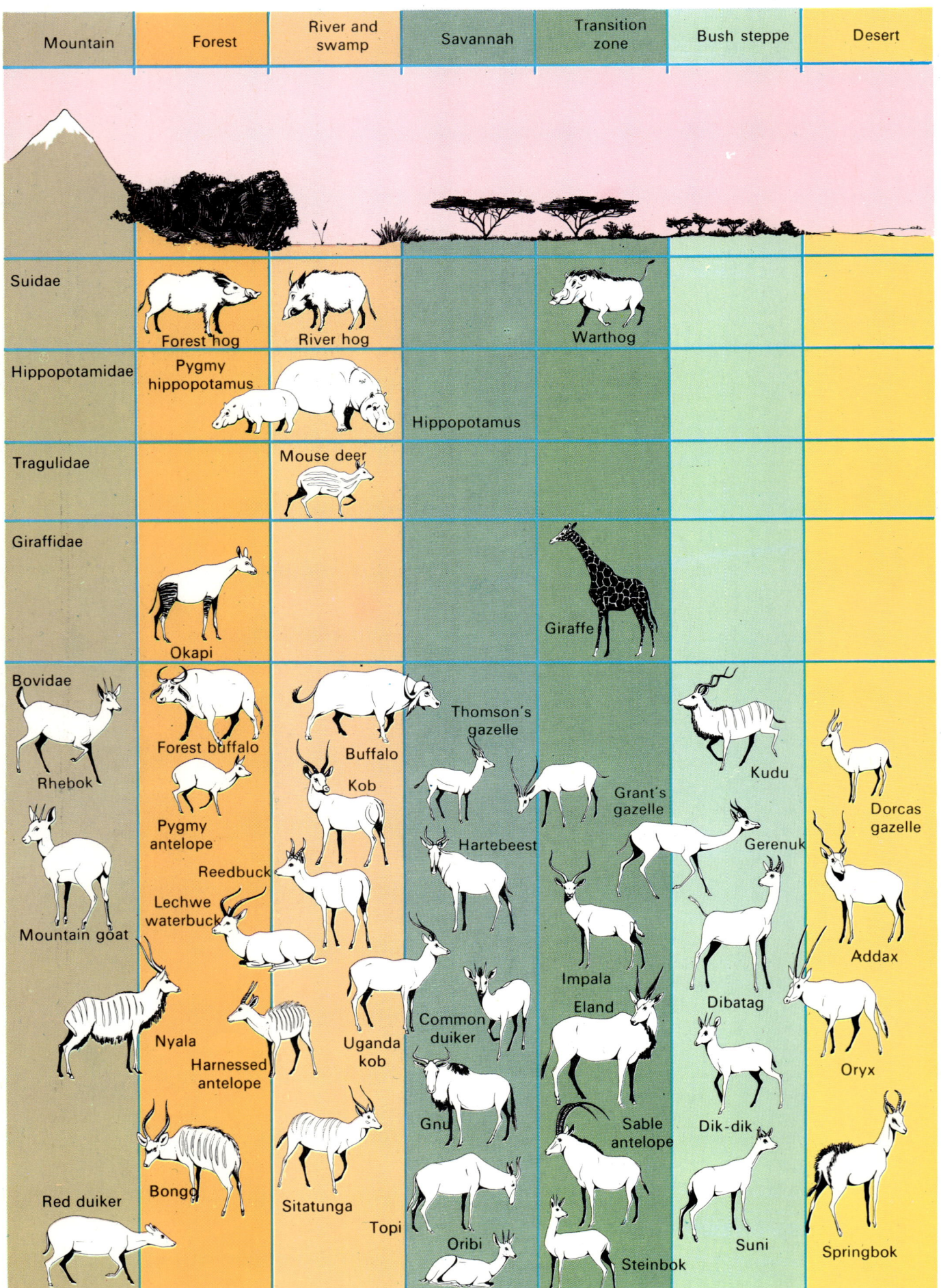

Mountain	Forest	River and swamp	Savannah	Transition zone	Bush steppe	Desert

Suidae — Forest hog, River hog, Warthog

Hippopotamidae — Pygmy hippopotamus, Hippopotamus

Tragulidae — Mouse deer

Giraffidae — Okapi, Giraffe

Bovidae — Rhebok, Mountain goat, Nyala, Red duiker, Forest buffalo, Pygmy antelope, Reedbuck, Lechwe waterbuck, Harnessed antelope, Bongo, Buffalo, Kob, Uganda kob, Sitatunga, Thomson's gazelle, Hartebeest, Common duiker, Gnu, Topi, Oribi, Grant's gazelle, Impala, Eland, Sable antelope, Steinbok, Kudu, Gerenuk, Dibatag, Dik-dik, Suni, Dorcas gazelle, Addax, Oryx, Springbok

buffalo (genus *Syncerus*), the various breeds of domestic cattle *(Bos)* descended from the long-extinct aurochs, the North American bison *(Bison)* and the small anoas which inhabit the Philippines and the Celebes islands (genus *Bubalus*).

The Alcelaphini are large antelopes with a long head, and withers that stand higher than the hindquarters. The flattened muzzle, though hairless, resembles that of a giraffe. Both males and females carry curved horns, the limbs are slender (with interdigital glands and lateral hooves well developed on the front legs but absent or rudimentary on the hind legs), and the tail, reaching to the hocks, ends in coarse hairs extending to the feet. Among these typical creatures of the African savannah are the hartebeest *(Alcelaphus)*, the bastard hartebeest *(Damaliscus)* and the two species of gnu *(Connochaetes)*.

Thomson's gazelle is primarily a creature of the plains where there is plenty of short, fresh grass. Living in small herds, it is found in many parts of East Africa.

The Hippotraginae are large antelopes with prominently ringed horns. The limbs are provided with lateral hooves and well-developed interdigital glands, and the long tail extends to the hocks. The sable antelopes *(Hippotragus)* are found from Transvaal up to Zambia and Angola in the west, and Kenya in the east. The genera *Oryx* and *Addax* are inhabitants of deserts and plains, mostly in North and East Africa.

The Reduncini are similar in size to the gazelles but somewhat heavier, living for preference in humid regions. Well-developed lateral hooves make them capable of keeping their footing in muddy ground, without risk of sinking. Only the males carry horns, which are ringed at the base. Among the many typical representatives of this subfamily are the kobs *(Kobus)*, the lechwes *(Adenota)*, the reedbucks *(Redunca)* and the rheboks *(Pelea)*.

The Antilopinae include the many graceful antelopes and gazelles with ringed horns, slender limbs, lateral hooves and interdigital glands. The tail is usually short but may in certain species reach the hocks and terminate in a black hairy tuft. The coat colour is generally fawn with reddish or greyish tints. The hairs on the belly are lighter, sometimes white, and many of these animals have a distinctive (often black) longitudinal band along the flanks, serving as a recognition or warning signal. The genus *Antilope* has only one species, the blackbuck *(Antilope cervicapra)*, of India and Pakistan, with spiral horns. Literally speaking, this is the only antelope proper but the term is of course more widely applied to the genus *Gazella*, broadly distributed over Asia and Africa, the genus *Aepyceros* – represented by the single species *Aepyceros melampus*, better known as the impala – the genus *Lithocranius* (gerenuk), the genus *Antidorcas* (springbok), and several other genera, found both in Africa and Asia.

The subfamily Saiginae boasts only a single species – the Saiga *(Saiga tartarica)* – an inhabitant of Russia and Siberia. This is a stocky horned animal with a prominently elongated snout which has developed into a short proboscis.

Finally there are the Caprinae – agile mountain animals whose limbs are provided with two toes, encased in stout hooves – ideal for clinging to and crossing rocky terrain. The head is convex and the horns often twist back on themselves in spectacular fashion. Of the numerous genera the two largest and most widely distributed are *Ovis* and *Capra*, and both contain wild and domestic species. The former, for example, comprises one wild and one domestic species of sheep. In the wild species the hairy tail is comparatively short but it is markedly longer in certain species of African domestic sheep. The genus *Capra* is even more diversified. Almost all of these animals have beards and although they do not have lachrymal glands they possess perianal glands secreting (in the males) a characteristic and powerful odorous substance. This well-known genus includes the wild goat of Asia Minor and south-eastern Europe *(Capra hircus)* – ancestor of the familiar domestic species – and the Alpine ibex *(C. ibex)*. Other genera of the subfamily Caprinae include the rocky mountain goat *(Oreamnos)*, the musk-ox *(Ovibos)* and the chamois *(Rupicapra)*.

Facing page : Artiodactyls are found throughout the African part of the Ethiopian region, though rarely in Madagascar. They inhabit mountains, forests, swamps and deserts, but are seen in greatest number in open terrain such as steppe and savannah. There is a far greater concentration of herbivores in the African savannahs than anywhere else on earth.

CHAPTER 5

Grace and grossness: gazelles and warthogs

Of all the animals of Africa none is as decorative as the gazelle– a creature equally well adapted to life in desert, steppe or grassy savannah. Well might the ancient poets of the Orient sing their extravagant praises, for what animals could compare with them in grace of movement, innocence of expression or subtlety of form and colour?

When the other migratory herds abandon the wide open spaces of the Serengeti for the summer–leaving behind them expanses of short grass now transformed into an arid wilderness and stretches of taller grass which are now fire-blackened or dust-blown wastes–two species of graceful antelope remain. So too do many carnivores who find they make delectable eating.

These two species–Grant's gazelle *(Gazella granti)* and Thomson's gazelle *(Gazella thomsoni)*–may look alike at a superficial glance. In mixed herds–more often found in the rainy season–their fawn coats and magnificent ringed horns are not easily distinguishable at a distance. But closer inspection shows that some are much larger and heavier, with a somewhat paler coat colour ranging from nut-brown to sandy-yellow. These are Grant's gazelles, the males weighing from 120 to 185 lb.

Size and weight are not the only distinguishing features. Grant's gazelle is a handsome, proud-looking creature with a stately gait. Its shining black horns are prominently ringed and proportionately larger in relation to body size than those of other antelopes. Thomson's gazelle, popularly known as the 'Tommy', is smaller and more thickset, with shorter legs and neck, the male weighing between 45 and 65 lb. In both sexes a large jet-black band adorns the flanks, distinguishing it from Grant's gazelle, although the female of the latter species shows

Facing page : This proud-looking Grant's gazelle, so slender and elegant, is one of the world's most beautiful animals. The female has a black band along the flanks– similar to that in both sexes of Thomson's gazelle–but this is not generally found in the male.

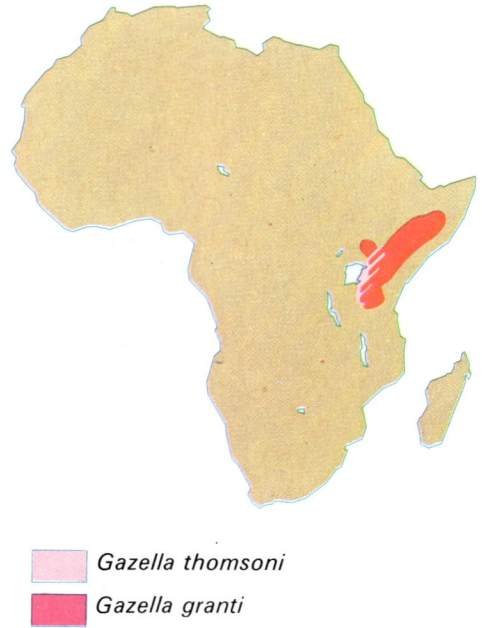

Geographical distribution of Thomson's
and Grant's gazelles.

THOMSON'S GAZELLE
(Gazella thomsoni)

Class: Mammalia
Order: Artiodactyla
Family: Bovidae

Height to shoulder: 22-28 inches (55-70 cm)
Weight: Males 45-65 lb (20-30 kg)
Females 35-45 lb (15-20 kg)
Diet: Grass
Gestation: about 190 days
Number of young: one, sometimes two

Adults
Smaller than Grant's gazelle, its head being
shorter and more rounded, the horns, similarly
ringed, also being shorter. A black stripe
extends from below the eye to the light-
coloured muzzle. The main feature of the
graceful, streamlined body is the black band
along the flanks, separating the reddish back
from the white belly. There is a white patch
on the rump, bordered by a black line, and the
black tail ends in a tuft of hair. As with other
gazelles the male is larger than the female and
the horns are more powerful, their maximum
length being 17 inches.

Young
The young gazelle has the characteristic black
band on the flanks as soon as it is born. Compared with the adult it has very long legs.

---

a similar band, but narrower and less clearly defined. Thomson's gazelle is a lively and amusing creature, ceaselessly swishing its tail, whether or not there are flies to be kept off. This may be a form of recognition signal for its companions, like the tail fluttering that goes on among certain kinds of gregarious birds.

The black band that decorates the flanks of Thomson's gazelle has provided a fruitful area of study for naturalists interested in the phenomenon of protective and warning coloration. Their conclusion is that the primary function of this pattern is to break up the outline of the animal in order to deceive predators. Viewed from a distance, against the yellow background of the savannah, the animal's shape is hard to see in its entirety, giving the illusion of several smaller and in-dependent bodily parts. Furthermore the experts suggest that the black stripe may also serve as an alarm signal, the animal using special muscles to vibrate its sides when it senses danger. So—as with the black and white stripes of the zebra—this is another example of coloration serving both for camouflage and recognition.

The gazelles, so perfectly proportioned and gracefully built, are without doubt the most beautiful grass-eating animals of the African savannah. Their elegance is all the more striking when they are seen in company with the cumbersome gnus or the hump-backed hartebeeste. But mere beauty is of little account in the wild and it is one of nature's cruel ironies that these delightful animals should fall victim, more frequently than any other herbivore, to the predators of the plains. Cheetahs and hunting dogs, leopards and lions (the last two particularly during the Serengeti's dry season), find them especially ap-petising, while hyenas, jackals, eagles and vultures show no mercy to the young.

The droughtproof antelope

The gazelles have a wide range of natural defensive equipment –sufficient, more often than not, to keep their enemies at a safe distance. Height, speed, powerful and flexible muscles, natural camouflage and keen sensory perception are adequate for self-protection in normal circumstances. They are not only long-sighted but are also able to distinguish moving objects very clearly. The combination of excellent vision and acute hearing usually gives them a head start on predators trying to track them at a distance, although they are more vulnerable to concealed attack from close quarters.

Both Grant's and Thomson's gazelles graze together during the rainy months, showing a preference for those regions where the grass retains a seasonal freshness. Yet even when the summer sun turns the savannah bare and dry, making life insupportable for most herbivores, Grant's gazelles in particular have no difficulty in surviving, for they require hardly any water. Thomson's gazelles are not quite so adaptable, needing to drink regularly; although they do not migrate they are obliged to gather together and wander off in search of pools and signs of sprouting new grass.

The degree to which they are dependent on water is of vital significance for all inhabitants of dry regions. Species that are well adapted to conditions of prolonged drought have a natural advantage over those that have to compete among themselves for water and hence run the risk of being picked off by lurking predators. Grant's gazelle is especially fortunate in this respect. Some authorities have suggested that the very pale colour of their coats (bright colours reflecting the sun's rays and dark colours absorbing them) play an important part in helping them to retain water and regulate their body temperature. But this has not been proved. What has been observed and seems more relevant is that the animals urinate very infrequently.

The smaller 'Tommy' is not so luckily endowed, requiring both water and short, fresh grass, whereas Grant's gazelle will browse quite contentedly on leaves. This preference for different types of vegetation has been strikingly demonstrated by Dr Estes in his surveys of the two species in Ngorongoro Crater. The local population of Thomson's gazelle tends to concentrate on the alkaline soil found on the floor of the crater near the lake, spreading out or closing ranks according to season, but never wandering far away. The Grant's gazelles, however, will stick to the hillsides where the grass is taller. But since there is no clear demarcation line between the zones it is not uncommon to find mixed herds, one species feeding on grass, the other on leaves.

Thomson's gazelle is distinguished from Grant's gazelle by its smaller size, black band along the flanks and black tail. The two species graze together during the rainy season but go their separate ways in the summer, Grant's gazelle not needing much water and able to endure drought.

Although this female Grant's gazelle appears to be reluctant she is in fact ready to receive the male's attentions. Her pretence of flight and withdrawal merely causes the male to redouble his efforts to court and mate with her.

Gazelles and their territories

Both species of gazelle are territorial animals, Thomson's more noticeably so than Grant's. Groups of females and their young—either separately or together—drift in the general direction of the adjoining pieces of grassland which the males have marked out as their own. The latter then attempt to entice the females inside the territorial bounds—a familiar pattern of behaviour. But these animals are in a minority. Most of the males, either still immature or having failed to win a portion of territory, tend to form 'bachelor' groups. They may make courtship gestures towards the females but since they are not involved in the breeding process such attempts are not serious and do not lead to actual intercourse or to fighting.

The territorial males trace out the boundaries of their property by establishing 'scent posts'—first urinating, then extending their hind legs to defecate on the same spot—something never done by the females or immature males. Thomson's gazelles also possess preorbital glands (in front of the eye-sockets) secreting a strongly-smelling substance which they deposit on the vegetation until it is thoroughly and visibly coated. Thus any would-be intruders receive a double warning—by means of sight and smell—that this is occupied territory and entry forbidden.

There are fights among the males of both species to conquer and defend territory. The small, wiry Thomson's gazelles approach each other cautiously, heads lowered and horns vertical, hoping to gain a bloodless victory by intimidation. This manoeuvre rarely works, however, and the ensuing battle consists of a series of head-on collisions, the opponents retiring to circle each other and then return to the fray.

Grant's gazelles prefer to put on a show of strength rather than come to actual blows. Each knows his own and his rival's capabilities from experience in the 'bachelor' herd. Necks stiff and cheeks puffed out, the two males place themselves some five or ten yards apart, eyeing each other warily, then shake their heads vigorously from side to side. As often as not that is the end of the hostilities; but should these tactics of mutual intimidation not succeed, the animals close in and circle each other, pausing from time to time to rub head or neck with a hind leg, to tear up a tuft of grass, to attack a bush with their horns, and to defecate or urinate. These are simple diversionary tactics to interrupt the flow of the contest and relieve tension, Thomson's gazelles, gnus and many other antelopes adopting a similar procedure. These methods may also fail to have the desired effect, in which case there is no alternative but to fight. Ears pricked, the two males nod their heads up and down and then interlock horns, each trying to twist the other's neck. The struggle usually ends with the headlong flight of the weaker animal, unless the rivals are unable to disentangle themselves, when death from exhaustion may result.

The sole purpose of these strenuous efforts to stake out and defend territory is to win possession of as many females as possible and then to hold them. The surrounding savannah is made up of a network of 'nuptial' zones, each dominated by a male bent on keeping rivals at a safe distance—about 800 yards in the case of Grant's gazelles and some 300 yards for Thomson's gazelles.

Generally speaking, the gazelle herds are less stable than those of gnus, the males having little time to do more than guard the females in their territory. The incessant activity of the male Thomson's gazelle reminds one of a sheepdog rounding up its flock.

Both species display similar sexual behaviour at courtship and mating time. When a female is on heat and ready to couple with a male she goes through an elaborate ritual. The male, drawn towards her by scent signals, is at first discouraged as she assumes a series of positions which restrict or prevent his approaches. It may be several days before she shows obvious signs of being attracted by him. When he advances to smell or lick her she retreats at a gentle trot, kicking out with her hind legs. This apparent unwillingness serves only to excite him further, but as the ritual of pursuit and retreat continues her pretence of withdrawal becomes increasingly half-hearted. Finally she comes to a standstill, raising her tail and slightly arching her back. The male is finally able to mount her, balancing himself on his hind legs and following the mare as she continues to move forward during the coupling itself. Once the act is completed, the male appears to lose all interest in his partner, leaving her and rejoining his male companions. The pregnant female will similarly remain with those of her sex until shortly before she is due to give birth. She will then separate herself from the rest of the herd and take up solitary residence on a small piece of grassland or, should that not be available, a portion of dry ground. It is there that she will give birth to one or sometimes two young.

The hazardous early days

As in most herds of herbivores, the majority of births take place in January and February, following the autumn rains and before the more prolonged and heavy spring downpours. But births may occur at intervals during other seasons as well. The female will drop one or more young in seclusion and provide all the necessary maternal attention and affection, eating the placenta immediately after the birth and licking the offspring clean. The latter is soon on its feet and suckling. Once it has drunk its fill it rests in a conveniently sheltered spot, under the watchful gaze of its mother. In due course she will rejoin the other females at pasture, taking careful note of the place where she has left her baby to await her return. This is the time when the offspring are at greatest risk where predators are concerned.

For the first few days of its life the baby gazelle will remain rooted to this same spot. Its natural instinct to flee when danger threatens is not yet developed and even if it were it would not be able to move far. At this stage the tiny legs are too fragile to permit free and rapid movement and any attempt to move would merely signal its presence to predators and seal its fate.

GRANT'S GAZELLE
(Gazella granti)

Class: Mammalia
Order: Artiodactyla
Family: Bovidae

Height to shoulder:
  Males 34-40 inches (85-100 cm)
  Females 32-38 inches (80-95 cm)
Weight: Males 120-185 lb (55-85 kg)
  Females 75-130 lb (35-60 kg)
Diet: Grass, leaves
Gestation: about 5 months
Number of young: one, sometimes two

Adults
Horns longer than those of other antelopes. Black, shining and perfectly ringed, they are lyre-shaped and in some males may measure up to 30 inches. The head is relatively short and the ears large. The upper part of the body is fawn, the belly, throat, chin and tail white, the last terminating in a hairy tuft. There is a black-bordered white patch on the rump. Two light-coloured bands extend from the root of the horns to the nostrils, above which is a black mark. On the flanks, where the fawn meets the white, there may be a black line, more clearly visible in the female and corresponding to the black band of Thomson's gazelle. It is absent in many males.

Young
The young gazelle, without horns, stands very high on its feet.

Subspecies
There are several subspecies characterised by geographical distribution as well as by the curvature of the horns. Thus *Gazella granti robertsi* has horns whose tips diverge instead of curving inwards.

Following pages : Three photographs of Grant's gazelles. *(Left)* herds of these gazelles are a common sight on the savannah, even in the dry season when other herbivores migrate. *(Above, right)* a male, in characteristic fashion, uses his hind leg to scratch his head. *(Below, right)* the young gazelle, horns just beginning to sprout, soon becomes independent.

So for the time being it is docile enough to be picked up and fondled, making no attempt to shy away. Remaining absolutely quiet and motionless is thus instinctive, given the circumstances and the animal's limited capabilities – nature's way of ensuring the survival of the species. Yet precisely because of this inbuilt mechanism of natural impulses, the baby gazelle – like other animals – finds it difficult to modify its behaviour to meet unforeseen contingencies. Thus the enforced passivity of the baby gazelles may prove fatal. For example, many of these animals have been burned to death by a savannah fire, either being unable to escape or leaving the decision until it is far too late.

Natural selection has brought into play yet another security precaution at such times. The mother will refuse to leave her young until it has urinated and defecated. She will then swallow the excrement so as to destroy all trace of odour which might otherwise guide carnivores, especially jackals, to the spot. Unfortunately the predators have other ingenious ways of overcoming the problem.

Enemies by land and air

Surveys carried out in the savannahs of East Africa have shown that Thomson's gazelle is regularly victimised by nine land predators, varying considerably in size and ranging from the

A newly-born Thomson's gazelle, its horns just beginning to sprout.

jackal to the lion. In Ngorongoro Crater, where lions are not normally found, the most determined hunters are common and black-backed jackals, hunting dogs, cheetahs and spotted hyenas. Each carnivore has its own chosen method of capturing prey, stalking it either individually or in packs. The options open to the victim are more limited and the utmost caution and watchfulness are essential at all times.

Jackals specialise in the capture of newborn gazelles and may often be seen at the appropriate seasons prowling round the antelope herds, either singly or in pairs. Hunting together often pays dividends, for the mother gazelle, instinctively giving chase to any animal threatening her young, can only cope with one at a time. Thus one jackal will create a diversion while the other snatches up the baby gazelle. Yet the carnivores' plans may be foiled by the arrival on the scene of other females – also with young hidden in the neighbourhood – ready to join in common defence.

The young gazelle's pattern of behaviour changes completely when it is a few days old. Its instinct when confronted by anything strange or menacing is now to take flight rather than stay put. Staggering to its feet, it sets off as fast as it can in a zigzag dash towards a new hiding place. Safely arrived, it will drop to the ground, hopefully giving enough time for the mother to come to its assistance.

Considering the many risks to which they are exposed, it is not to be wondered at that a large proportion of newly-born gazelles are killed during the first few days of their life. Those

The black band decorating the flanks of Thomson's gazelle is clearly visible at close quarters, but at a distance becomes blurred and serves to break up the animal's outline. This optical illusion helps to deceive predators.

that manage to get through this critical period, however, have little to fear any longer from jackals.

Spotted hyenas also pose a serious threat to the newborn gazelles but on the evidence of their droppings they seem to prefer young gnus or zebras, given the opportunity.

Leopards, concealing themselves in the thick undergrowth near watercourses, may frequently take a gazelle unawares. Yet they also perform a service by killing the occasional jackal. During his study of gazelle behaviour in Ngorongoro Crater Dr Estes watched the activities of a female leopard. In a single month the wild cat was observed to kill and hoist up into the tree where she habitually hid two male Grant's gazelles and no fewer than eleven jackals.

The fiercest and most relentless enemies of adult gazelles are the cheetahs and hunting dogs. When menaced by these predators the gazelles are unable to conceal their fear and nervousness, snorting loudly and beating the ground with their fore feet. If the danger is imminent Thomson's gazelles will contract the skin of their flanks, causing a rippling movement of the lateral black stripes and their white borders. Then, with tail stiffly erect and limbs at full stretch, they leap into the air and make off at top speed. This very distinctive behaviour on their part is believed by some zoologists to be a specific reaction to the presence of hunting dogs, the most persistent of their many predators.

Grant's gazelles, lacking the black stripe, have their own alarm signal. They also leap high into the air but bristle the white hairs which form a wide black-bordered patch on their rump.

Although most of the animals in the herd will gallop off at speed, the obstinate behaviour of the territorial males and the females with young makes them particularly vulnerable to the attacks of hunting dogs. As the pack draws near, the proud males, not daring to infringe upon the domains of their neighbours, stay where they are, while the females equally stubbornly refuse to abandon their young, hidden in the grass. Even when the males decide to take flight they do not run in a straight line but gallop about in circles, reluctant to journey too far from their home territory. Such tactics are predictably disastrous, ending in certain capture and death. Once the females are caught, their orphans too have little hope of surviving.

Gazelles also have to be on the alert against aerial attack. Newborn animals, hidden and motionless in the long grass, are fairly safe from birds of prey such as eagles, dependent on sharp vision to guide them to their prey. But as soon as the young gazelles begin straying away from their mothers they are open to frequent attacks from the skies.

The members of a zoological expedition on their way from the Serengeti to Ngorongoro Crater one February morning witnessed such an attack. From their open Landrover they watched a Verreaux eagle swoop down on a small group of Thomson's gazelles grazing in the short grass about 500 yards from the road. Perched firmly on the ground, the huge bird whirled around twice and then, with wings half-extended,

Facing page (above): A small herd of Thomson's gazelle at a waterhole. Unlike Grant's gazelle, this species requires a good supply of water in the dry season. *(Below)* a splendid male Grant's gazelle with shining black horns. These natural weapons are generally used only to intimidate other males during the mating season.

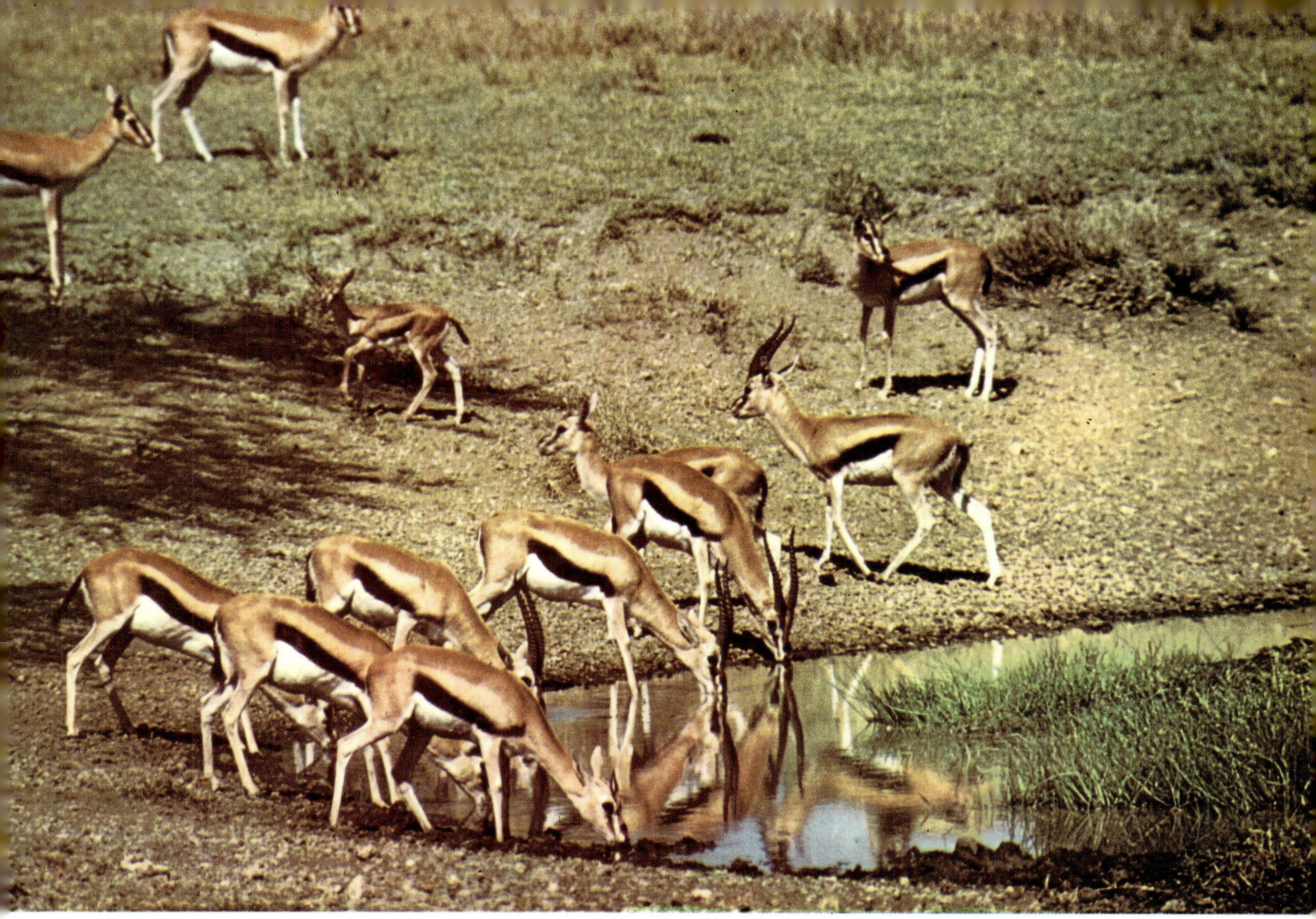

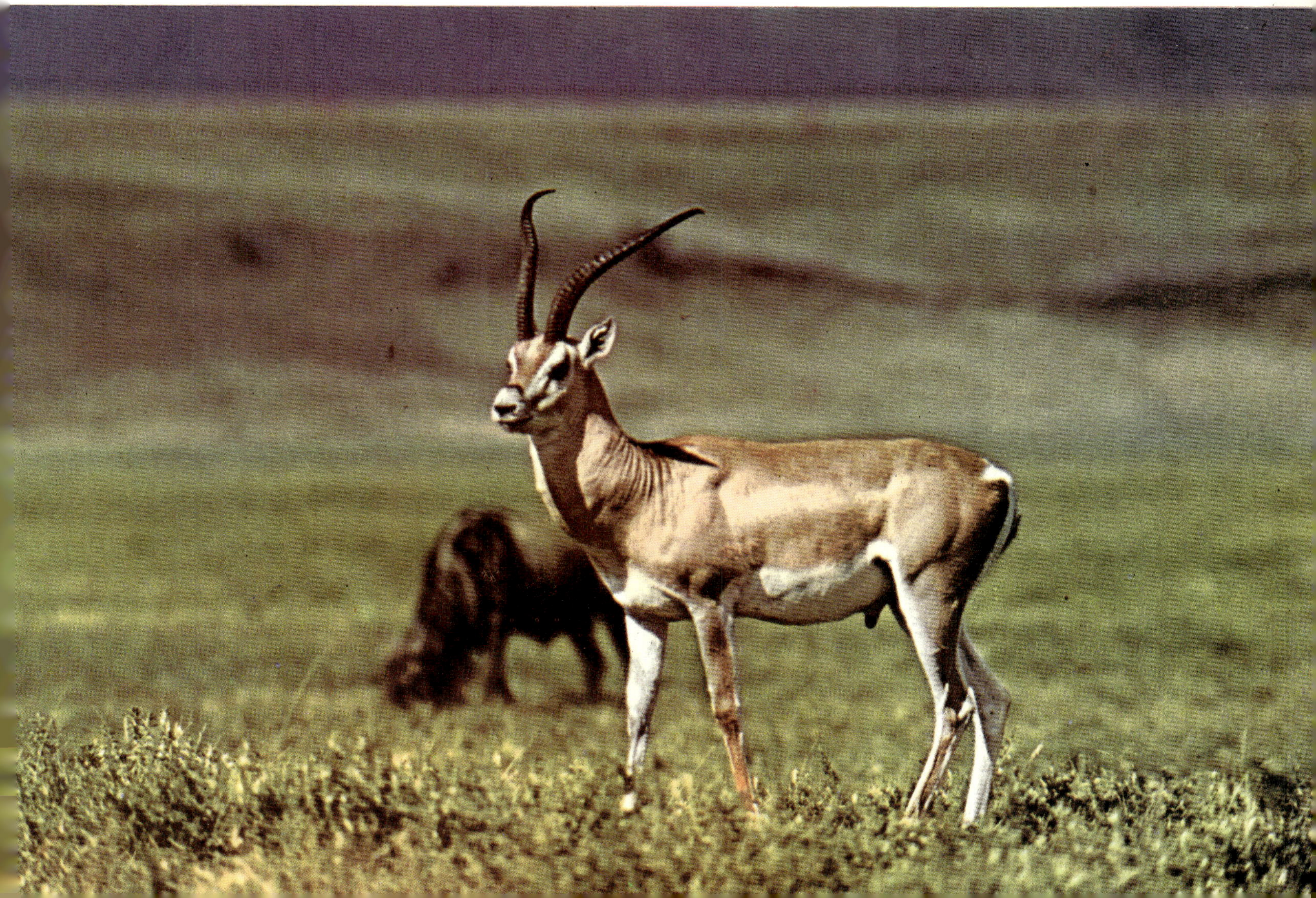

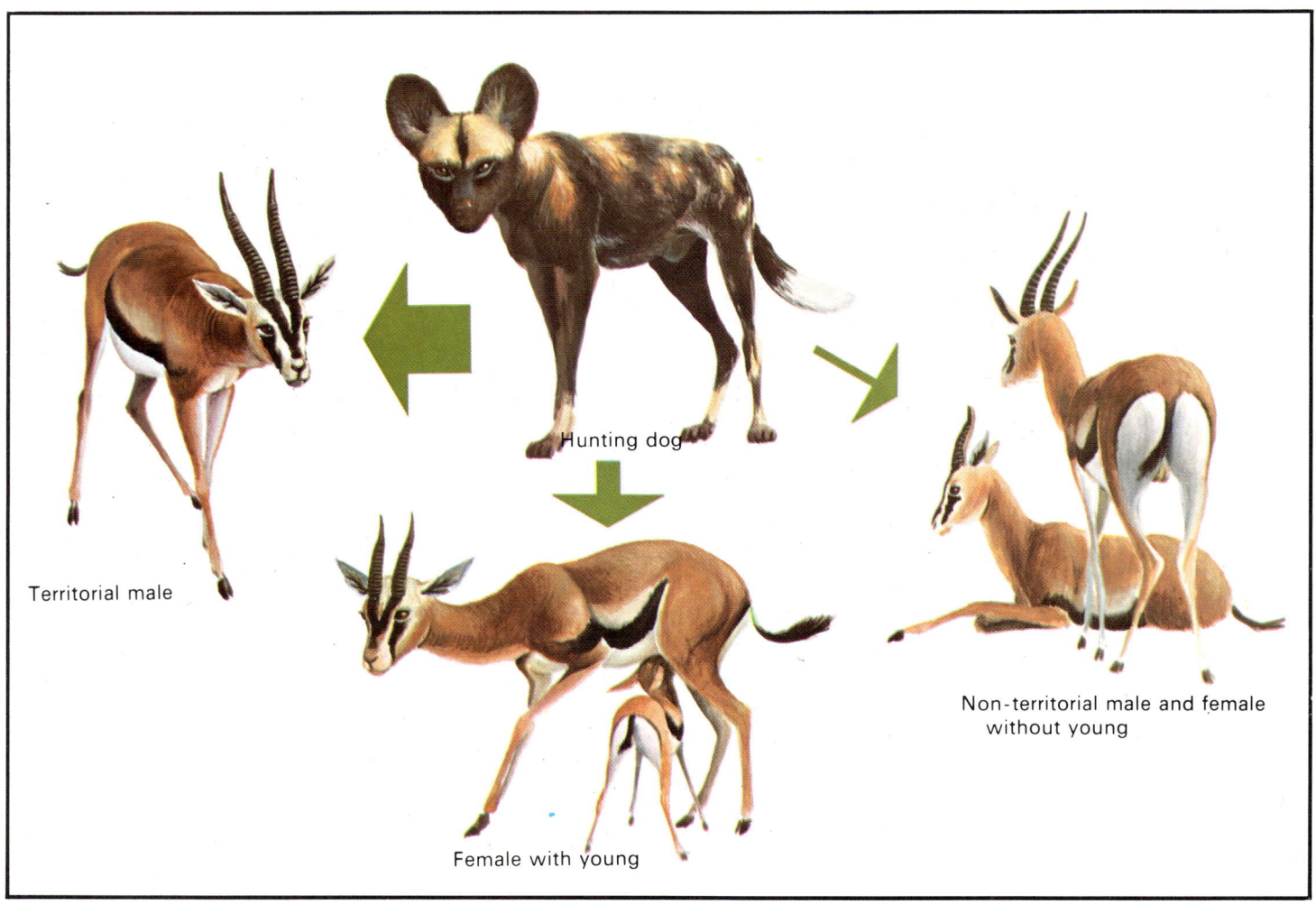

Territorial male

Hunting dog

Non-territorial male and female
without young

Female with young

This chart shows (according to the comparative thickness of the arrows) how much more vulnerable some Thomson's gazelles are than others to the attacks of African hunting dogs. The territorial males, refusing to abandon their domains, are easily picked off one by one, and the females, anxious to save their young, are likewise frequent victims, the latter then being at the mercy of the savage carnivores. The non-territorial males and unaccompanied females, tending to group together, are less often attacked.

fastened its talons on one of the younger animals. A female, presumably the victim's mother, charged the bird, swinging it round in a desperate effort to loosen its hold. Then a male with splendid horns joined the attack, hurling itself at the eagle, which did not await the impact but immediately released its victim. There was to be no happy ending, however, for no sooner had the baby gazelle struggled to its feet than a great Nubian vulture plummeted down and proceeded to devour the tiny animal with incredible despatch, quite unperturbed by the frantic attempts of the adult gazelles to intervene. In the end they gave up the hopeless fight and galloped off, while the frustrated eagle soared up into the sky. When the zoologists arrived on the scene a few minutes later, the vulture also took wing, leaving the remains of its meal—just the head and the two front feet.

The Verreaux eagle will also attack older gazelles, dropping on them from the branches of an acacia and strangling them on the ground with its powerful claws.

In spite of their numerous natural enemies, winged and otherwise, the African gazelles—and there are several others in addition to the two discussed here—have successfully increased their numbers. In many areas of steppe and savannah where unrestricted hunting and uncontrolled rearing of domestic livestock have driven out or destroyed large herds of ungulates, the gazelles have managed to survive. But there is no room for complacency, knowing from experience how precariously balanced the ecology of any given region can be; and indeed the gazelles sometimes strike the only cheerful note in parts that once boasted a much more varied fauna.

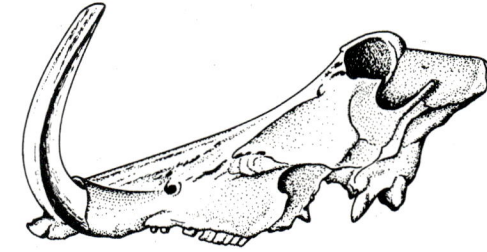

The grotesque warthog

Gazelles and antelopes are familiar creatures of the open plains of East Africa, while giraffes, buffaloes and even rhinoceroses are not uncommon sights in the savannahs. What is rather more unexpected is the appearance of a herd of wild boars – traditionally animals of woods and forests – in the middle of a broad plain practically denuded of shrubs or bushes. Most wild pigs, it is true, prefer a wooded habitat, but the warthog *(Phacochoerus aethiopicus)*, though often seen on bush-covered steppes and forest fringes, is quite at home in the open savannah country as well.

No less strange than the very presence of these creatures is their hideous appearance. The huge elongated head ends in a flattened snout. Two pairs of powerful canine tusks protrude from the mouth – the lower pair comparatively short, the upper ones long and upward-curving. Four unsightly protuberances or warts on the face, two below the eyes and two above the snout, give the animal its common name. The skin is almost naked, apart from some scattered whitish bristles, similar to those of elephants; and a mane of long, bristly grey-brown hairs extends from the top of the head down the back to the root of the tail.

The warthog is indisputably an ugly beast yet its behaviour belies its looks. It is, for example, a surprisingly fast runner and a devoted parent. In order to feed it will often go down on its knees and entire herds, adults and young alike, may sometimes be seen in this odd position. The adults are well equipped to defend themselves against a wide range of enemies and the

The warthog, with its formidable tusks, is the African equivalent of the European wild boar. But whereas the latter is a forest creature, the warthog is equally adapted to life on the savannah.

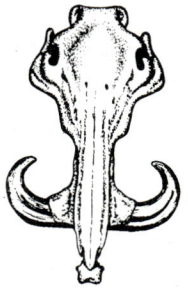

Skull of warthog.

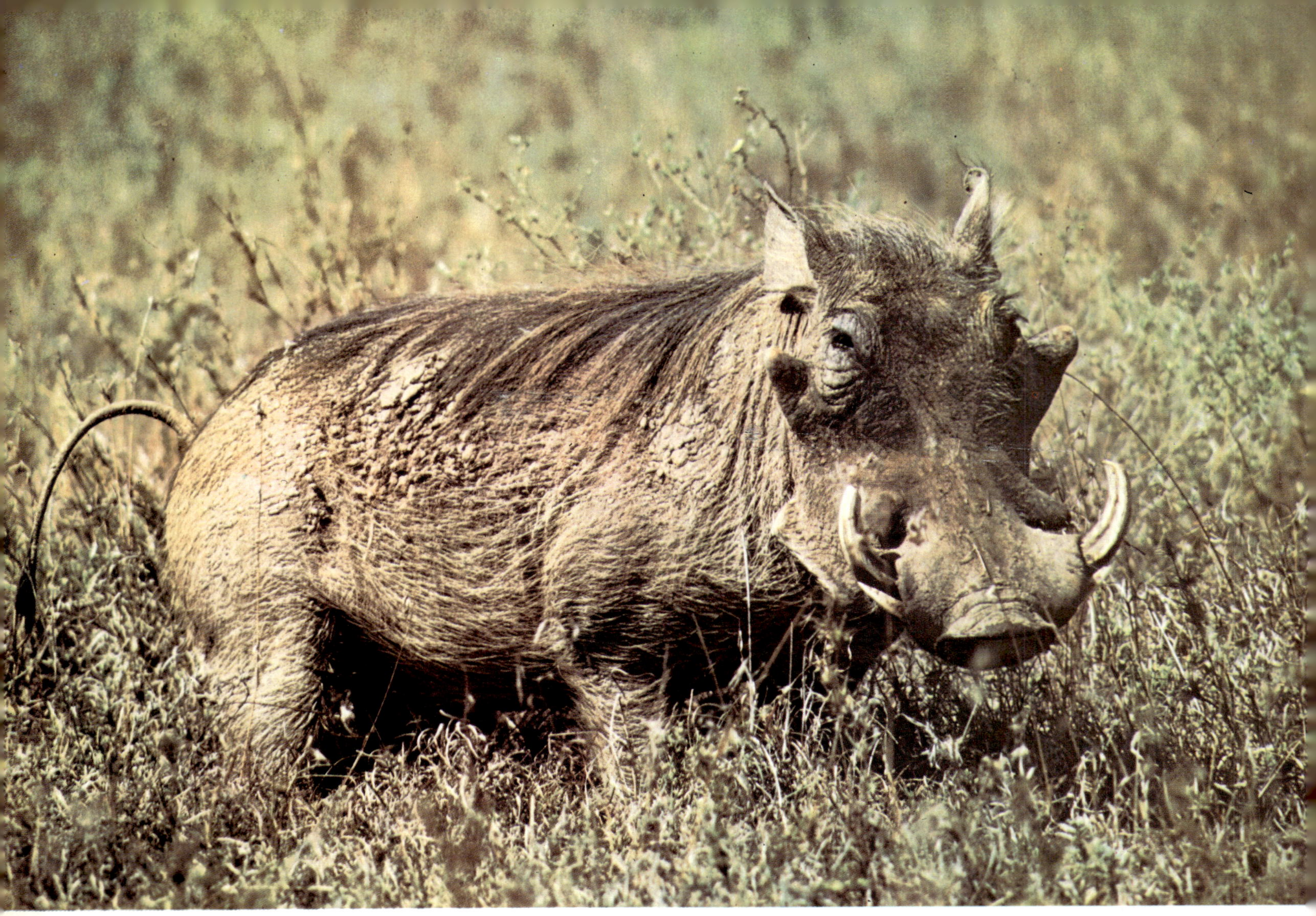
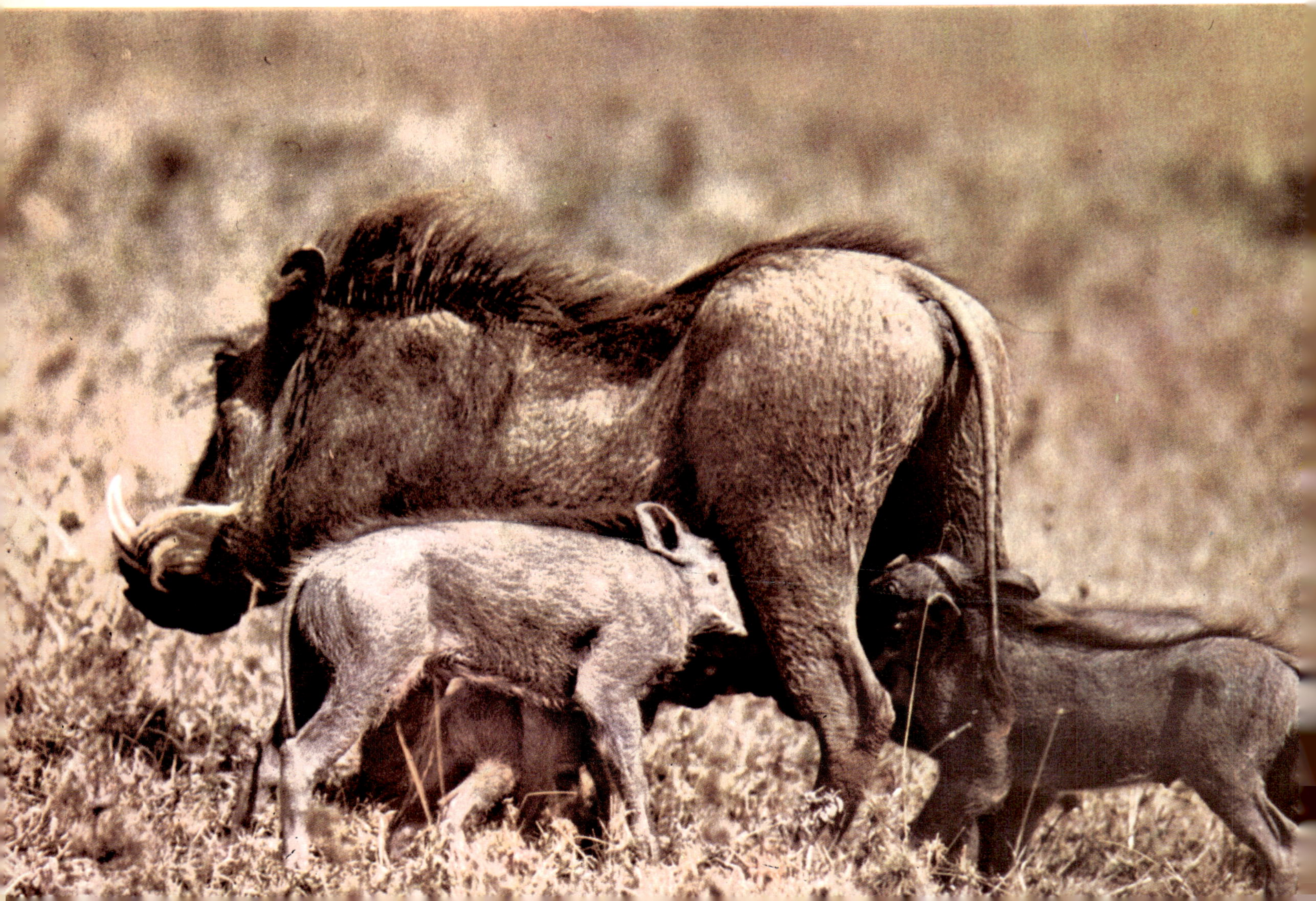

young prudently remain close to the side of their elders. Should a cheetah launch an attack on a baby warthog, for example, the adult will charge in fury, its momentum and the menace of its sharp tusks being enough to put the predator to flight. Meanwhile the young boars will scuttle about seeking refuge under their mother's belly. If the enemy is a lion or a leopard, the warthog may adopt the more sensible course of making for its lair—often the abandoned burrow of an animal such as an aardvark, the space being enlarged by scooping out earth with the tusks. The rush to safety is usually led by the young boars, followed closely by the adults, all with tails held high. They retreat backwards into the lair, the smaller animals without difficulty, the adults, especially the males, making rather heavier weather of it, but with tusks at the ready to ward off the pursuing predator. Naturally they do not always come off victorious in fights against carnivores but they give a good account of themselves; certainly there is no animal of similar size that can fight so fiercely and effectively, particularly when it is in a tight corner.

The mating season is characterised by bloody contests in defence of territory and for possession of females. Gestation lasts from 22 to 24 weeks and a litter consists of from three to six young. The latter, very weak during the first few days, remain in the lair and are attentively cared for by their mother, who suckles them and protects them from intruders. Even when they venture out into the open—kneeling down to grub for worms and sample the vegetation—they do not wander far from her side, gambolling about in an abandoned manner and uttering little excited grunts.

Warthogs are sedentary creatures, only straying short distances. Their diet is basically herbivorous but in fact they will eat almost anything, including roots, berries and seeds, using their tusks as tools to dig food from the ground or rummage in thickets.

Although some males choose, at certain times, to take up a solitary existence, warthogs usually live together in family groups, joining to drink on the banks of rivers or waterholes but dispersing to their separate lairs when alarmed.

Adult warthogs, because of their size and weight, do not need to stand in awe of many predators. But their numbers may be seriously affected in other ways. They, like other wild pigs and many ruminants, are highly susceptible to the disease known variously as murrain, cattle plague or rinderpest. Before effective vaccines were developed there was a grave risk of their spreading the disease to domestic animals.

Although by human standards they appear repulsive, warthogs are in fact likeable creatures, well adapted to a difficult life in highly inhospitable terrain that is shunned by all their other wild relatives. They possess both speed and stamina, defend themselves courageously and display a deep sense of family loyalty and responsibility. Most predators will think very carefully before coming to grips with this powerful and well-armed animal which has proved itself to be uncommonly tough and resourceful.

Facing page (above): Despite its grotesque and ferocious appearance, the warthog is a friendly creature with close family attachments. *(Below)* after the young have been weaned they will adopt an omnivorous diet, including grass, roots, seeds and berries.

WARTHOG
(Phacochoerus aethiopicus)

Class: Mammalia
Order: Artiodactyla
Family: Suidae

Length of head and body:
  Male 50-60 inches (125-150 cm)
  Female 42-56 inches (105-140 cm)
Height to shoulder:
  Male 26-34 inches (65-85 cm)
  Female 22-30 inches (55-75 cm)
Weight: Male 130-330 lb (60-150 kg)
  Female 110-165 lb (50-75 kg)
Diet: Omnivorous—grass, roots, berries, seeds, etc.
Gestation: 150-175 days
Number of young: three to six

Adults
Common name derived from prominent warts on either side of face. Flattened muzzle terminating in a snout. Canines, especially those of upper jaw, protrude in pairs from the mouth. Small, pointed, flexible ears; relatively long legs. Body covered with thick wrinkled skin, with occasional bristles. A mane extends from the head down the length of the back. Fore limbs have callouses already present in fetus. Tail terminates in tuft of stiff hairs. Maximum recorded length of upper tusk is about 24 inches but average is 10-12 inches.

Young
Unlike in other wild pigs the coat has no longitudinal stripes.

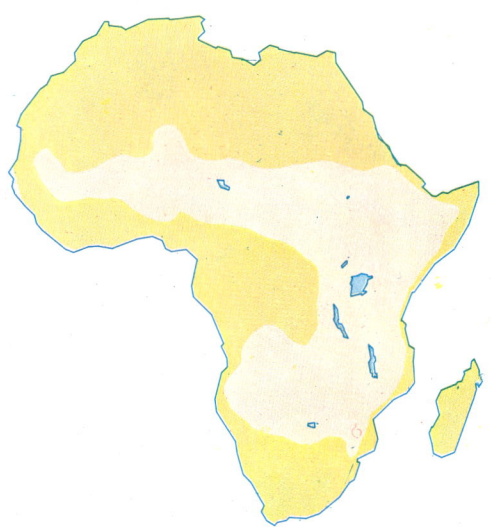

Geographical distribution of the warthog.

FAMILY: Suidae

Unlike other hoofed mammals the pigs (Suidae) are not grass eaters but omnivores, feeding on roots, fruit, mushrooms, insect larvae, small vertebrates and—in the case of the wild boars—snakes (to whose venom they seem impervious).

Their dentition is as one might expect for such a varied diet, though it differs according to species. The domestic pigs and wild boars have six incisors, a pair of canines and seven teeth in each half of each jaw for chewing, the premolars having assumed the form and function of the molars. Other members of the family have fewer premolars and sometimes—as in the warthog—fewer incisors. The canines are often extended into tusks which protrude from the mouth; those of the wild boars and the warthogs become extremely sharp through constant rubbing and serve as formidable defensive weapons.

The flattened muzzle is prolonged into a snout, from the end of which open the nostrils. Two special bones help to make this snout exceptionally tough and resilient for rooting out food from the ground.

Compared with the thickset body the shortish limbs are surprisingly slender, making for a minimal area of contact with the ground, and hence for rapid movement—useful for escaping from predators. Each limb terminates in four toes (with the exception of the peccary, which has only three functional toes on the hind legs), and of these four only the two central toes grip the ground. But on soft terrain, including swamps, the two less developed lateral toes are brought into use and help to prevent the animal from sinking and possibly drowning.

Only the warthog feeds during the day on open savannahs; most other Suidae are forest lovers, preferring to remain in the shade by day and to emerge only at dusk or by night. Their sensory organs are all well developed, the eyes being small, the ears large and flexible. The senses of hearing and smell are particularly keen; the scenting powers of domestic pigs are invaluable for rooting out truffles. In the wild both senses help to give warning of predators.

The stomach of most Suidae consists of a simple sac but shows signs of division into two chambers. This is especially marked in peccaries and babirusas, though the differentiation is not sufficient to permit rumination. Some authors suggest that the relatively simple gastro-intestinal structure of the Suidae gradually evolved into the far more complicated digestive system of the other ungulates.

Whereas the females of most hoofed mammals give birth to only one, or more rarely, two young at a time, sows are much more prolific, with litters of up to twelve in some wild species.

The Suidae comprise five genera, the most familiar of which is genus *Sus* with some fifteen species. They include the domestic pig *(Sus domestica)* and the European wild boar *(S. scrofa)*. Domestic pigs are probably descendants of wild boars, a theory which seems to be borne out by the fact that if they are allowed to revert to the wild state they soon become indistinguishable from their wild relatives.

The warthog *(Phacochoerus)* is a creature of the African savannah, but the solitary forest hog *(Hylochoerus)* has more conventional tastes, as does the bushpig or river hog *(Potamochoerus)*. Both are African genera, the latter including the widely ranging *Potamochoerus porcus*, and *P. larvatus*—confined to the island of Madagascar. The slender-legged Babirusa *(Babirusa)*, with its long curved tusks, is an inhabitant of the Celebes and other Indonesian islands.

Finally there are the South American wild pigs or peccaries belonging to the family Tayassuidae—fairly small and with internal characteristics more resembling those of ruminants.

CLASSIFICATION OF SUIFORMES

Family	Genus
Suidae	Sus
	Potamochoerus
	Phacochoerus
	Hylochoerus
	Babirusa
Tayassuidae	Tayassu

The Suidae are represented in both the Old and New World. Genera include the domestic and wild pig, the bushpig, the warthog, the forest hog, the babirusa and the peccary.

Wild boar
(Sus)

River hog
(Potamochoerus)

Warthog
(Phacochoerus)

Babirusa
(Babyrousa)

Forest hog
(Hylochoerus)

Collared peccary
(Tayassu)

CHAPTER 6

The ostrich: a bird that cannot fly

The ostrich *(Struthio camelus)*, that curious flightless bird with long legs and serpentine neck, is a familiar sight in many of the East African national parks – such as Nairobi and the Serengeti – sharing the same pastures as zebras, gnus, hartebeeste and other antelopes.

Although the majority are nowadays protected in reserves, it is not long since they roamèd freely in large numbers all over the African steppes and savannahs – and even deserts. No bird is better adapted to a life on the open plains. Although they are omnivorous – including insects, reptiles and rodents in their diet – they prefer vegetable foods, particularly tough plants which they tear up and swallow, roots and all; a specialised digestive system enables them to retain the greater part of all ingested material.

The 8-foot-tall ostriches are the sentinels of the savannah, their large dark eyes, protected from the dust by lids fringed with thick hairs, scanning the distant horizon for predators. Keen eyesight is one form of natural protection; speed is another. There are two toes on each of its powerful legs, but only the inner one, protected by a flat nail, is used when running. Thus equipped, the ostrich is capable of speeds of up to 40 miles per hour, and it can keep going for considerable distances.

Exceptional vision and remarkable speed are of course invaluable for survival in regions renowned for their number and variety of predators. From the ecological standpoint ostriches fall into the same category as many hoofed mammals, for their diet consists to a large extent of the same kinds of herbaceous plants. They are therefore liable to be attacked by the same carnivores and form defensive alliances with the grass-eating animals. The far-ranging vision of the birds

Facing page : Ostriches dig out a shallow depression in the sand to make a nest. In this are deposited the eggs of several hens, each laying 6-8, which are then incubated for 42-48 days by the cock and a single hen. When the chicks are hatched they are immediately exposed to the perils of the savannah. Helpless for the first two days, they soon acquire a measure of independence, though they are carefully tended by both parents.

combines with the keen hearing and smell of the antelopes and other mammals to constitute a highly effective alarm system.

Although the ostriches' wings are useless for flying they do serve a purpose in permitting rapid change of direction and sudden braking. They also come in useful for chasing off flies, just as other animals swish the insects away with their tails. The breastbone lacks the distinctive keel to which the muscles of flying birds are attached and has been transformed into a kind of flat shield, protecting the bird from the burning sand when it is lying down or from the kicks of enemies when fighting.

Fine feathers

Throughout history man has been intrigued and fascinated by this huge flightless bird—and has exploited it ruthlessly for his own profit and entertainment. Stylised outlines of ostriches, scratched by prehistoric hunters more than 5,000 years ago, have been found on the walls of caves in the Erongo mountains, at a height of over 7,000 feet. Chinese chronicles of great antiquity refer to ostrich eggs being offered as gifts to the rulers of the Celestial Empire. The Assyrians regarded the ostrich as a sacred bird, and according to the evidence of frescoes it was accepted, together with other animals, as a form of tribute by the Pharaohs of Ancient Egypt. There are also frequent references to this unusual bird in the Bible.

During the imperial period of Rome, ostriches imported from Africa and Syria were paraded before awestruck spectators at festivals, and even graced the dinner tables of the nobility. Aelius Lampridius tells how the Emperor Heliogabalus, who reigned in the 3rd century A.D., had 600 ostrich brains served at a banquet. Julius Capitolinus, writing at about the same period, describes a hunting expedition at which some 300 ostriches, all painted red, were set loose for the amusement of the emperor Gordian III and his court; and the Egyptian tyrant Firmius, also in the 3rd century, ordered ostriches to be harnessed to his chariots.

In the Middle Ages both men and women set a popular fashion of wearing ostrich feathers to ornament their clothing. Hats were freely decorated with these large gaudy plumes and many a knight flourished one proudly on his helmet. Such a craving doomed the birds in their thousands. Arabs and Tuaregs, for example, regularly hunted them on horseback, beating them to death when they dropped from exhaustion and making a meal of them after plucking out the valuable feathers. These found ready buyers in Europe.

The South African Bushmen still hunt ostriches—as they have for centuries past—but they have an additional reason. Decorating themselves with ostrich feathers, they stalk their prey by imitating its strutting gait and kill it as soon as they are within arrow range. Then they take the eggs (the largest of any living bird though small in relation to its stature), scoop out the contents and refill the shells with water. These 'bottles' are buried at strategic points so that when they go hunting there is a permanent supply of fresh drinking water—life and death in the Kalahari Desert with its suffocating droughts. The women also paint the shells, turning them into colourful and authentic works of art, though too fragile for more than local display. It is interesting to note, also that drinking utensils made of the shells of ostrich eggs were also used in ancient times by the Assyrians, Egyptians and Greeks.

Although the Bushmen always looked on the ostrich as fair game they did not kill them in sufficient numbers to affect the population balance. But when the white man arrived on the scene from Europe with his deadly firearms the story was very different. The ostriches, together with the zebras, rhinoceroses and antelopes, were slaughtered in their hundreds; and when, during the 19th century, ostrich feathers once more came into vogue in fashionable circles the situation became desperate. Hunts were now organised by professionals and what began as a sport ended as a massacre. The victims were numbered in thousands, then in millions. To satisfy the growing demand ostrich farms were established, first in Algeria, later in California and South Africa. The passing trade developed into a booming industry. The feathers were taken from both the male and female birds, those of the former being more showy until artificial dyeing processes made the distinction unimportant. They were cut for eight months of the year, a halt being called

The female sits on the eggs during the day, her brownish plumage blending well with the savannah background. The male takes over the responsibility by night when his black plumage is also relatively inconspicuous.

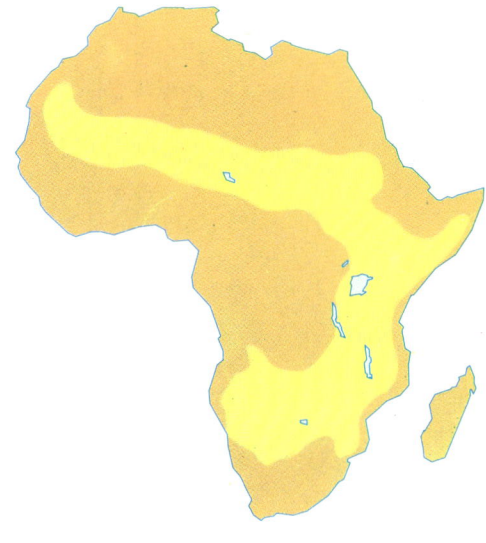

Geographical distribution of ostriches.

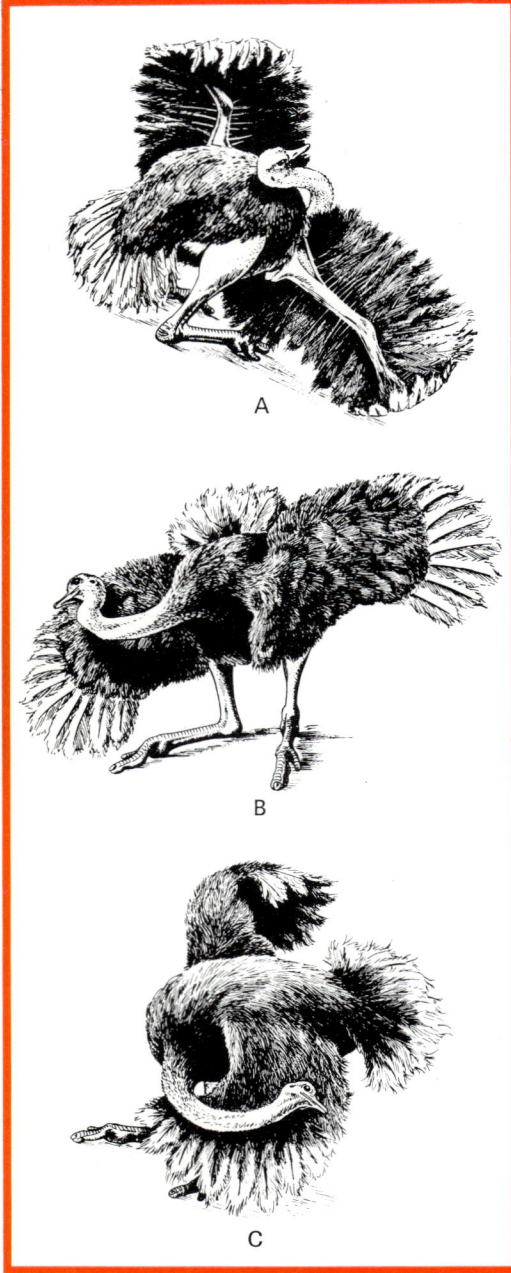

The courtship ritual of the male ostrich is most elaborate. He kneels before his partner, supporting himself on the long tarsi, and extends his wings and tail. Then he rests his head against his back (A) and moves it rapidly from side to side. At the same time he beats his wings rhythmically (B, C) and flutters the tail up and down to coincide with this movement.

during the breeding season to allow them to regrow. This procedure, which apparently causes no pain to the birds, is still used today.

The idea of rearing ostriches in captivity was not new. Pierre Belon, back in the mid-16th century, reported that both the Libyans and Numidians were raising these birds for commercial purposes; and two centuries later the French naturalist Georges Buffon described how certain tribes in North-east Africa captured and raised troops of ostriches for precisely the same reason.

The 19th-century craze for ostrich feathers soon passed—as all fashions will. When the demand dwindled some optimistic businessmen tried to popularise ostrich racing, harnessing the birds to two-wheeled carts or saddling them to be ridden like horses by jockeys. But the sport never attracted the crowds and was soon abandoned. The ostrich farms too began to close their doors. Nowadays the only large-scale farms are in South Africa's Cape Province, the birds being raised not only for feathers but also for their skins—made up abroad into soft leather for gloves, purses and similar articles. The South African trade, still very profitable, paradoxically helped to save the ostrich from total extinction—a happy feature of an otherwise dismal story.

Nevertheless ostriches were exterminated in many regions—and for obvious reasons. A steady worsening of climatic conditions killed off large numbers of birds, while the wholesale collection of eggs by local inhabitants further reduced the numbers. Indiscriminate hunting, with ever more efficient long-range weapons, speeded up the destruction. By the first half of the 19th century ostriches had vanished from the Libyan Sahara and by the end of the century from southern Algeria. Not many years afterwards the last ostriches had disappeared from other parts of the central Sahara and from southern Morocco.

Modern weapons and the extensive use of land transport were to bring about a tragedy in more recent times. During the second world war the entire race of Arabian ostriches—most northerly of all the subspecies, ranging through the Syrian and Arabian deserts as far as southern Iran—was exterminated before naturalists even had the opportunity of studying their behaviour in detail.

These birds, quite capable of standing up to the attacks of such natural predators as lions, leopards, cheetahs and hunting dogs, were powerless against the brutal ingenuity of man himself. Although ostrich chicks had always been vulnerable to the raids of carnivores, once they had grown sufficiently they had little to fear. But the white man's rifle could pick off adult birds at long range and soon entire communities were being destroyed. Happily the more southerly subspecies were able to find protection in national parks and nowadays flourish in reasonable numbers. Elsewhere they are similarly protected by local farmers who complain, however, that their voracious appetites tend to reduce the amount of pasture available for their domestic herds.

The courting season

As the sun sets on the savannah, bringing cool, fresh night air, nature seems to come alive. The weaver-birds twitter in the acacia branches and a variety of animal noises echo through the darkness. One sound in particular may puzzle the inexperienced visitor. It is quite unlike the bark of a jackal, the howl of a hyena or the mewing of a cheetah. If anything it bears some resemblance to the throaty roar of the lion—a single cry, repeated regularly and rhythmically. It sounds like 'mbuni' and this is the word that the Swahilis use for the bird that makes it. For this is the call of the male ostrich at mating time.

The curious cry is sometimes heard at sunrise as well as at the end of the day, and its purpose is to attract females to the nesting sites and to warn other males of the intention to mark out a piece of territory for the occasion. Any young or inexperienced bird rash enough to approach will be driven off by a volley of violent kicks. As night falls, the mating call may become more harsh and strident, indicating that an enemy—

Ostriches generally live in small troops consisting of several males and a larger number of females. Exceptionally keen vision provides good warning of predators but should danger be imminent the group scatters, each bird relying on speed and stamina to make its escape.

Although the ostrich's wings are useless for flying they do help to balance the bird when it is running. The outspread wings also provide much-needed shade for the chicks when first hatched, protecting them against the scorching rays of the savannah sun.

Facing page (above) : The adult ostriches are constantly on the alert for possible danger during the days and weeks following the hatching of their tiny chicks. The lives of the young birds depend on their parents' vigilance and tactical skill in evading and deceiving predators. *(Below)* while the hen incubates some eggs, others, further developed, are merely guarded and allowed to be warmed by the sun. Thus all the eggs eventually hatch at about the same time, permitting the chicks to be reared together by the parents.

perhaps a solitary prowling leopard—is in the immediate neighbourhood.

During the mating season the ostriches adopt two clearly distinguishable routines. In the day both sexes mingle harmoniously in troops but when night falls the males stand guard over their territories and clamorously invite the females to join them. The rival males often fight each other, displaying surprising ferocity, the kicks from their powerful feet sometimes sending an adversary reeling back many yards. The claws too are equally likely to inflict deep wounds, both on rivals and enemies, but as with many species of birds and mammals, serious fights among the males are usually avoided wherever possible.

The birds also engage in extraordinary courtship rituals. The male drops on folded legs in front of his intended partner, then extends his wings, flapping them backwards and forwards, with tail lowered. Reclining the head against the back he proceeds to nod it rhythmically from side to side, rapidly fluttering the tail up and down. This curious ritual lasts about ten minutes, after which the bird gets up and approaches the female with wings outstretched. The procedure is then repeated, together with ground stamping, while the female looks on with every appearance of being thoroughly bored. Once mating has taken place, with the eggs laid and the chicks safely hatched, the troops break up and the birds settle down to family life. The procedures of preparing the nest, laying the eggs and incubating them are fascinating.

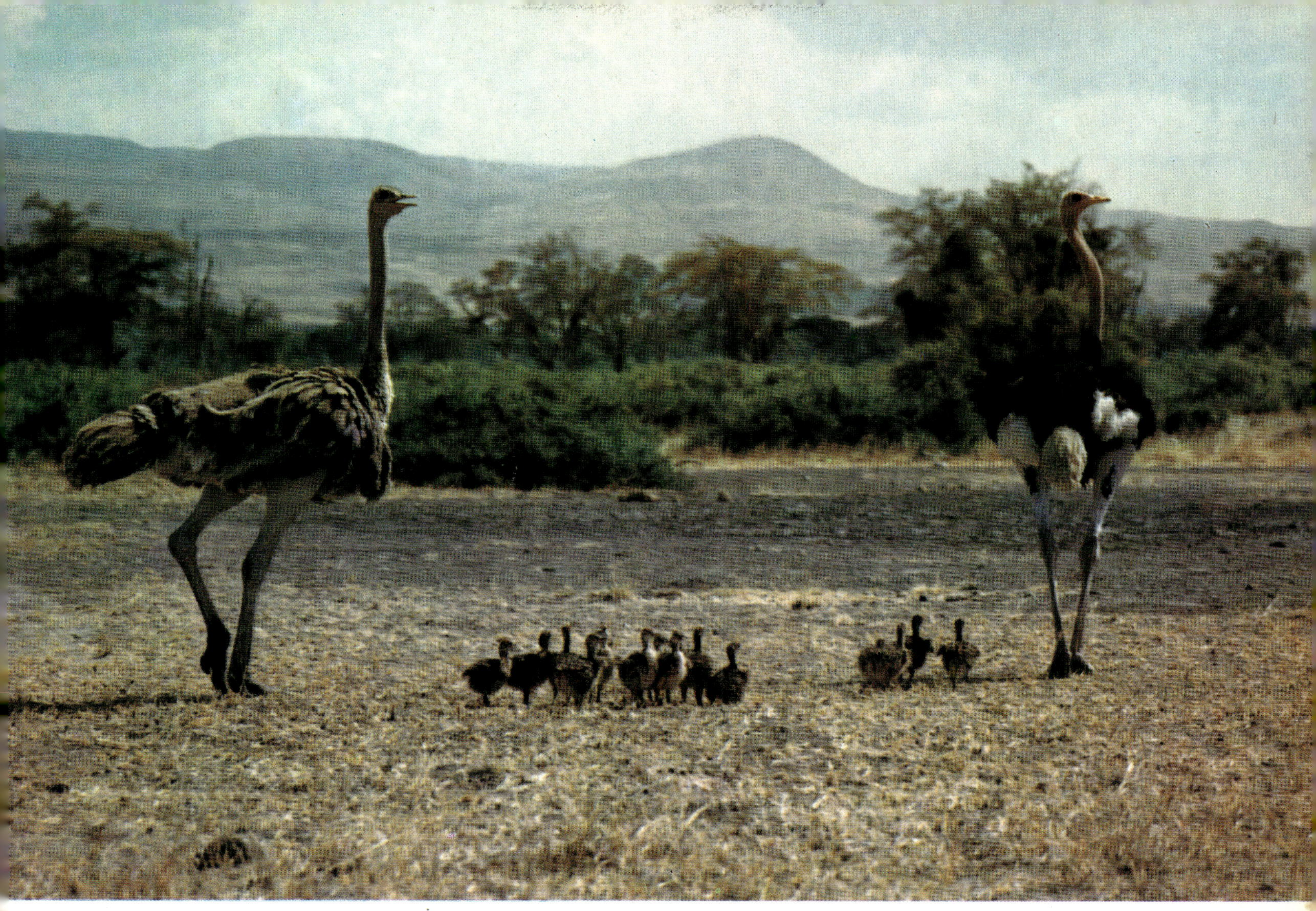

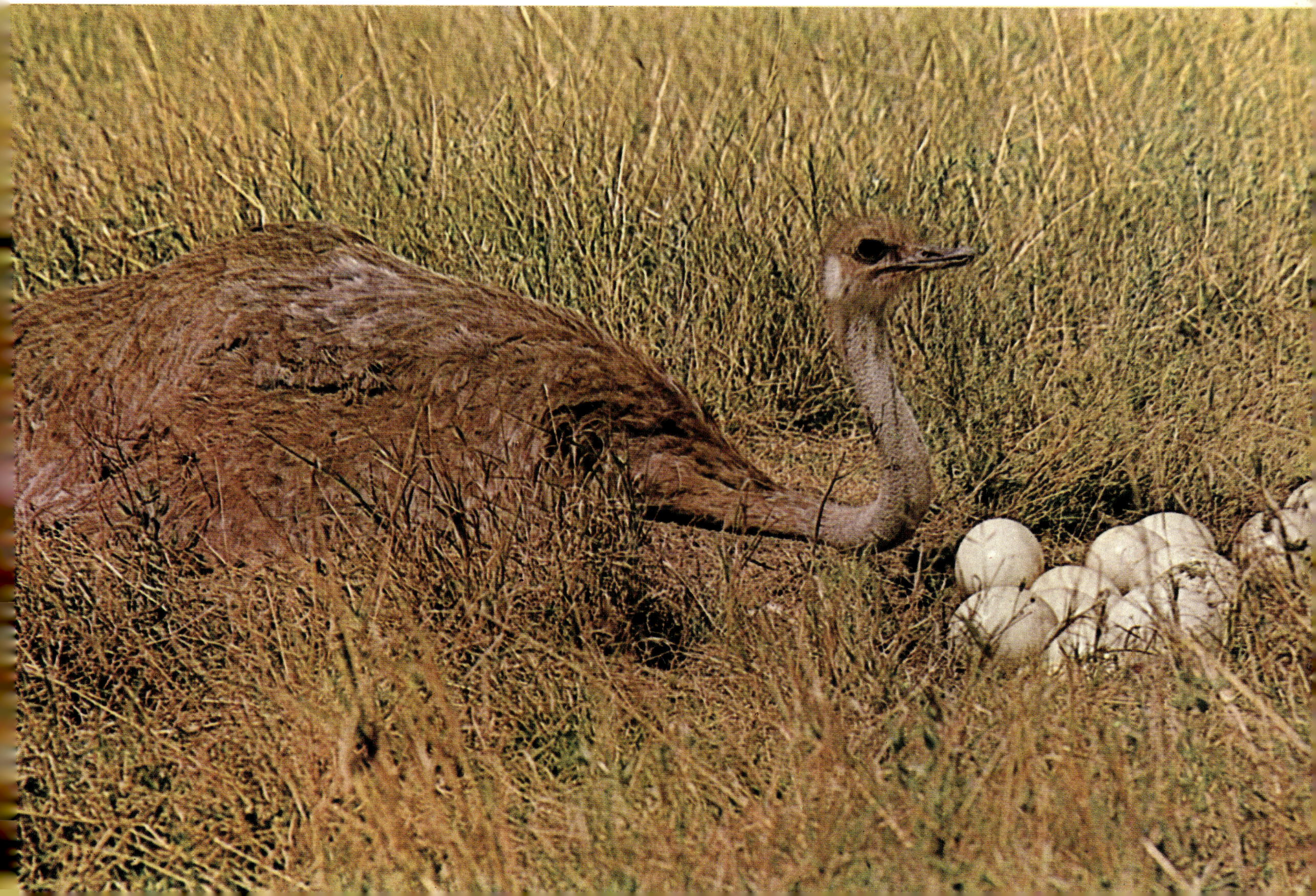

The newly hatched ostrich chick looks more like a hedgehog than a bird, with spiny feathers covering the upper part of its body.

Although the ostrich's egg is the largest of any living bird–6 inches long by 5 inches wide–it is the smallest in relation to the bird's size and weight.

The new generation

When the ostriches have selected a suitable site for laying their eggs they stir up the earth or sand, or clear an area of grass–some 10 feet across–by trampling it, tearing it up by the roots or eating it. The nest consists of a simple hollow, about a foot deep and 3 feet in diameter. From it issue well-worn tracks, used by the parent birds for coming and going.

Each breeding cock partners several hens, sometimes as many as 5 or 6, but more generally 3. Each of these hens lays 6-8 eggs, often more, which are deposited in the communal nest at regular intervals every other day. The cock, however, only forms a close liaison with a single hen and together they will later be responsible for the rearing of the chicks. The other hens seem to have no other function but to lay eggs, after which they leave the nesting site. This is the conclusion reached by several authorities, though not definitely established. On the other hand, there is no confirmation of the popular belief that the ostrich is a polygamous bird.

The time of egg-laying depends on the rainy season and varies considerably from one region of Africa to another. Over a large part of the continent, however, it occurs between September and November. In very dry years it is probable that ostriches living in areas badly affected by drought do not succeed in breeding. This of course is an additional reason for steadily decreasing numbers, and a strong argument for making sure that they are protected.

The oval eggs are grey-white or ivory, the shells being hard and glossy but slightly pitted, this varying according to subspecies. They measure 6 inches long and 5 inches wide, weighing approximately 3 lb, these figures differing of course in individual cases. Though the largest eggs laid by any living bird–they are equivalent to about 25 ordinary hens' eggs–they are the smallest of any bird species in relation to body size and weight.

Naturalists have sometimes been surprised to come across ostrich eggs laid in hollows some distance from the main nest. These were at first thought to be supplementary eggs, designed eventually to provide food for the chicks, but subsequent studies showed that this was not the case, for such eggs were clearly more advanced in development than the others. This appears to be due to the fact that the adults begin to incubate some eggs before others are laid, the complete egg-laying cycle lasting some two weeks. Furthermore, the heat of the sun in these latitudes is so intense that incubation of the eggs continues even when the parent birds are not in attendance. It would therefore seem logical for the first eggs to hatch before the last–staggered in such a way as to make it difficult or impossible for the parents to rear the chicks. Yet in some subtle, instinctive way the adult ostriches prevent this happening. Careful study has shown that when the embryo has reached a certain stage of development the parents hollow out a patch not far from the nest, depositing in it each egg that is near to hatching, allowing the sun and the embryo itself to complete the process. This ensures that all

eggs hatch more or less simultaneously, enabling the parents to rear the chicks normally. A succession of staggered hatchings would pose insuperable problems.

Some authors claim that even before incubation commences the adult ostriches cover their eggs with sand and keep a watch on them from the nearby nest. If danger threatens they flatten themselves against the ground in the tall grass; but should the enemy come too close—say within 6 feet—they will make their escape, returning only when the coast is clear. Some eggs and chicks may thereby be sacrificed.

The incubation period lasts between 42 and 48 days. The cock assumes the responsibility of sitting on the eggs by night, when his black plumage provides effective camouflage, while the hen takes over during the day, when her drabber plumage makes it equally hard for predators to spot her, because she looks like a dried bush. Both parents, however, run the risk of continuous attack during this period. Fortunately they are proficient in hiding from their enemies—stretching their necks and flattening their bodies against the ground, remaining absolutely motionless. Their reactions, however, vary according to the identity of the attacker. Confronted by a lion or a human, even the cock will flee. During the final stages of incubation the cock may show aggressive tendencies and even challenge an intruder, but these are little more than displays of intimidation and should they fail he will abandon the nest. Smaller predators, however, are

Two ostriches keep a lookout for predators in Ngorongoro Crater while a gnu grazes in the background. Ostriches and antelopes often form mixed herds—eyesight, hearing and sense of smell all combining to provide both birds and mammals with a highly effective advance warning system.

often driven off with hefty kicks. Imminent danger is signalled by the male, which emits a loud hissing sound.

The newborn chicks are hardly able to lift their heads or walk during the first few hours of life and take no food for 24 hours. They soon gain strength, however, and after a couple of days are able to run alongside their parents. At this age they look rather strange, the upper part of the body being covered with a coarse down while softer feathers grow on sides and belly. The chicks are ash-grey or reddish, with dappled black markings, especially on neck and head. Their immediate instinct when danger looms is to run as fast as they can, and if closely pursued they will flatten themselves against the ground—shamming dead—where their coloration makes them surprisingly difficult to detect.

At such times the cock will try to divert the attention of the enemy, while the hen seeks safety. He will sometimes run directly into the predator's path, zigzagging from side to side and drooping one wing as if it were broken. The hen also attempts to fool the aggressor by pretending to be wounded, though in less ostentatious fashion. Should the carnivore pursue her she will resume her normal gait and try to lead her chicks as swiftly as possible to a sheltered spot where all can hide. Once the danger is past the parents rejoin each other, nervously flapping their wings, and then round up any stray chicks, which are easily located by their loud piercing cries.

As with most animals, the baby ostriches are at greatest risk immediately after they are hatched. All kinds of predators—lions, leopards, cheetahs, hyenas, jackals, snakes and birds of prey—are waiting to strike. The German ornithologists Mr and Mrs Sauer, who made a special study of breeding ostriches, once saw a huge crow kill a chick and then fly off with it after devouring its intestines. But the birds may be endangered even before they are hatched. George Adamson watched a hyena prowling about 400 yards from a nest containing 17 eggs, which were being incubated by one parent bird. A few hours later only the shells were left. Adamson also saw Bateleur eagles and other birds of prey keeping a close watch on a nest where both parents were present, waiting for a convenient opportunity to swoop on the chicks.

As they grow the young ostriches gradually become more self-reliant and less vulnerable, so that eventually the only enemy they have to worry about is man himself. At about two months they begin to lose their strange spiny covering of down. This is replaced by plumage similar to that of the hens and these feathers are retained by both sexes for approximately two years. The cocks then take on their characteristic shiny black plumage, with white feathers in the tail and on the wing tips. But they are not fully developed and sexually mature until they are three or four years old. They will have grown 2-3 inches every week until about 6 feet tall, the growth rate then slowing down to give an eventual height of nearly 8 feet.

Ostriches are relatively long-lived birds. In captivity they have been known to live until they are about 50 but their average longevity is 30-40 years.

The ostrich has a small head compared with the rest of its body. The short beak is deeply cleft to a point just below the large black eyes.

Food and drink

Because of its excellent eyesight and running ability the ostrich normally lives on open plains where predators can be detected at a distance and sufficient time afforded to escape from them. Some subspecies inhabit bush-covered steppes, one of these being *Struthio camelus molybdophanes*. At breeding time the male boasts a magnificent cobalt-blue plumage on legs and neck, blending perfectly with the black feathers on the rest of the body. It is amusing to see these birds rearing their heads like periscopes above the dwarf acacias. Neither they nor other subspecies venture into denser vegetation where leopards and other predators might lurk.

Although not a migratory animal, the ostrich has to move about, especially to find drinking water. But given plenty of fresh grass and certain water-retaining bulbous plants it does not need to visit pools or rivers too often. It has been suggested that it can quench its thirst adequately on morning dew. In any event its modest water requirements enable it to live contentedly not only in tall-grass savannah but also in the bush, on mountain-sides and even in deserts. This wide range proves it to be a remarkably adaptable bird.

Ostriches lead an uneventful life. In the morning they all troop out in the same direction, heads lowered, pecking away

A pair of ostriches in the Serengeti. The birds are omnivorous, attracted by any unusually shaped or coloured objects. Although their basic diet consists of grass and plants they also eat insects, lizards and other small animals as well as inanimate objects such as pebbles, shells and pieces of wood.

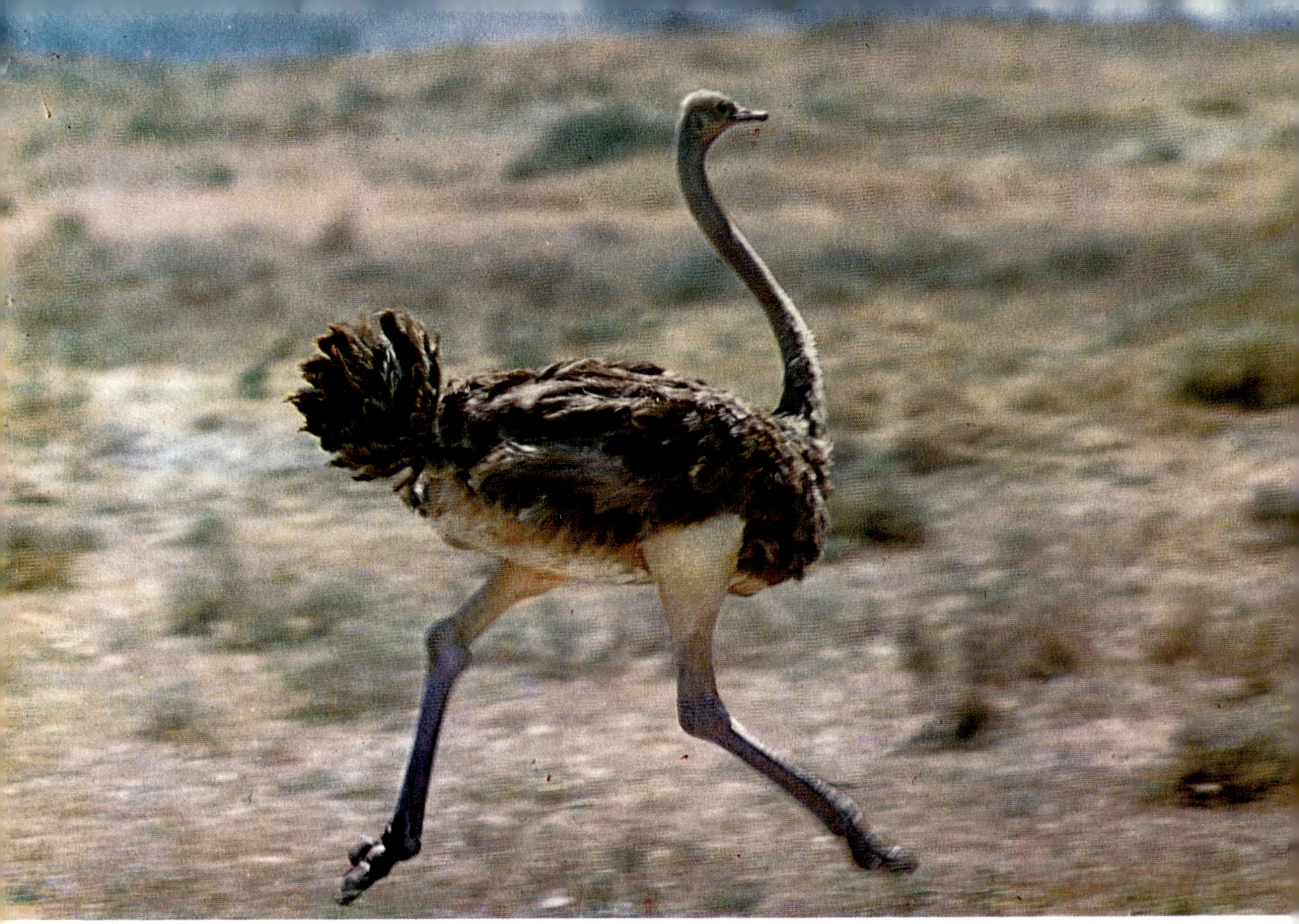

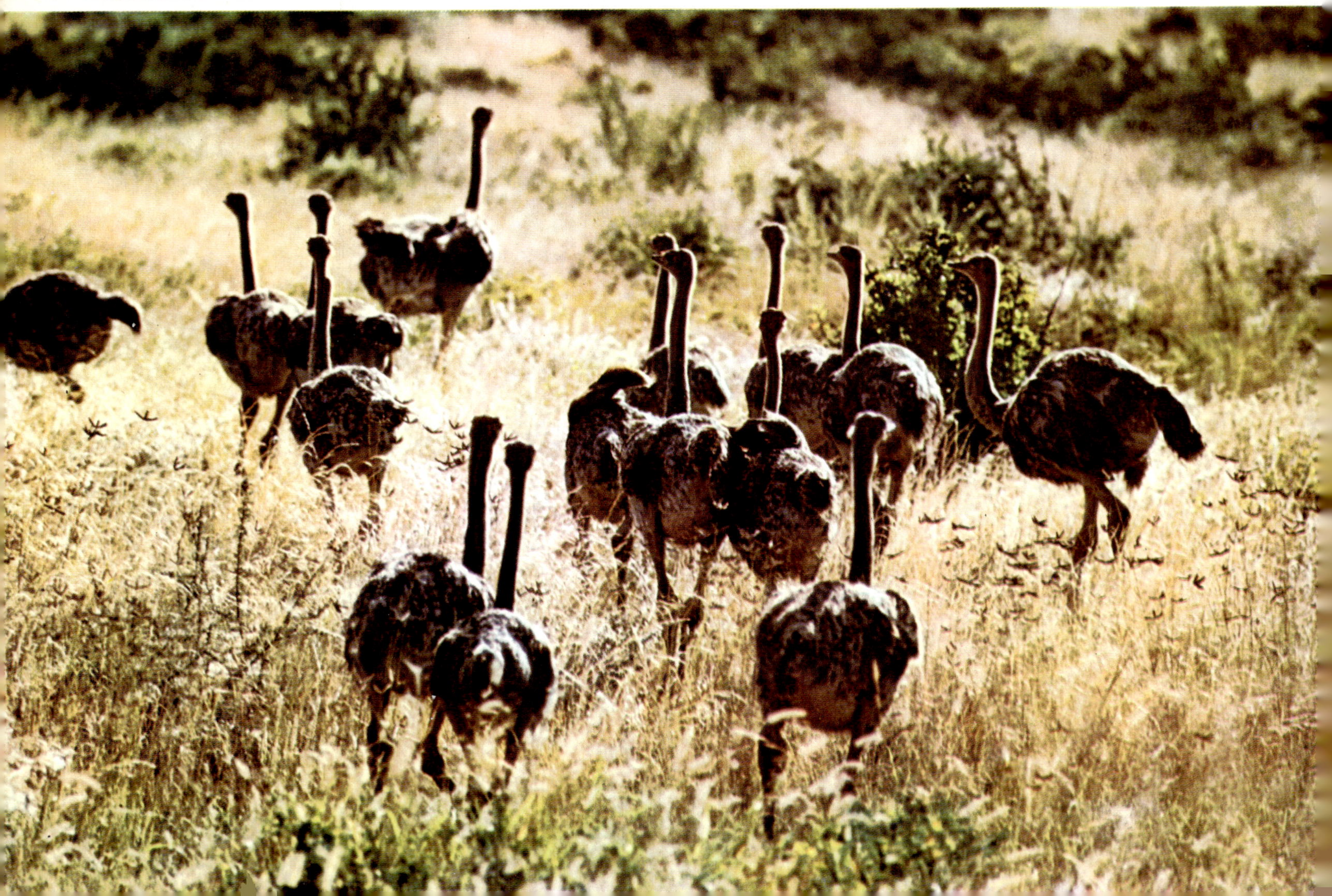

without interrupting their leisured walk. Their widely cleft beak is an all-purpose tool, enabling them to tear up grass, swallow seeds and fruits, and catch insects, lizards and other small vertebrates. They uproot small plants but will also browse happily on the stems and branches of shrubs or swallow sharp thistles, as is proved by the examination of their stomach contents.

Ostriches in captivity are renowned for eating a wide range of unusual objects and in the wild they will swallow anything that because of its shape or colour happens to catch their eye. Dr Valverde, who studied their food habits in the Sahara, found about 13 lb of assorted materials in the gizzard of one adult male, including about 2 lb of oval-shaped pebbles, probably to help crush tough vegetable matter. Most of the residue consisted of plants, mainly in the form of thistles, while there was also a small quantity of bean-like seeds of a plant of the genus *Crotolaria*. Finally he found more than 1 lb of cricket larvae and nymphs.

Visitors to zoos are usually requested not to feed strange objects to ostriches. The experiment may be entertaining, but even the strong digestive system of the ostrich has its limits. Some metal objects may be ground down with other material, but those with sharp points and edges may be harmful.

Fastest animal on two legs

Olympic runners who break 10 seconds for the 100-metres dash or who clock under 4 minutes for the mile can be assured of hitting the headlines in the daily newspapers. Yet the sprinters are running at only a little over 20 miles per hour while the long-distance runners' rate has dropped to near 15 miles per hour. By animal standards—the cheetah, for example, can reach almost 70 miles per hour—their efforts are puny. Of course the Big Cats and other fleet-footed animals such as the hare and the greyhound have a natural advantage in possessing four feet. Yet the ostrich, provided with only two, does not seem to find this a handicap; and in a straight race it would leave the finest of our runners flat-footed at the start.

Its legs are of course exceptionally strong. Even a month-old bird can run at 35 miles per hour while the adult ostrich can keep up a speed of about 40 miles per hour for some considerable distance. It moves forward without any apparent effort, taking giant 6-foot strides, wings slightly extended, head erect—easily the fastest creature on two legs.

Speed and stamina are the ostrich's natural defences and best means of survival on the African savannah. Although some carnivores are faster over short stretches none—except perhaps the formidable hunting dogs—can touch the huge flightless bird in distance running. This is one reason why it has managed to to survive in an environment that was surely not designed for it. The ostrich, having triumphed over so many natural handicaps and hazards, is indeed unique among the animals of the African continent and merits the fullest possible protection in its last remaining refuges.

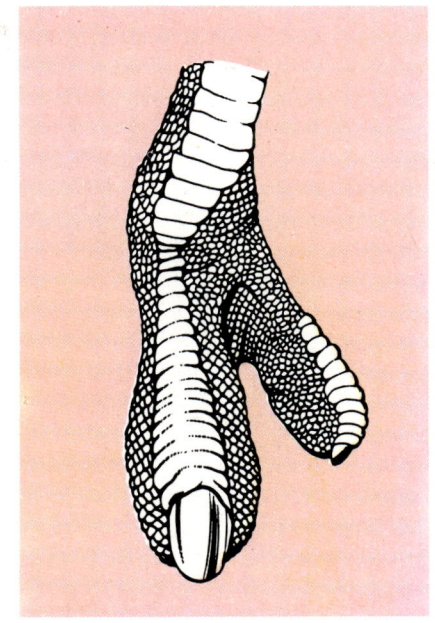

The foot of an ostrich. All animals that rely on speed have limbs and extremities which are well adapted for running, with a minimum of contact between the feet and the ground. Thus the ostrich only has two toes, the inner one being provided with a tough, sharp nail. This is also a formidable weapon for the cocks in their ritual mating fights.

Facing page (above): The ostrich takes long strides when it runs, keeping its head high and its wings slightly outstretched. At full speed it can move at 40 miles per hour. *(Below)* a troop of ostriches on the savannah. Although these birds were widely distributed throughout Africa in the 19th century, their numbers are now much reduced and they depend on the protection afforded by national parks and game reserves.

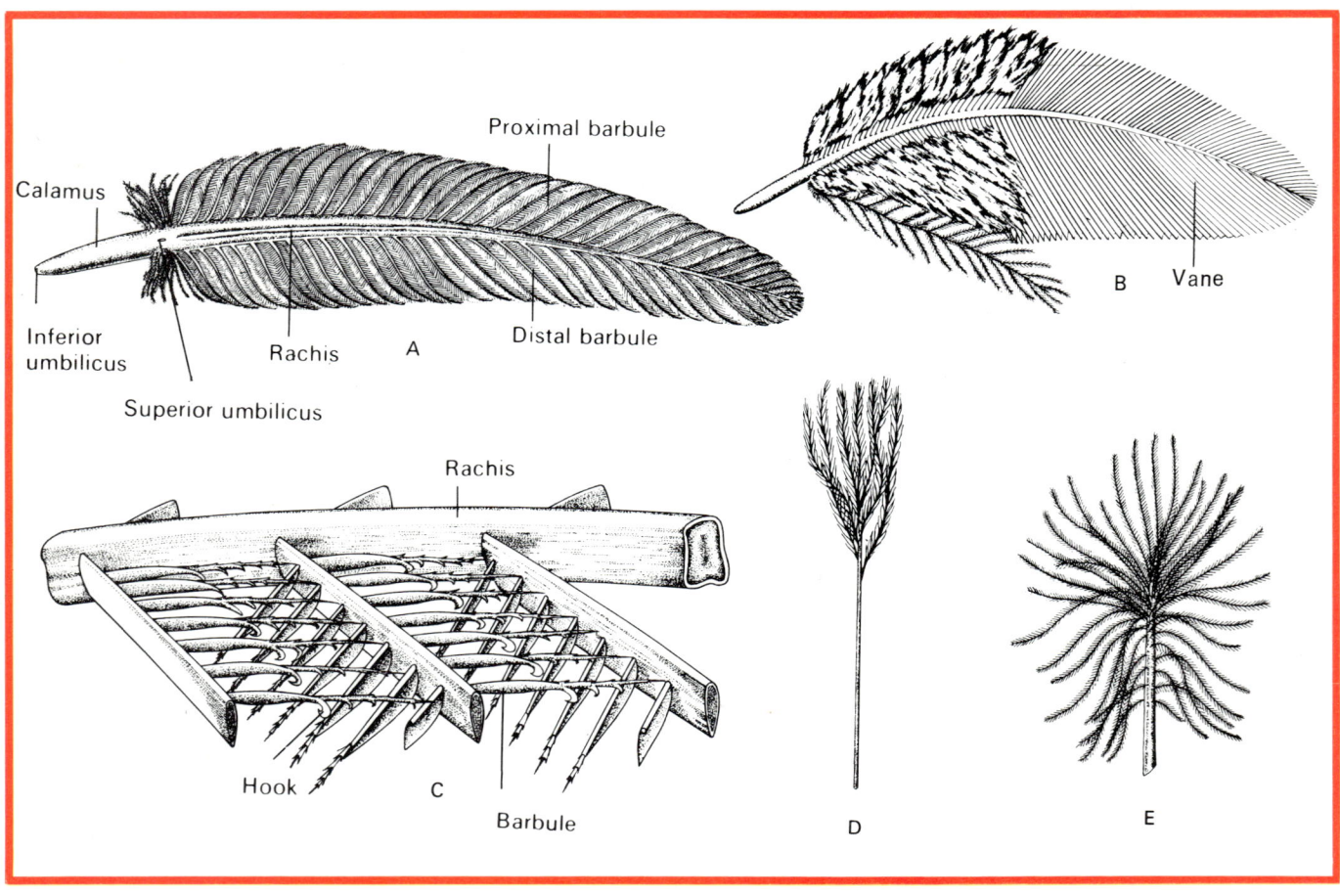

Calamus

Proximal barbule

Inferior umbilicus

Rachis

Distal barbule

Superior umbilicus

A

Vane

B

Rachis

Hook

C

Barbule

D

E

CLASS: Aves

In the late Jurassic period there lived a very strange animal. It was about the size of a pheasant, had bird-like feet and fore-limbs in the shape of wings, each with three digits terminating in claws, which were presumably used for clinging to branches. Yet anyone looking first at the head, the powerful jaws and the sharp teeth might be justified in thinking it a reptile, and closer examination of the skeleton, with its heavy compact bone structure, keel-less breastbone and large number of movable caudal vertebrae, would reinforce this opinion. However, this extraordinary creature was no reptile for it had one anatomical peculiarity that placed it firmly in quite a different category. Its body, wings and tail were covered with feathers. It was in fact the ancestor of the modern birds.

This animal was clearly not able to fly in the manner of true birds. It probably spent most of its time in trees, gliding from one to another, or from tree to ground, rather like a modern flying squirrel. It may have moved about on dry land too, though perhaps not rapidly enough to escape the many predatory reptiles of the time. One such predator apparently trapped one of these prehistoric birds, forcing it from the comparative safety of the branches into a swamp below, where it sank and drowned. The outline of its body was preserved in the thin layer of mud that formed its shroud. About 150 million years later, in 1861, paleontologists near Langenheim in Bavaria found its fossil remains embedded in Jurassic limestone. They gave it the name of *Archaeopteryx lithographica*. The skeleton was later bought and exhibited in London's Natural History Museum.

The discovery caused a sensation in the scientific world. For *Archaeopteryx* (literally 'ancient wing'), though it showed unmistakable bird-like features, was equally clearly reptilian in origin. It was a triumphant confirmation of the hotly-debated theory of evolution advanced only two years previously by the English naturalist Charles Darwin in his momentous book *On the Origin of Species by Means of Natural Selection*. In 1877 a second skeleton was unearthed near Eichstätt, also in Bavaria.

Here was incontrovertible proof that birds were descended from reptiles. But with which reptile group was *Archaeopteryx* linked? Science has still not provided a certain answer. Logic would seem to demand that it be allied with the pterosaurs–flying (or at least soaring) reptiles dating from the Early Jurassic, with beak-shaped jaws, teeth, long tails and wings (some had a wingspan of 20-30 feet). But these creatures, extinct for 60 million years, are not thought to have been the ancestors of birds. Paleontologists are more inclined to see a remoter link with the thecodonts, small carnivorous reptiles that lived in the Early Triassic about 190 million years ago. Yet if this theory is correct there is still a missing link and some scientists have tried to bridge the gap between the flightless reptiles and *Archaeopteryx* by speculating on the possible existence of an intermediate creature to which they have given the name *proavis*.

The acquisition of feathers was a vital step in the evolution of creatures destined to master a new environment–the air. Feathers are in fact epidermal (skin-like) structures, similar to and derived from the scales of reptiles. A typical contour feather–so called because together they define the bird's streamlined body shape–consists of a tapering axis or shaft, the hollow base of which is the quill or calamus and the extension of this the rachis. Attached to the latter on either side is a series of parallel rods or barbs. Each barb has two rows of barbules, one on the distal (away from the body), one on the proximal (towards the body) side. The barbules overlap one another and also have two rows of smaller barbicels, some ending in tiny hooks, which enable the barbules to interlock and form a strong but delicate vane or web. Where the quill and rachis meet there is a tiny pit called the superior um-

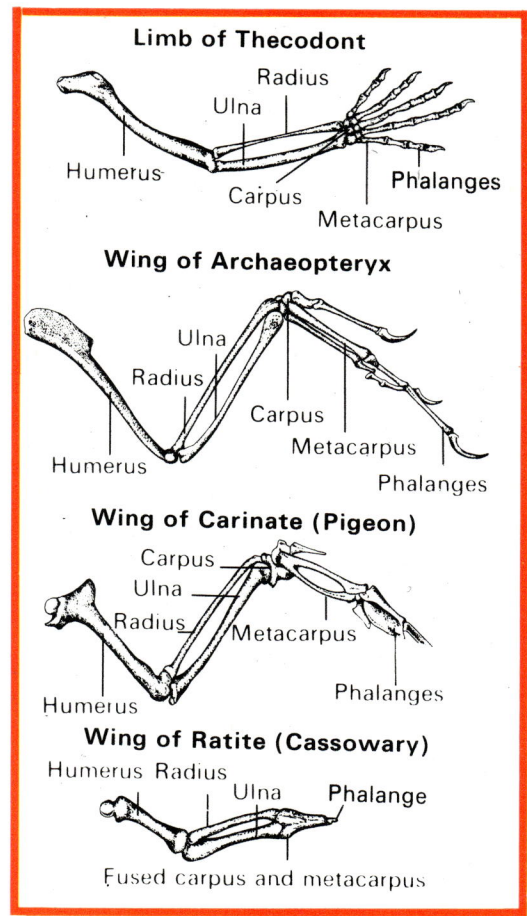

During the course of evolution the fore-limbs of birds have gradually developed in order to support their wings, which have become longer and more powerful with the passage of time. Among the progressive changes which took place between the age of reptiles such as the thecodonts (some 200 million years ago) and the emergence of true flying birds were the atrophy of the functional fingers and the fusion of the carpal and metacarpal bones. In the flightless ratites the wings also became atrophied.

Facing page (above): Two seagulls in flight. *(Below)* the typical structure of various forms of bird feathers. A. Remige or flight feather. B. Contour feather. C. Detail of interlocking barbule hooks. D. Filoplume. E. Downy feather or plumule.

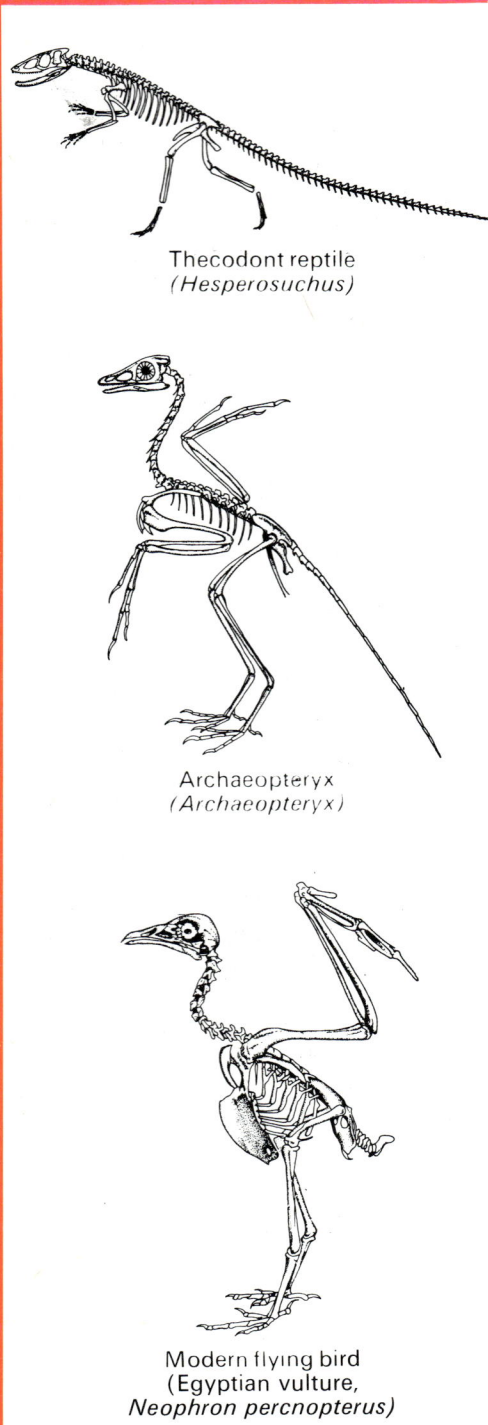

Thecodont reptile
(*Hesperosuchus*)

Archaeopteryx
(*Archaeopteryx*)

Modern flying bird
(Egyptian vulture,
Neophron percnopterus)

As birds took to the air their skeletons underwent numerous changes. In this chart—which does not necessarily show a direct line of descent from reptile to bird—the main features to note are the gradual shortening of the tail, the lengthening and specialised function of the fore-limbs and the appearance of a breastbone keel to which are attached the powerful flight muscles.

bilicus out of which sometimes grows the hyporachis or aftershaft, varying considerably in size and number according to species.

The contour feathers which form the bird's normal covering are not distributed evenly over the entire body but are found in defined areas known as feather tracts (pterylae) with naked areas of skin (apteria) in between. But the fluffy contour feathers extend over these latter zones to provide insulation for the whole body. The penguin is exceptional in having contour feathers evenly distributed over the whole body.

Concealed by the contour feathers are downy feathers or plumules. These have fluffy barbs and provide an additional lower layer of insulation. Scattered among the contour feathers are also hair-like filoplumes with a few barbs and barbules at the tips.

The most important of the contour feathers are the wing or flight feathers (remiges) and the tail or steering feathers (rectrices). The larger portion of the wing surface is made up of overlapping remiges which differ in number and kind. The primaries, attached to the digits of the hand, are well developed and vary from 9 to 12 according to species. The secondaries (varying more widely from 6 to 37) are attached to the ulna of the forearm; and the tertials, when present, are attached to the humerus of the upper arm. Smaller overlapping feathers called coverts (tetrices) cover the bases of the flight feathers. The tail, which is used for steering, is formed by the long retrices, attached to a terminal bone (pygostyle) and these too vary, according to species, from 4 to 16 pairs.

In the long course of their evolution the birds underwent anatomical and physiological changes which made it possible for them to grow feathers. The fore-limbs gradually increased in length while the fusing of certain bones gave additional strength. This lengthening of the limbs can already be seen in *Archaeopteryx* but the creature still had three fingers with claws suited to its arboreal life. Today only the chicks of the South American hoatzin show similar primitive features in the shape of claws on the wing tips, enabling them to cling reptile-fashion to branches, and these disappear later. The ends of the metacarpal bones of the second and third fingers of *Archaeopteryx* were not joined as they are in true flying birds, where they form a solid shaft to which are attached the primary flight feathers.

The tail of *Archaeopteryx* was comparatively long and flexible, like that of a lizard, but could not have been used for steering. In later species it gradually shrank, thus providing proper support for the tail feathers needed for changing direction during flight. But wings and tail alone, though indispensable, were not in themselves sufficient for aerial flight; a light body was also necessary. *Archaeopteryx* had a heavy skeleton. Modern birds have light, hollow bones while a series of internal air-sacs also help to lighten the body weight. Their flight structure is very varied and adaptable, dependent mainly on the well-developed pectoral muscles attached to the breastbone by means of a deep keel. The latter is such a characteristic part of a bird's anatomy that the entire subclass of flighted birds is known as Carinatae (from *carina*—keel), as distinguished from the subclass of Ratitae—keel-less and flightless birds—such as the ostrich and the kiwi.

The outer framework of a bird's body—comparable to an aeroplane fuselage—must be strong enough to support the wings, to activate the powerful flight muscles and to protect the internal organs from injury. To this end certain bones (clavicles, ribs, pelvic girdle and some vertebrae) are fused rigidly. The muscular power necessary to initiate and maintain flight is aided by a rapid rate of blood circulation and high arterial pressure. The heat generated by the energetic flight action would lead to a significant increase of body temperature were it not for the natural cooling system provided by the air-sacs. These are long tubes with pouches extending to various parts of the body, including the viscera and many bones. They are also linked by a single opening with the lungs. During flight the fresh air inspired by the bird enters the lungs and passes through to fill the air-sacs, and is circulated to regulate the body temperature. The system of lungs and air-sacs also results in a more efficient use of oxygen, some being taken in when entering the lungs and more after air is returned by the air-sacs to the lungs, prior to exhalation.

The cerebellum—that part of the brain governing motor co-ordination, by which the bird responds immediately and correctly to all manner of complex stimuli—is large; so too are the optic lobes and nerves, providing excellent vision. The sense of hearing is also well developed—some nocturnal birds managing to orientate themselves by sound alone—but the sense of smell appears to be relatively poor in most species, the olfactory lobes being comparatively small.

Judging by the evidence provided by fossils we know that by the end of the Cretaceous, about 70 million years ago, birds had already acquired most of the afore-mentioned characteristics. The only significant developments after that period were the return of the flightless Ratitae to a terrestrial mode of life and of the flightless Spheniscidae (penguins) to a semi-aquatic existence.

The birds—capable of travelling rapidly from one region to another and extraordinarily adaptable to differing environments—have populated every part of our earth, from the frozen wastes of Antarctica to the equatorial jungles. So diversified are they in shape, size, coloration and behaviour that scientists have classified approximately 8,600 living species.

Most birds are monogamous, forming pairs for the rearing of the brood, and a number are known to mate for life. Some are polygamous and a few are polyandrous, the female mating with several mates. The extent to which the partners of a pair share parental duties differs widely. In some species there is almost complete equality, but in others the male takes only a minor role, such as helping to feed the young, or, as in birds of paradise, takes no part at all. There are a few species, notably phalaropes, in which the roles of the sexes, apart from egg-laying are almost completely reversed. In these species the female is dominant, usually larger than the male and more brightly coloured. In six families of which the Old World cuckoo is the best documented, the females are parasites laying their eggs in the nests of other birds.

SUBCLASS: Ratitae

The birds belonging to the subclass Ratitae are incapable of flying and have instead succeeded in adapting themselves to running on the ground. All save one of these species are unusually tall. The ostrich may stand nearly 8 feet high and weigh over 300 lb, but other ratite birds, now extinct, were even bigger. The elephant birds *(Aepyornis)* of Madagascar and the moas *(Dinornis)* both outstripped the ostrich in size. The latter were the largest birds ever known, some of them towering 13 feet high and weighing 500 lb.

Nobody knows how the ratites originated. If in fact they were descended from creatures that never adapted themselves to aerial flight, they can be ranked as the most primitive birds in existence. On the other hand, if they are simply descended from carinates, having somehow lost their powers of flight—and most experts studying fossil remains, embryology and comparative anatomy are inclined to believe that this is the case—they have evolved in a unique manner. The carinate birds that most closely resemble the ratites are the tinamous (Tinamiformes), South American ground birds which ornithologists find difficult to classify. Their keeled breastbone is similar to that of ordinary flying birds whereas other anatomical features, including an unusual palate structure, seem to link them more closely with the ostriches and rheas.

Storks live in the African tropics and are noteworthy for their boldly contrasted coloration. This is the saddle-billed stork *(Ephippiorhynchus senegalensis)*.

CLASSIFICATION OF AVES (BIRDS)	
Subclass	**Order**
Ratitae	Struthioniformes
	Rheiformes
	Casuariiformes
	Apterygiformes
Carinatae	Tinamiformes
	Sphenisciformes
	Gaviiformes
	Podicipitiformes
	Procellariiformes
	Pelecaniformes
	Ciconiiformes
	Anseriformes
	Falconiformes
	Galliformes
	Gruiformes
	Charadriiformes
	Columbiformes
	Psittaciformes
	Cuculiformes
	Apodiformes
	Coliiformes
	Trogoniformes
	Coraciiformes
	Piciformes
	Passeriformes

Two typical representatives of the subclass Ratitae. (*Above*) an emu and (*below*) a cassowary.

Most ratites have soft feathers—the barbs not being hooked—distributed uniformly over most of the body and not just in the pterylae. Their skeletons show many unusual features, some peculiar to themselves, others—more primitive—also found in prehistoric reptiles. What is immediately obvious is the complete absence of a breastbone keel. Other characteristics are the rudimentary or non-existent clavicles (collar-bones), the absence in most species of the bony joint (pygostyle) to which the tail feathers of the carinates are attached, and the unorthodox palate structure (found in adult tinamous but only in the embryonic stage of certain other carinates).

The wings are rudimentary and non-functional, used only for lightening the body weight and balancing the bird when running. But a comparison of the different ratites shows considerable variation in the size of these wings. The rheas, for example, have long wings with traditional primary and secondary flight feathers, while those of the kiwis are so small as to be practically useless and hardly visible because they are hidden under the thick plumage.

The feet are strong and ideally suited for running on hard ground. Ostriches, uniquely, have two toes while most of their relatives have three. The kiwi alone has an additional rudimentary, backward-pointing toe (though embryos of emus show a similar feature.)

The ratites are usually divided into four orders—Struthioniformes, Rheiformes, Casuariiformes and Apterygiformes. All of these categories include extinct forms. Living ratites are more simply and conveniently divided into Struthioniformes—notable for their large size and short beaks—and Apterygiformes. The former order comprises the ostrich and its taller relatives, the latter only the kiwi.

Reverting to the wider classification to take in extinct as well as living ratites, the Struthioniformes consist of the Struthionidae (ostriches) and the fossil Eletherornithidae, from the Middle Eocene. The Rheiformes comprise one family only, the Rheidae (rheas). These are birds of the Argentinian pampas, north-eastern Brazil, Peru and Patagonia. The Casuariiformes include the Casuariidae (cassowaries) and Dromiceidae (emus), the former from New Guinea and north-eastern Australia and the latter from both northern and southern Australia. Also in this order are the extinct Dromornithidae, creatures of the Pleistocene. The Apterygiformes consist of the single family Apterygidae (kiwis). The kiwi, a resident of New Zealand, is a much smaller bird, shy and nocturnal in habit, and for these reasons not often seen in the wild. It is about the size of a barnyard cockerel and whereas the taller ratites have short beaks with nasal openings about midway down, the kiwi has a long beak with nasal openings at the tip. It also has four toes, of which one is rudimentary, facing backwards and completely raised off the ground.

Until modern times there were also living representatives of two other orders—Dinornithiformes and Aepyornithiformes. The former included the family Dinornithidae or moas of New Zealand and adjacent islands. Since they figure prominently in Maori folklore they were probably still found in the islands when the first Maori settlers arrived from other parts of the Pacific, but they would seem to have disappeared about 600 years ago. The family Aepyornithidae or elephant birds lived in Madagascar and their immense eggs, about six times the size of ostrich eggs, are sometimes still uncovered, astonishingly well preserved. But the birds themselves vanished from the earth at about the same time as the moas.

The ratites, being flightless, have naturally fallen prey to many terrestrial carnivores and this partially explains why they are nowadays less plentiful than they once were in Africa, South America and Australasia—their traditional habitats. In islands where such predators were less abundant or late in arriving, they have survived longer, but there is good reason to be concerned for some subspecies which are finding it increasingly difficult to contend with the various pressures of modern civilisation.

CHAPTER 7

The African lion, king of the wild

Preceding page : A lioness and her cub are silhouetted against the evening sky. Once widely distributed through Asia as well as Africa, the lion is now numerous only in the reserves and national parks of Central, South and East Africa.

Facing page : Two magnificent lionesses at rest on the African savannah. It is the females who are responsible for hunting the prey while the males are concerned with the defence of territory.

In the African tropics the transition from day to night is spectacular but of short duration. The stifling heat and blinding glare are relentless, seeming to continue without any prospect of respite; but eventually the fiery red ball of the sun plunges—and it happens so rapidly that the word is apt—below the horizon. The evening sky is tinted with soft golden light and in the refreshing coolness of the air nature comes alive again.

Nowhere are the sounds of renewed activity so pronounced as in the Serengeti. The diurnal animals—birds, small rodents, and antelopes—make the most of these last precious moments of the dying day, while the nocturnal hunters rouse themselves from their torpor and begin preparing for the grim but necessary business of the hours of darkness that lie ahead. Among these hunters the most vociferous is the lion, justly called 'king of beasts'.

In olden times the lion's range extended even into Europe. Cave paintings and fossil remains testify to its existence in Spain and other parts of Europe during the Pleistocene period. Because it clearly spent much of its life in caves paleontologists refer to it as the cave lion. It was the most powerful carnivore of its time, more than a match for other predators such as the cave bear and hyena.

Although we think of the lion as a typical animal of Africa *(Panthera leo leo)*, the Asiatic lion *(Panthera leo persica)* was once just as widely distributed from Asia Minor (Turkey) southwards into Arabia and westwards through Iraq, Iran and Afghanistan down into parts of the Indian peninsula. From the many references to lions in the Bible and in Greek mythology it is apparent that they roamed at will through Palestine and Macedonia. Nowadays, however, the Asiatic lion population has

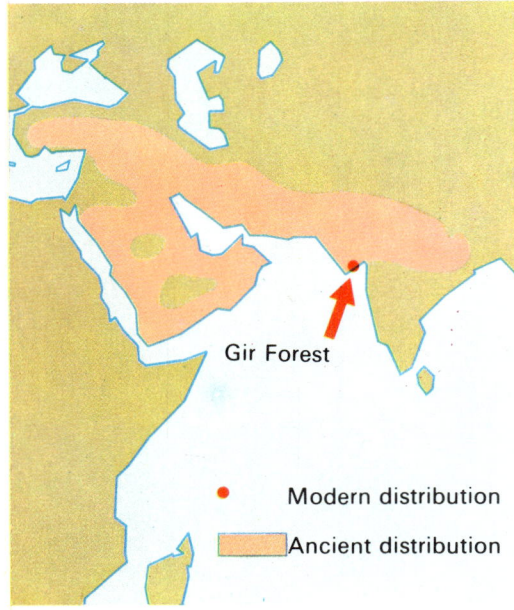

Gir Forest

●   Modern distribution

▭   Ancient distribution

Since the end of the 19th century the range of the Asiatic lion has been drastically reduced. Once roaming freely through the Near and Middle East as far as India, the last few hundred representatives of the species are now protected in the Gir Forest sanctuary. Even in this last refuge they are hunted by the owners of domestic livestock grazing in the reserve.

Facing page (above) : The African lion is a creature of the open steppe and savannah. Fearing no rival it may often be seen basking during the daytime, without making any special effort to conceal itself. *(Below)* both lionesses and cubs, seen here in relaxed mood, have the streamlined contours of typical feline hunters, and both show similar spotted coat markings.

vanished from almost all these regions. The last survivors—only about 250 animals—are protected in the Gir Forest wildlife sanctuary on the Kathiawar peninsula of India. Happily the African lion ranges over a wider area, though also in need of protection. Most of them are found in the 3½-million-acre Serengeti National Park.

The ideal conditions enjoyed by the lions of the Serengeti could be matched, barely a century ago, in many other parts of the dark continent, from the Mediterranean down to the Cape of Good Hope—with the exception of the tropical forest and desert regions. At that time the lion was still undisputed lord of the savannahs, steppes, plains, hills and mountains. Wherever the vegetation cover was not too dense and the water supply adequate, the African night was rent by the proud roar of the 'king of beasts'.

The decline of the lion population of Africa began in the north—in Tunisia, Algeria and Morocco, countries where they had been plentiful in the Middle Ages. The handsome black-maned lion was forced to abandon the Atlantic and Mediterranean coastal plains and to seek refuge in the Atlas Mountains. Among contributory causes were road-building activity, the encroachment of cultivated land and the brutal extermination of the North African hartebeeste and other creatures that formed part of the lion's normal diet. When it had to resort to the desperate expedient of preying on domestic animals, farmers and local authorities declared war. Between 1873 and 1883 the Algerians proudly announced that 202 lions had been 'officially' slaughtered; the last recorded killing was at Souk-Ahras in 1891. In Morocco the lions found more security in the forests of the Middle Atlas, where they survived until 1922. But eventually the combination of industrialisation and population pressure put an end to the reign of the lion in North Africa, just as the agricultural expansion policies of the Boers and the British decimated their numbers in South Africa. Today their last refuges are the huge reserves and parks of Central and East Africa, where they are no longer exposed to the 'civilising' activities of men and governments.

King of beasts

Lions are basically nocturnal creatures, resting by day and hunting by night. Darkness is their proper element, where they can make most efficient use of their senses of vision, hearing and smell and where the stealthy approach on padded paws goes undetected even on open, unsheltered ground. For this reason sunrise and sunset or a clear moonlit night are ideal times for studying their behaviour.

Even by day, however, lions may be seen lazily taking their ease on the plains of the Serengeti, usually in family groups. The cubs are predictably more active than their parents, energetic and playful in spite of the heat. The chances are that they will be of varying ages. The youngest may be little more than three months old, still with the spotted marking on their coats that makes them so difficult to see among the bushes where the

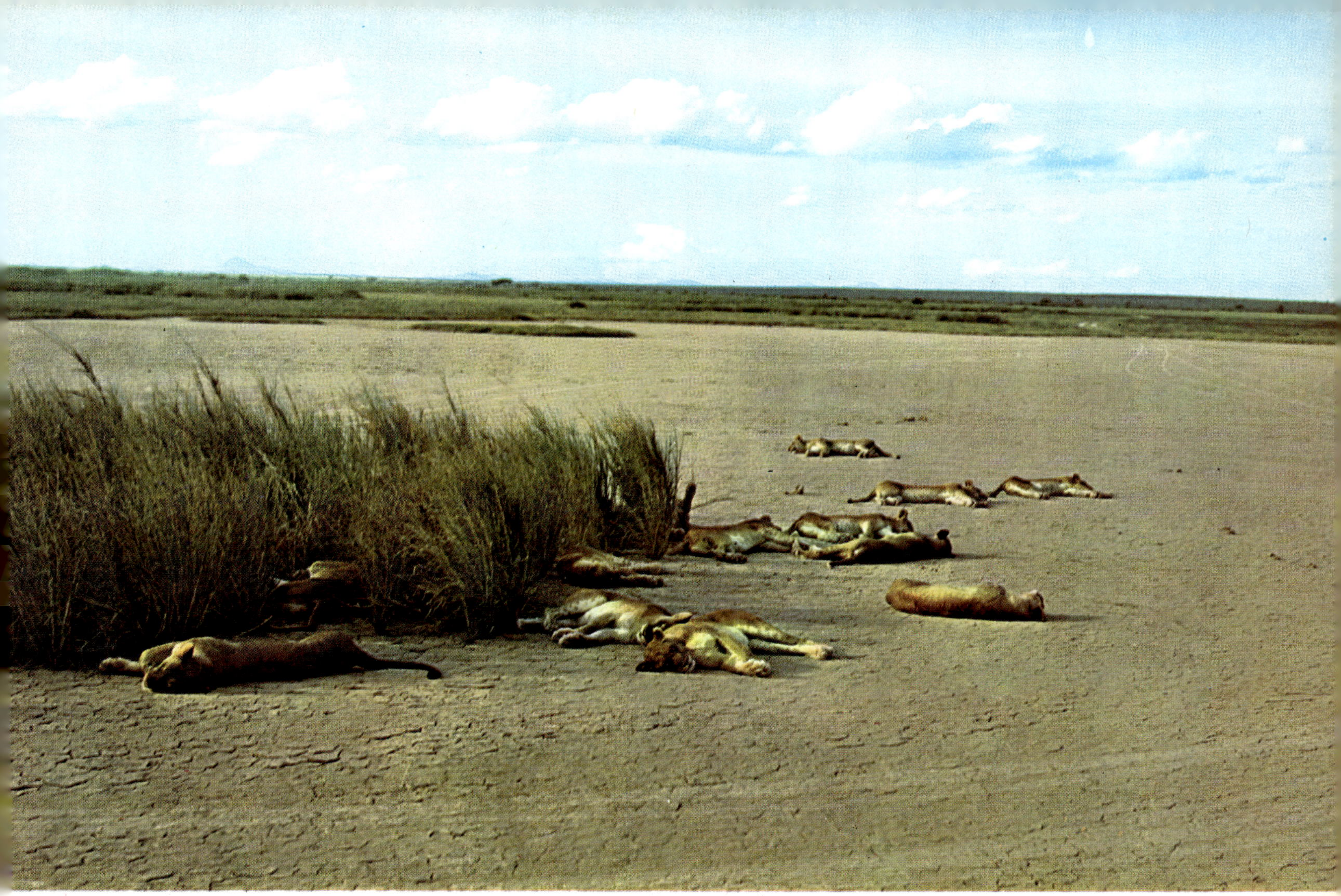

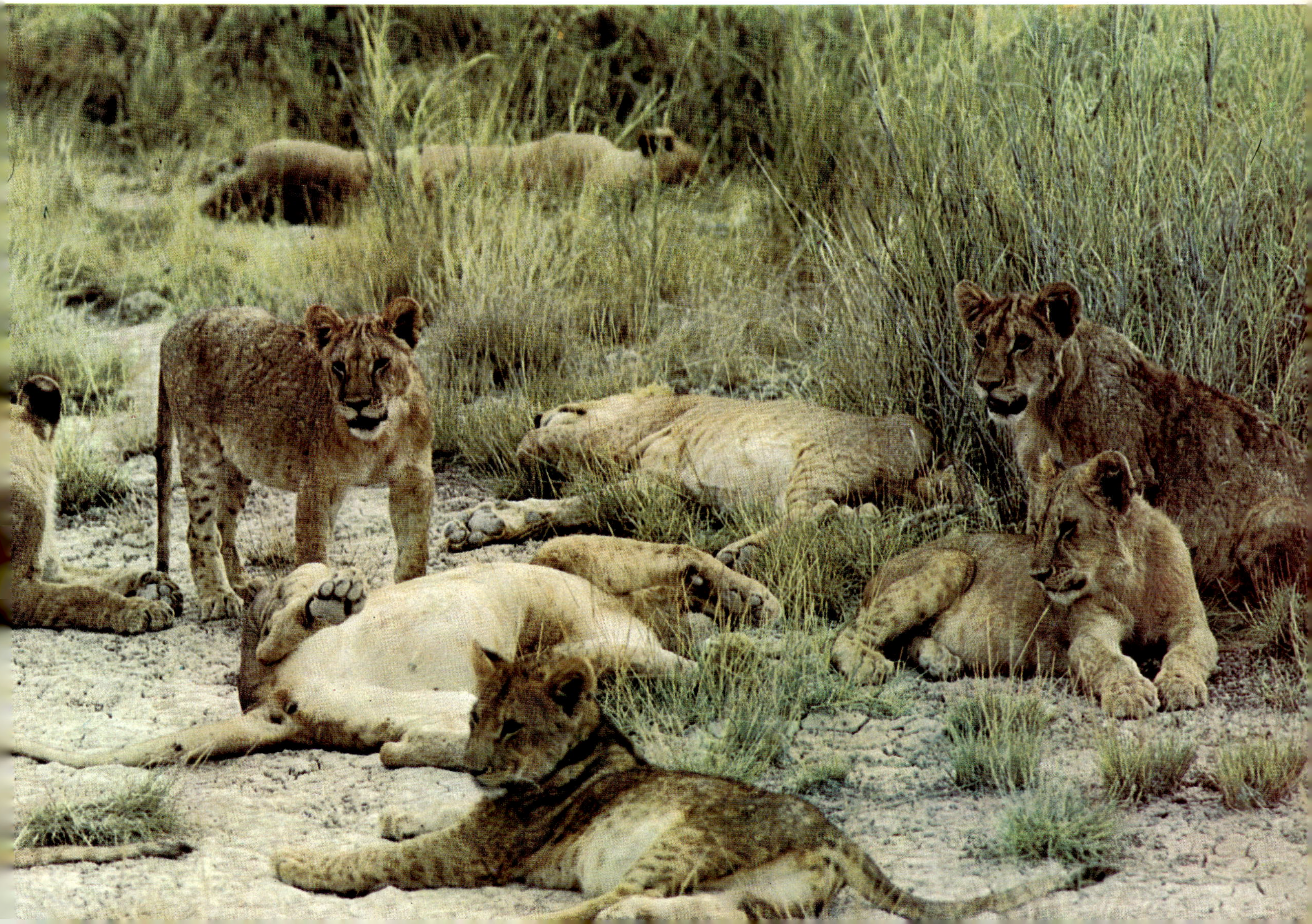

Tiger
(*Panthera tigris*)

Jaguar
(*Panthera onca*)

Clouded leopard
(*Neofelis nebulosa*)

Leopard
(*Panthera pardus*)

Lion
(*Panthera leo*)

Puma
(*Puma concolor*)

Ounce or snow leopard
(*Panthera unica*)

Forests and woods

Savannahs and steppes

Hillsides and rocky plateaux

High mountains

The world's wild cats are all carnivores, adapting their hunting techniques to a number of different environments. The tiger, leopard, clouded leopard and jaguar are all creatures of forest, jungle and bush, the first three being inhabitants of the Old World, the last confined to the New World. The lion roams the open steppes and savannahs with sparse vegetation, while the puma or mountain lion, another wild cat of the American continent, is a hunter of the hills, deserts and mountains. The little known snow leopard lives thousands of feet up in the mountains of Central Asia.

mother will leave them in order to go hunting. Some, nearer a year old, have already assumed the uniformly golden colour on their back; and others—between fifteen months and two years of age—show numerous mottled marks on the inside of the legs (marks that the females do not lose and which may be throwbacks to the days long past when they were probably forest animals rather than creatures of the plains).

The large amber-coloured eyes of the lion cubs are delightfully expressive, reflecting all the curiosity, innocence and mischief of infancy, yet with a hint of the wariness and pride that characterise the adults.

The lioness in the wild is a picture of serene elegance and hidden power—a far cry from the bored, under-exercised creature caged up in so many zoos. The unwavering intensity of her stare serves clear warning that anyone daring to come closer to her cubs does so at his peril. She will fight fiercely in their defence.

The slender, streamlined shape of the lioness is in sharp contrast with the altogether more solid, heavy bulk of the lion. For whereas she is the huntress he is the warrior. The male is a truly majestic creature with his magnificent mane covering head, neck and shoulders. There are few more impressive sights than the silhouette of a lion against the horizon nor many sounds more spine-chilling than his majestic roaring as night falls over the savannah.

The male's mane begins to grow at about two years and by the time the animal is five or six it may be 10 inches long. But not every lion possesses a mane and its length and colour may vary enormously according to individual and location. The manes of lions inhabiting regions of shrub-covered steppe are sparse, almost non-existent, while those of the lions of the Kalahari or Ngorongoro are abundant and flowing, the colours ranging from light chestnut-brown to deep black. These dissimilarities may be partly hereditary, partly accidental–influenced, for example, by such simple actions as continual rubbing against thorns and rocks. The only general conclusion is that the manes of young lions are invariably lighter in overall colour than those of the adult males.

Short white hairs cover the chin and surround the throat and eyes of both males and females. Perhaps because of the highly developed masticatory muscles, the eyes are slanting, with the colour of the iris changing–according to the angle of the light or depending on the age of the individual–from yellow to various shades of brown.

The long tail ends in a black pointed tuft of hair, tipped by a horny spur. The waving of the tail is a sure sign of bad temper or displeasure.

Hunting the hunter

Although it is a fairly common experience in East African parks and reserves to see lions in a family group and to come across them hiding in the undergrowth or devouring a freshly killed herbivore, there are not many opportunities of watching a lioness actually hunting her prey. Long hours of very patient observation may still be unrewarded by such a spectacle. Any description of the habits and behaviour of lions and lionesses in quest of prey must therefore be based to a large extent on the specialised publications of men who have spent years studying the subject. In recent years there has been much learned literature about the African lion, and several outstanding books designed for the general public which, because they are based on personal experience, are not only highly informative and readable, but also imbued with a deep sense of sympathy and genuine understanding.

Lions are known to be territorial predators, proof of which lies in the ferocious fights that take place among the males and the loud roaring by which they make their presence known. Within their extensive domains it is the females that are responsible for stalking game, raising the young and teaching them in their turn to become hunters and warriors. The males are the defenders of territory against intruding lions from other troops, feeding on the prey secured by the lionesses. They are polygamous, the number of mates depending on the prevalence of prey and the size of the troop (or pride). Zebras, gnus and other large antelopes are regularly hunted, giraffes occasionally and buffaloes infrequently. But there is no herbivore which is not in some measure affected, directly or indirectly, by the activities of these carnivores.

These are well-known and comparatively superficial items of information. Other problems are very much more intriguing.

The lioness has the build typical of the hunters of the family Felidae, a low-slung muscular body well adapted to the chase or to sudden attacks from ambush.

The impressive flowing mane of the lion serves to intimidate enemies and rivals, its colour and texture varying according to the individual and its locality. Although it furnishes protection for the head and neck during a fight, this visible means of recognition helps to avert rather than initiate such contests.

How extensive are the lions territories and just how are their frontiers delineated? Are they sedentary hunters or do they follow the herds of ungulates in their seasonal migrations? What happens to the young males and do they fight for domination within the troop or are they simply thrown out? Are fights between lions ritualised—fought, as in so many animal societies, without any intention of shedding blood—or are they duels to the death? Do lions attack only sick and stray herbivores or do they not bother to discriminate among their victims, showing themselves opportunists rather than planners?

Such questions can be answered, objectively and scientifically, thanks to the opportunities afforded for studying lions in their natural surroundings in parks and reserves. By now the animals are so accustomed to human presence that naturalists can examine them, without undue risk, at close range. But even under such favourable conditions some questions can only be resolved after many years of study, often involving uninterrupted observation by day and night.

Dr George B. Schaller, the American zoologist already known for his work in India on the tiger and in the Congo on the mountain gorilla, was invited by John Owen, director of Tanzania's national parks, to carry out a thorough survey of the social behaviour of lions and their impact on the herds of grass-eating animals. The survey, which was begun in 1966, was financed by the Zoological Society of New York and the US National Foundation of Science.

Dr Schaller started by marking the ears of 150 lions with metal plaques. This was done under anaesthetic by means of hypodermic syringes fired from rifles. In this way, assisted by a colleague, he was in a position to follow them wherever they roamed and to take detailed notes of their activities. The information arising out of Dr Schaller's survey, coupled with the published work of other distinguished zoologists, provides answers to some of the questions raised.

Most lions live in prides, some of which are small and fixed, with only four or five individuals. But in the Serengeti larger troops are often found, typically composed of, say, two adult males, a dozen females and about twenty cubs. In the Masai-Mara reserve prides of more than twenty lions are common, while in South Africa's Kruger National Park prides of up to forty animals have been seen.

Some males habitually roam quite freely across a large area of savannah, but the majority prefer not to wander far beyond the bounds of their own territories. The size of these domains depends on the concentration of game in the district. Where there are plenty of grazing herbivores the hunting territory will be confined; but if the wildlife is scarce the predators are forced to enlarge their circle of activity and to embark on a semi-nomadic existence.

The males mark out the limits of their territory, just as other species do, by establishing scent posts. They urinate over bushes, tufts of grass and other selected landmarks. The urine contains a strong-smelling secretion which constitutes a scent barrier—unmistakable to other lions and so persistent as to be easily detected even by the human nose.

Strangers—keep off

Reinforcing this chemical signal—unseen but effective—are other warning signs, both auditory and visual, which do not go unheeded by would-be intruders. The shattering roar and the flowing mane are the two most obvious examples of such supplementary warnings.

The terrifying echoes of the male lion's throaty roar can be heard for miles around, and it has now been established by scientists that one of its main purposes is to proclaim the exclusive right to a piece of territory. The American naturalist Jim Fowler conducted an interesting experiment to check this theory. He placed a tape recorder in the centre of a region frequented by a large group of lions. When he switched it on the silence was shattered by the prolonged and mighty roaring of a 'strange' lion. A scene of utter confusion resulted as a group of lionesses approached the machine, casting uneasy glances in every direction. They were followed by several large males, showing obvious signs of anger and eager to expel this bold intruder which was daring to defy them in the midst of their hallowed territory.

The two territorial lions reacted to this 'invasion' of their private domains exactly as predicted. Any unaccustomed sound which might represent a threat to communal security would have elicited the same response, but the challenge of a rival lion is

The curving canines of the lion are powerful weapons in the wild, often used to put the finishing touches to the deadly work begun by the sharp claws.

Following pages : In the parks and reserves of East Africa the branches of an acacia offer lions relief from the scorching sun and respite from the onslaughts of insects. The males, with their heavier bodies, usually leave the higher branches to the lionesses and the cubs.

LION
(Panthera leo)

Class: Mammalia
Order: Carnivora
Family: Felidae
Length of head and body: 90-115 inches
(225-285 cm)
Length of tail: 30-35 inches (75-85 cm)
Height to shoulder:
Males 38-44 inches (95-110 cm)
Females about 34 inches (85 cm)
Weight: up to 475 lb (225 kg)
Diet: zebras, gnus, antelopes, gazelles, etc
Gestation: 105-112 days
Number of young: 2-3
Longevity: 16-20 years, up to 34 in captivity.

Adults
Large massive head, short rounded ears, yellow
eyes with round pupils, triangular muzzle. Cleft
upper lip, sometimes light marking above eyes,
with long whitish hairs. Relatively slender tail,
ending in tuft of dark hairs. Short coat, brownish
or fawn, but varying according to individual and
locality. Males larger than females, with mane
formed of long darkish hairs extending over head,
neck, upper part of chest and down to top of fore-
paws.

Young
Cubs have a large, somewhat disproportioned
head. Coat spotted during first three months and
adult colour only evident at two years of age.
Some, especially females, retain the spots into
adult life. In males the mane begins to grow after
two or three years.

potentially more dangerous than the mere presence of another species.

The roar of a lion serves both as a warning to adversaries and a friendly reassurance to companions. Some authorities believe that in addition to announcing the male's whereabouts it may also be intended to intimidate prey and drive them in the direction of the lurking lioness. But the roar is only one of many sounds made by the huge beast to express a wide variety of emotions and intentions. Growls, purrs, coughs and grunts are all familiar noises and depending on their volume and intonation the lion can convey a tremendous range of contrasting feelings from authority, anger and fear to contentment and desire.

If roaring represents the lion's audible assertion of rights, the mane is a visible reinforcement of that claim. In some ways it would not seem to be an obvious boon to its owner. It harbours parasites, it is uncomfortably hot and it would appear to provide ample advance warning to a potential prey. But scientists have concluded that the mane's significance arises from the very fact that it is so conspicuous against the background of yellow plains and green grass—a means of announcing the owner's identity and presence to a potential rival or enemy, even when some distance away.

There are many territorial animals—the fishes of tropical coral reefs and the insect-eating robin redbreasts are striking examples on a smaller scale—that draw attention to themselves by distinctive shape or coloration. Other animal species give out unmistakable sound or scent signals for the same reason. Lions are no exception to this rule, emitting similar warning messages to others of their kind. Yet such signals are not intended to be provocative—rather the contrary. They are designed primarily to prevent fights taking place between healthy members of their own species—an absolutely vital way in which to safeguard their survival.

Even if a fight proves unavoidable lions will adopt a variety of ritual attitudes and movements designed to give the weaker animal a 'let-out' and perhaps save its life. Certain postures reminiscent of infancy are clearly intended to indicate submission, acknowledging the superior strength of an opponent and providing an opportunity for escape. Such tactics—substitutes for aggression—are employed by nearly all species of powerful, gregarious animals, though rarely by solitary predators. Nor are they needed by species whose natural weapons are not deadly enough to inflict serious or lasting injury on others of their kind.

Dr Schaller has stated that when an intruding lion is surprised on territory that does not belong to him he may be chased off by the rightful owner—sometimes for several miles—until he acknowledges submission. But even in such cases a simple warning is sufficient. The pursuer seems to measure his pace by that of the retreating intruder, again trying to avoid a critical encounter. Thus the angry roars, the swishing tail and all the other fierce-looking movements may be seen as parts of an elaborate pattern of ritual behaviour, intended to prevent rather than provoke fighting.

Duels to the death

Although comparatively rare, duels of life and death do sometimes take place. The wardens of Ngorongoro once came across a huge black-maned lion, aged more than fifteen years, with an injured fore-paw. During the night terrible roars echoed through the crater. Next morning, when they drove up to the spot where they had seen the wounded beast, they found nothing but some scraps of skin, tufts of mane and parts of the body which had been literally ripped up and half devoured by hyenas. Among the black hairs they discovered several tufts of reddish hair, obviously belonging to lions with lighter-coloured manes and labelling them quite indisputably as the sole authors of the crime.

Mervyn Cowie, in his monograph *The African Lion*, also tells how one night he listened to the roaring of two lions fighting for hours on end not far from his house. At sunrise he set out towards the place from which the sounds of battle had come, and found the horribly mutilated carcase of a male lion. Flanks and belly had been torn open by the claws of its opponent, while neck and head were disfigured by deep fang marks. Not far away sat a lioness, probably its mate, quietly contemplating the remains. Grass, bushes and dwarf acacias had been torn up by the roots or flattened as if by a bulldozer. Cowie followed the tracks left by the second lion and soon came across its dead body lying near a river, evidently as a result of wounds inflicted by its vanquished rival.

Evidence shows that when the leaders of a troop of lions seclude themselves with their females during the mating season, neighbouring lions—perhaps attracted by the scent of the rutting lionesses—sometimes intrude upon their privacy, with tragic consequences. Quarrels between rival males may also take place over title to a prey. George Schaller, during an expedition in the Serengeti, spotted the lacerated corpse of a male lion, the surrounding area being littered with tufts of hair. Nearby a lioness was eyeing the body of her mate with evident signs of affection, while a short distance away lay a recently killed zebra. Dr Schaller concluded from the clear tracks leading from the scene of the crime that the culprits were in all probability three adult males from another group, using superior weight of numbers to assert their will.

The huntress and the warrior

It has long been known that the adult male lions in a pride do not—except on rare occasions—take part in any active hunting. Furthermore when they are particularly hungry they will even deny the females and the cubs a share of the freshly killed prey. Judging by such behaviour one might conclude that the typical African lion was a pretty despicable creature—arrogant, lazy and selfish—growing fat at the expense of his family. But this would be a hasty and totally unjust verdict, based on applying human standards to animal behaviour. We still know too little of the animal world, where things are seldom as they seem.

The adult male lion is responsible for defending territory and safeguarding the troop (or, as it is usually called, the pride), prerequisites for the lioness's hunting and cub-rearing activities.

Geographical distribution of the African lion (based on Dr George Schaller's work).

Recent surveys by zoologists have shown that the male lions have a vital role to play in ensuring the survival of the pride. It is they who, by sheer weight of numbers and superior strength, provide the territorial security which the lionesses need in order to hunt freely and rear their cubs without risk of outside interference.

Dr Schaller made a study of one pride consisting of two males and thirteen females. One of the males happened to be killed by three lions of another pride and although the surviving male was a powerful animal he was quite unable on his own to safeguard his little community. A few days later a couple of lions from the neighbouring territory launched a surprise attack and, despite the resistance of the solitary chief, succeeded in breaking through to a clump of bushes where a lioness had left three cubs sleeping. The intruders showed no mercy, taunting the helpless cubs cruelly before slaughtering them. One of them was eaten on the spot, another dragged off between the fangs of its killers as a trophy.

When the lioness returned from hunting to the place where she had left her cubs, Dr Schaller witnessed a remarkable scene. The mother approached the corpse of the third cub, sniffed it, then calmly sat down and devoured it.

This kind of behaviour is fairly frequent among predatory animals. Although the interplay of impulses that stimulates cannibalism among wild creatures is very complex, it would seem that the movements, the purring, the smell and even the breathing of the young all act as a barrier to the so-called carnivorous instinct. Thus the appealing behaviour of all young animals, their tender expressions and affectionate movements, may well be designed—subconsciously—to hold in check the aggressive impulses of their parents, enabling the latter to distinguish between their offspring and a legitimate prey or foe. But the absence or cessation of such signals causes confusion. Hence the dead body of a cub will be regarded as nothing more than a hunk of edible flesh, as appetising as any other, no matter what the source.

Dr Konrad Lorenz and his colleagues conducted an experiment related to this phenomenon. It proved that turkeys that had become deaf as a result of an operation killed their newly-hatched chicks with blows of their beaks. The adult birds were able to recognise their chicks as soon as they uttered their first cries, but this was their sole means of distinguishing them from the various intruders threatening the nest during the incubation period. Deprived of their hearing, the adult turkeys had no way of identifying their offspring and proceeded to kill them one by one. This is virtually what may happen in the case of a lioness suddenly and unexpectedly confronted by a dead cub, even when it is her own.

Dr Schaller made another interesting observation about the pride that lost one of its male protectors. Out of twenty-six cubs born in a two-year period, only two managed to survive, whereas in a neighbouring pride protected by three males, twelve out of twenty cubs born during a similar period benefited from the additional security to reach adulthood.

The lioness is a creature of contrasting moods, merciless and grimly determined when pursuing prey, patient and sociable when she adopts a maternal role. Wariness, however, is always the keynote.

The conclusion must be that in the world of lions there is a highly effective division of labour between male and female. The lionesses hunt, feed and raise their cubs, and the lions defend their territory, often at the price of their own life. This exempts them from any duties that might interfere with their warrior role. They are entitled to eat whatever prey is killed by their mates and they display no interest whatsoever in the upbringing of the cubs. Any group deprived of its quota of protecting males is seriously weakened and is likely in due course to disintegrate, the males from rival prides being quick to take advantage of such a significant change in the social structure.

The very striking physical differences between the sexes may well have arisen as a result of this clear-cut variation in activity. The superior size of the male, the mane that enlarges his face and protects him during a fight, and the powerful roar are all designed to impress and intimidate enemies. The lioness's sleek, unencumbered body is far better suited to her role as huntress—lying in ambush, chasing and capturing prey. She looks much more like a typical feline, obviously a close relative of the tiger, the leopard and the jaguar. Strictly speaking, it is she and not her mate who takes a place alongside the other wild cats as one of the most active of predators; yet it is he who can legitimately claim to make it all possible.

Male lions leading a nomadic existence often form alliances, co-operating instead of coming to blows over rights to territory, females and prey. Even those content to lead a sedentary life find it essential to avoid rivalry within the pride, two or more protecting males usually being needed to provide stability and security.

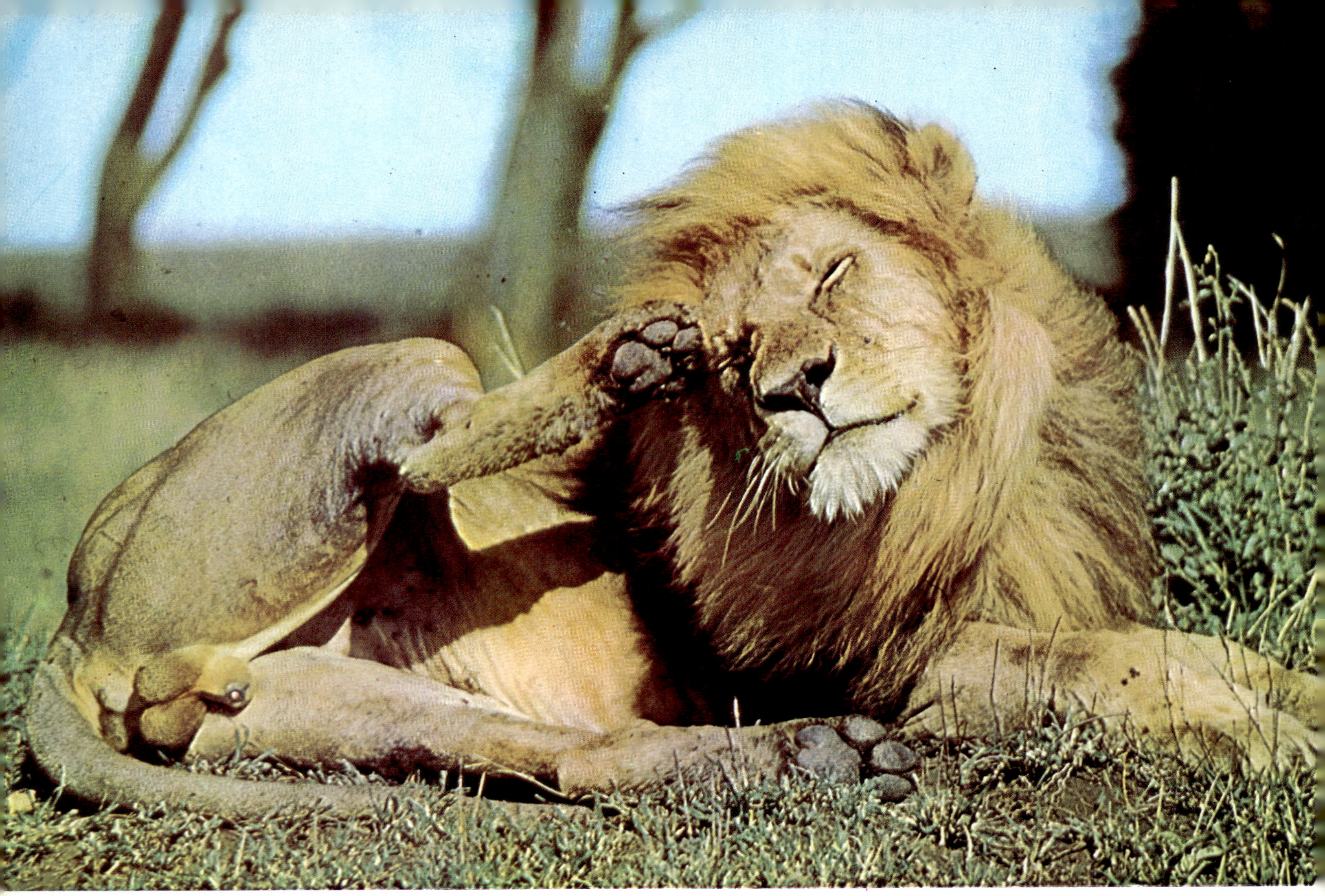
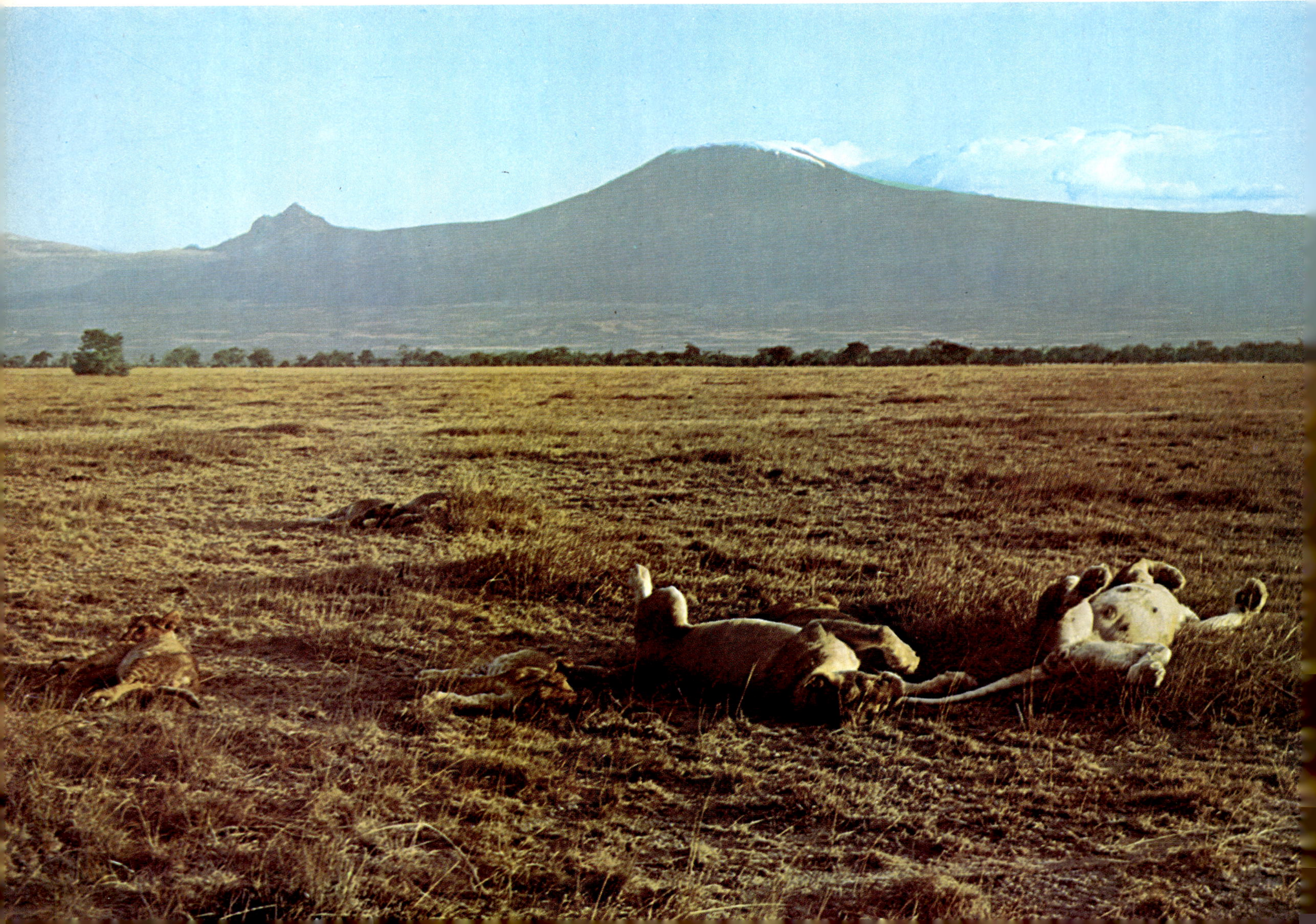

The outsiders

The great herds of gnus and zebras set out on their migrations as soon as the dry season returns, leaving the parched dusty plains for the lower-lying, well-wooded country to the north. When the rains complete the seasonal cycle the herbivores trek back to the high savannahs and the new growths of grass. Nobody was really certain whether the lions followed the zebras and antelopes when they embarked on their annual travels or whether they patiently awaited their return at the beginning of the rainy season. Dr Schaller's survey with his marked lions showed conclusively that the majority of lions in the Serengeti stay where they are, modifying their feeding habits according to circumstances. In the absence of zebras and gnus they turn their attention to Thomson's gazelles, hartebeeste and buffaloes.

Nevertheless some lions lead a nomadic existence all year round, wandering the length and breadth of a region some 900 square miles in area, and straying into territory that does not belong to them. These nomadic groups generally consist of young or adult males unattached to any more stable troop, and occasionally include females that have abandoned their regular companions due to shortage of food. In the lion community it is quite common for males over the age of three years to be expelled by the dominant males and condemned to wander like outlaws until they succeed in getting themselves accepted by a new group. They may achieve this either by taking over the place vacated by the death of another male or by vanquishing the leader of a troop in open combat.

In regions such as the Serengeti where lions live in absolutely ideal surroundings the whole park is made up of different prides and there is little room for newcomers. Yet exceptions are sometimes made. Wandering lionesses, for example, are often welcomed by dominant males though viciously rejected by other females; conversely, a strange lion may be accepted by the females of a pride, though shunned by other males. Zoologists admit, however, that there is still much research to be done on this fascinating question.

The transistorised lion

More than 200 lions in the Serengeti—unable to establish fixed territories because of the shortage of suitable prey for one half of the year—follow the migrating herds across the empty plains and back again. To study their behaviour Dr Schaller and his colleagues selected a few individuals and trailed them day and night. They found it comparatively simple to keep watch on them when the nights were clear, for as soon as the sun went down the animals showed obvious signs of confidence and friendliness, even coming close enough to nibble at the tyres of the Land Rover. By moonlight their brownish forms took on silvery overtones, beautiful to watch as they made their slow, stealthy approach. But the barking of a jackal, the howl of a hyena, the whinnying of a zebra or the distant roar of another lion would bring them instantly to a standstill. Turning their heads in the

Facing page (above) : As is the case with so many large mammals, insects are a continual nuisance. This lion is trying to get rid of them by rubbing a paw against his forehead. *(Below)* Amboselli game reserve. Kilimanjaro in background.

direction of the strange noise they would immediately slink off and disappear into the thickets.

Dark moonless nights, on the other hand, posed considerable problems for the animals could not then be clearly distinguished from the surrounding murky vegetation. Dr Schaller hit upon a brilliant idea. One large male was anaesthetised with the aid of a hypodermic arrow shot from the Land Rover and its neck fitted with a tiny electronic transmitter. Taking turns with his companion, William Holz, to listen to the bleeps of the receiver inside the truck, Schaller was able to follow the animal uninterruptedly, day and night, for three weeks, recording all its movements and activities.

The lion, a strong, healthy individual, had joined with another male to claim a piece of territory about 75 square miles in area. Here they both shared a number of females. Thanks to his remote control system Dr Schaller kept a constant watch on the movements of his subject throughout a normal day. At 8 o'clock in the evening, for example, it accompanied its companion on a short stroll of a mile or so to steal the remains of a zebra from a scavenging hyena. After the meal both animals ambled back in the opposite direction to rejoin one of the lionesses and her two cubs. They stretched themselves out beside her with their hunger temporarily satisfied and slept soundly for the rest of the night.

Next day the two males made a tentative effort to do some hunting on their own but succeeded only in wearing themselves out. They then spent the greater part of that day and the following night resting or sleeping. The pattern was not greatly different on other days, with about twenty hours devoted to rest and only four to more active pursuits. When not asleep the transistorised lion would spend hours on end listening to the sounds of the savannah. The howling of a hyena pack would bring him instantly to his feet and bounding off in the expectation of snatching away their meal. On one such expedition he managed to steal a half-eaten gazelle from a leopard; another night he shared an eland with the lioness that had killed it. In the entire three weeks when he was under observation he lived entirely on prey taken by other animals. His only attempt to catch his own food—a dik-dik—ended in total and abject failure, due partly to inexperience, partly to lack of determination.

A life of ease

Some of Dr Schaller's observations confirmed those of other zoologists. The fact that lions generally spend the greater part of their time sleeping was of course well known. This is apparently not just the result of boredom (as would seem to be the case with lions in captivity) for lions in the wild appear to be equally indolent. Since they fear no threat from any other creature they do not trouble to seek refuge in thick vegetation and have an air of complete relaxation, the appearance of nonchalance being reinforced by frequent, voluptuous yawns. They may often be seen in the middle of an empty stretch of savannah, surrounded on all sides by hundreds of other equally impervious grazing

The lioness, although an indomitable huntress, will rob a scavenger and feed on carrion should the opportunity arise. Her ferocious aspect is usually enough to keep any rival at bay but a pack of hyenas may prove more than a match.

Gnu

Giraffe
and other prey

Zebra

Buffalo

Large antelopes
excluding gnu

Carrion

Gazelle and warthog

animals. Even the approach of human observers will not disturb their languid siesta. They generally sleep on one side but may sometimes be seen on their back, paws in the air. At other times they may take their ease in the shade of an acacia, leaning comfortably against a tree trunk; certainly wherever shade is available they will make use of it. Although they can endure heat they do not crave sunlight for its own sake.

The phenomenon of two adult males harmoniously sharing the same piece of territory—as reported by Dr Schaller—is not an isolated or unusual occurrence. Two, even three or four, males may decide to band together in common interest and establish their own pride on a permanent basis. If for any reason they are compelled to separate temporarily, their reunions are accompanied by many evident signs of joy—heads coming together, rubbing of noses and other ritual movements indicating mutual affection. The lion is not by nature a solitary animal and here is further evidence that even those leading a nomadic life seek safety in numbers.

Lions prey on a variety of herbivores, especially gnus, zebras and antelopes. They resort to gazelles, warthogs, young buffaloes and giraffes only when unable to satisfy their hunger in other ways. The width of the arrows indicates the comparative frequency with which they take different types of prey, the main point of interest being that carrion is an important supplement to their diet.

Lions will seek shade if available rather than endure the broiling heat of the sun. But well-wooded glades such as this are few and far between in their customary habitat of steppe and savannah.

Predator versus scavengers

We have seen that the male lion relies almost exclusively on his mate for a continuous supply of food. But the lioness, assiduous hunter though she is, will not miss the chance of relieving hyenas of their prey nor turn up her nose at carrion. When the need is acute both sexes will turn to food which they would normally disdain. A troop of lions has been observed devouring a warthog in an advanced state of putrefaction, having painstakingly recovered its water-swollen corpse from the bed of the river into which it had fallen.

In 1968, during the rainy season, Dr Schaller observed that out of 121 animals eaten by lions in a certain district only about a half had been killed by lionesses, roughly a quarter having been caught by hyenas and the remainder by other unidentified predators.

The Dutch zoologist Dr Kruuk, who made a special study of hyenas in Ngorongoro, was astonished at the number of adult gnus and zebras that these celebrated carrion eaters had themselves killed. He confirmed Schaller's findings that the lions in the reserve relied to a large extent on the marauding hyenas for their sustenance. It had been commonly assumed that the situation was exactly the reverse.

One night a group of naturalists driving through Ngorongoro Crater were startled by a diabolical sound which they soon recognised as the sharp, irregular howling of a pack of hyenas. Picking their way through swampy terrain and illuminating the path ahead with their powerful headlights, they followed the hubbub until they swept their beam over a confused dark seething mass which eventually resolved itself into a group of more than 20 hyenas busily devouring the carcase of a gnu. About 50 yards away the headlights picked out the form of a lioness which was loping off, belly close to the ground, in the direction of a clump of acacias.

The lioness had apparently left her cubs hidden in some bushes at sunset and gone off to kill a gnu. This is the time of day when the 200 hungry hyenas living in the reserve habitually emerge from their lairs in quest of food. On this occasion they were quick to catch the scent of fresh meat on the breeze. Soon they were on the prowl—a long column of sinister ghosts—all heading towards the site of the kill and forming a shadowy circle round the lioness which had barely found time to sink her fangs into the still-warm flesh. Minutes later, about 20 of the scavengers moved forwards. Snarling, with ears flattened against her head, the wary lioness uttered a deep roar and began sidling away from her prey. As if obeying an unheard signal the hyenas raised their hairy tails, lifted their voices in an unearthly howl and hurled themselves at the retreating predator. Yard by yard she gave way before her attackers, lashing out now and then with a paw. But eventually she was forced to yield her prey to the voracious carrion eaters. As she crawled off into the safety of the surrounding thickets, the hyena pack converged on the spot to claim their bleeding trophy. The naturalists watched them pick the carcase clean without further disturbance.

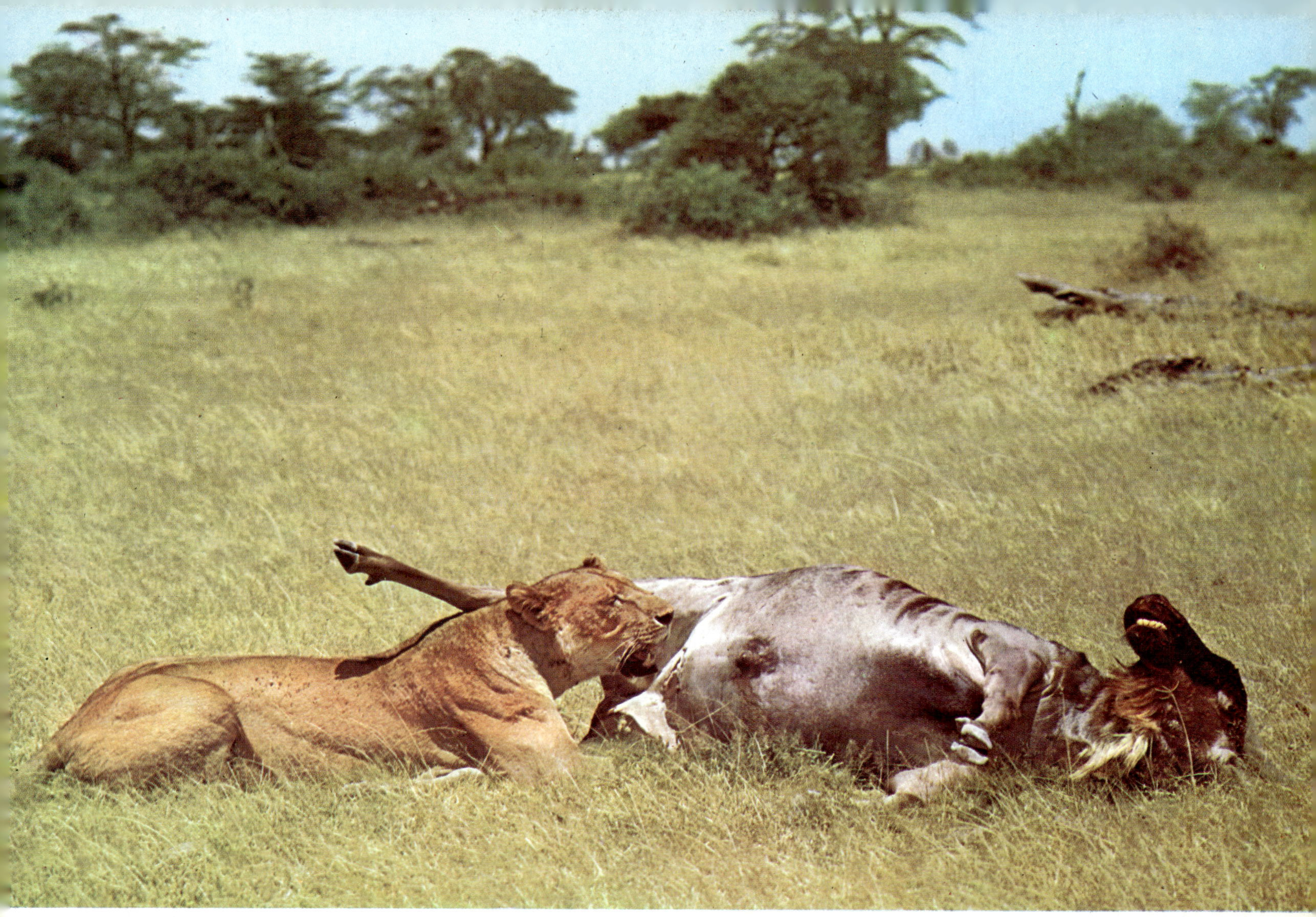

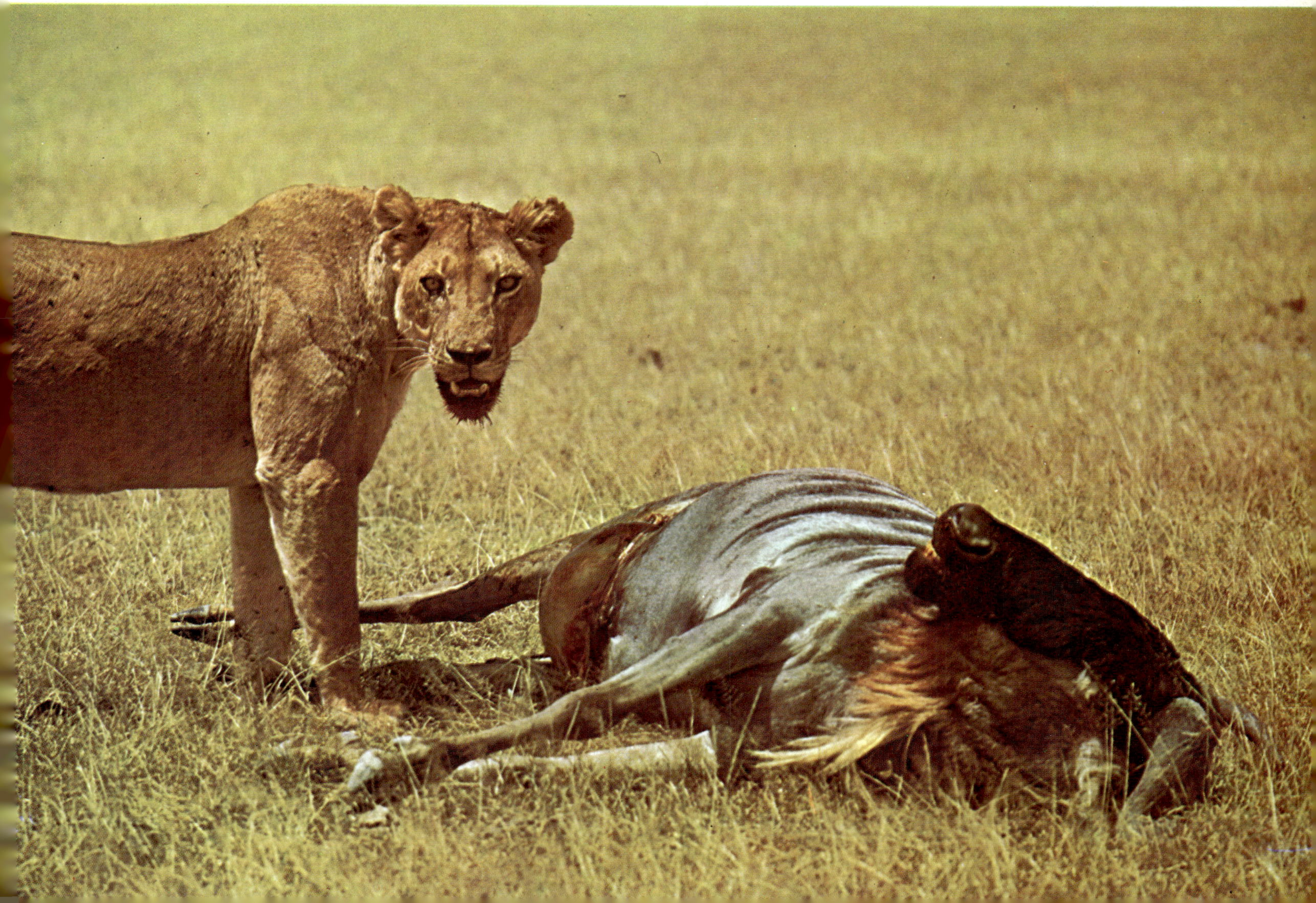

The footprints of a lion are clearly outlined in the wet mud.

Preceding page : A lioness has killed her favourite prey–a gnu. Her problem now is to drag the carcase into the shade of the bushes where the other members of the pride await her.

Fang and claw

The first fact to bear in mind when examining the hunting technique of lions is that they are felines. Although they can call upon vast reserves of energy in a final pounce, they lack stamina, tire quickly and are unable to keep up a high speed for any distance. An attack from ambush suits them better than direct pursuit. This method allows the hunter to remain completely motionless, hidden up to the moment when the prey comes sufficiently close, the attack being launched by a mighty leap and perhaps a lightning sprint across open ground. The carnivores of the bush, such as the leopard, tiger and jaguar, all adopt this technique and so, in its savannah habitat, does the lion. The speeds attained in the course of such attacks are not high in comparison with those of antelopes in full flight. Some say a lion can reach 50 miles per hour; George Schaller puts it more conservatively at a maximum of 35 miles per hour.

The lion's aim is to hurl itself through the air with all the weight of its heavy body, accelerating as it goes. Thus it literally gets off to a 'flying start'. Even the most fleet-footed of animals scarcely has time to collect its wits in the face of such a whirlwind assault. So the herbivores have to profit as far as possible from the known weaknesses of the predator, keeping well away from areas of tall, thick grass or from rocks and bushes. This is why gnus, zebras and Thomson's gazelles are generally found grazing on open savannahs and plains where they are best protected from carnivores and stand a good chance of outrunning them. The lions, in return, do everything they can to entice the herbivores into their own territory. To achieve this they have to adopt the only technique that can assure their survival in open country with a minimum of tree or bush cover–hunting in groups, so that ambush, pursuit and attack can be perfectly co-ordinated. The reason why lions are commonly found in community groups stems from their food requirements and hunting methods.

Some authors, including Mervyn Cowie, contend that the males collaborate actively in these group hunting activities, playing the role of beaters by terrifying the prey with roars and driving them towards the spots where the lionesses are posted. A distorting effect, somewhat like that achieved by a ventriloquist, makes it difficult for the hunted animals to locate the male's position. Other authorities claim that the lionesses deliberately place themselves downwind of their prey, so that they cannot be scented, while the lions make a wide detour until they are upwind of the herbivores. When the latter catch the scent of the lions they panic and rush headlong into the trap set for them. Support for this theory comes from one eye-witness account of three male lions which were spotted towards evening advancing in a straight line, some fifty yards apart from one another, towards a herd of grazing gnus. Their objective seemed to be to drive the gnus in the direction of some high grass where presumably the lionesses lay concealed. They caused a virtual stampede and for the whole night the loud roaring of the lions echoed across the plain. Their tactics would seem to have been successful for next morning revealed the lionesses gathered round the remains of a newly slaughtered gnu.

Schaller, however, argues that there is no proof whatsoever that lions roar in order to frighten their prey. He is equally convinced that many group attacks fail precisely because the chosen prey scents the predators. Only once has he seen two males actually collaborating with a lioness, and his opinion is that the females alone act as both beaters and hunters.

Dr Schaller tells how he once watched five lionesses engaged in a communal hunt. Step by step, almost creeping through the dry grass, they drew steadily closer to a herd of Thomson's gazelles. Suddenly, when about 40 yards away, one lioness sprang forward like a flash of lightning, scattering the gazelles in every direction. One of them rushed past the spot where another lioness crouched in waiting. Two powerful paws were unleashed like springs, cutting the gazelle down in full stride. The other lionesses then joined her in the feast.

This organised form of hunting makes good sense when one considers the problems faced by a solitary lioness seeking food. In the beautiful Amboseli reserve—where Mount Kilimanjaro rises spectacularly in the background—a single lioness was seen hunting near a watercourse. Her two cubs were hidden among the bushes. She was using every deceptive technique in the book—hiding herself among the dead stumps and high grass bordering the swamp, gliding silently from one vantage point to another, almost flattened against the sand. Her every movement revealed an extreme state of tension, the taut muscles rippling with hidden reserves of power, the eyes unblinking and alert to any strange activity. Yet her successive attacks on a zebra, a hartebeest and two gnus were all failures. A further unobserved attack at night was apparently more successful for the following day she was seen with her cubs feeding on a gnu while a flock of two dozen vultures waited patiently in a nearby acacia for a share of the remains. Faced with the double problem of her own survival and the care of her cubs she had evidently made repeated attempts to surprise the herbivores as they came down to drink. These attacks had all but exhausted her. One animal, probably the weakest, had finally been caught. Failure to kill any prey would surely have doomed the cubs to death by starvation.

The solitary lion is thus constantly confronted by difficulties; but even in a group there may be problems. George Schaller has pointed out that on the open plains, whether the lions hunt singly or together, only one attempt in ten is likely to succeed. Where there is no cover they are limited to nocturnal hunting for by day their movements are quickly detected by ever-wary herbivores. Where there is tall grass cover they have a better chance of daytime success and it is possible that at such times the black marks behind their ears serve as mutual recognition signals.

Although lions are heavy, powerful creatures their natural prey—zebras, gnus, hartebeeste, topis and other large antelopes—are capable of defending themselves with surprising courage

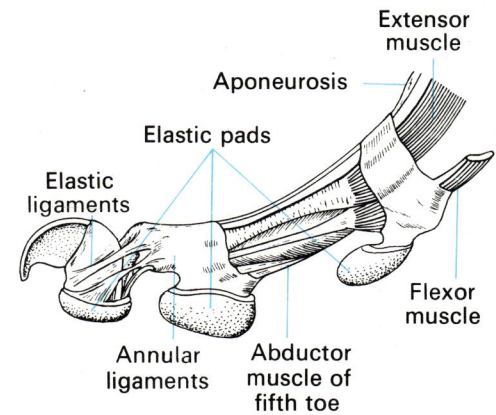

Anatomy of front paw of lion, showing the muscle mechanism of the retractile claw (based on sketch by H. Böker).

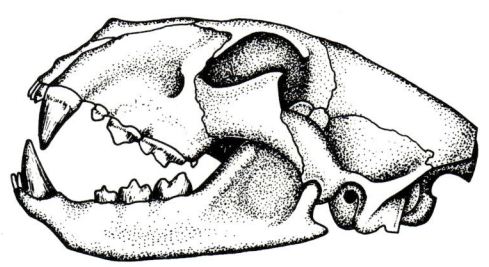

1: $\frac{3}{3}$ C: $\frac{1}{1}$ PM: $\frac{3}{3}$ M: $\frac{1}{1}$

Skull and dental formula of lion (see page 20 for key).

and tenacity. Trying to catch a Grant's zebra from a moving Landrover, for example, is a formidable task. It needs a lasso attached to a long pole and the concerted efforts of three or four strong men to bring the sturdy animal to its knees and even then they have to keep well out of range of its powerful hooves. Gnus have an additional defensive weapon in the shape of their tapering horns and antelopes such as oryxes, whose horns are as slender and pointed as daggers, can also put up a brave fight when cornered.

It is not sufficient for a lion simply to creep up on its prey, surprise it and leap at its throat. It has to be able to bowl it over, immobilise it and then kill it in as short a time as possible. This is not how the hunters of the 19th century and the writers of adventure stories always describe the process. They talk glibly of lions hurling themselves at their victims' necks and breaking the vertebral column with their teeth. This is not correct. Having first caused the herd to stampede, the lion singles out its prey (usually an animal separated from its companions) and goes straight for it—not in leaps and bounds, for this would not permit a rapid change of direction, but running level with the ground.

The shock of the ensuing impact is normally so great that one blow from the lion's mighty paw is often enough to send the reeling victim to the ground. The predator can then bear down with the full weight of its body, at the same time sinking its powerful jaws into the fallen animal's throat. The deep wound caused by the fangs cuts through the windpipe and severs the esophagus and part of the vascular tissue. Meanwhile the lion grasps its prey firmly between its sharp curved claws and increases the pressure until the unfortunate creature dies of strangulation, generally within five or ten minutes. This method of killing is typical of all the wild cats. On some occasions, however, the victim may die immediately on impact with the ground, of a fractured skull. Lionesses also vary the procedure at times by gripping the victim firmly by the muzzle until it suffocates.

The Masai warden Maison-Ole-Lepore, who has lived since his childhood in the Masai Mara reserve, took us one morning to see the remains of a lions' banquet. The victim on this occasion was an adult giraffe and all that now remained of it were the hooves and a few bones. We had already seen giraffes killed by lions in other districts and wondered how they went about toppling such an immense creature, 18 feet tall and liable to weigh about one and a half tons. While we gazed at the pathetic remains our guide explained to us how the lions tackled their lofty opponent.

'Twiga (the Swahili name for the giraffe) is very much afraid of Simba (the lion). When Twiga walks through the tall grass, away from the cover of trees, it looks carefully around in every direction. If a lion tries to launch an attack on open ground Twiga will probably fracture its skull with a single kick. Twiga is usually quite safe because it spends much of its time near acacia trees and does not need to lower its head when it eats. But Twiga must move about to drink and as it

Facing page (above) : Lionesses feed on the carcase of a small antelope killed during the dry season in the absence of gnus and zebras. It hardly furnishes enough meat to feed the entire family so the cubs are likely to go short. *(Below)* a lioness sinking her fangs into the carcase of a Thomson's gazelle, one of the species remaining behind in the Serengeti when the larger herbivores migrate.

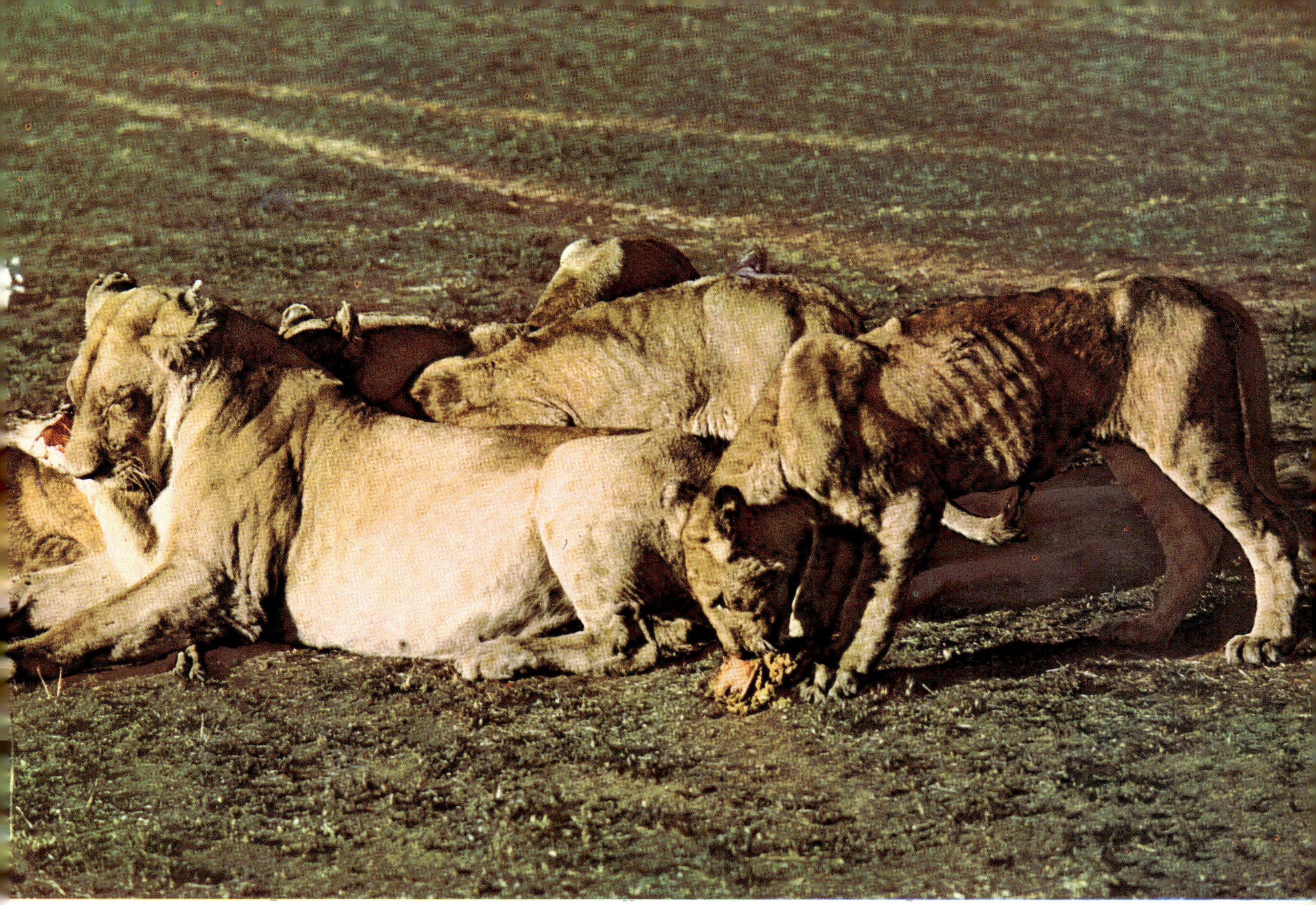

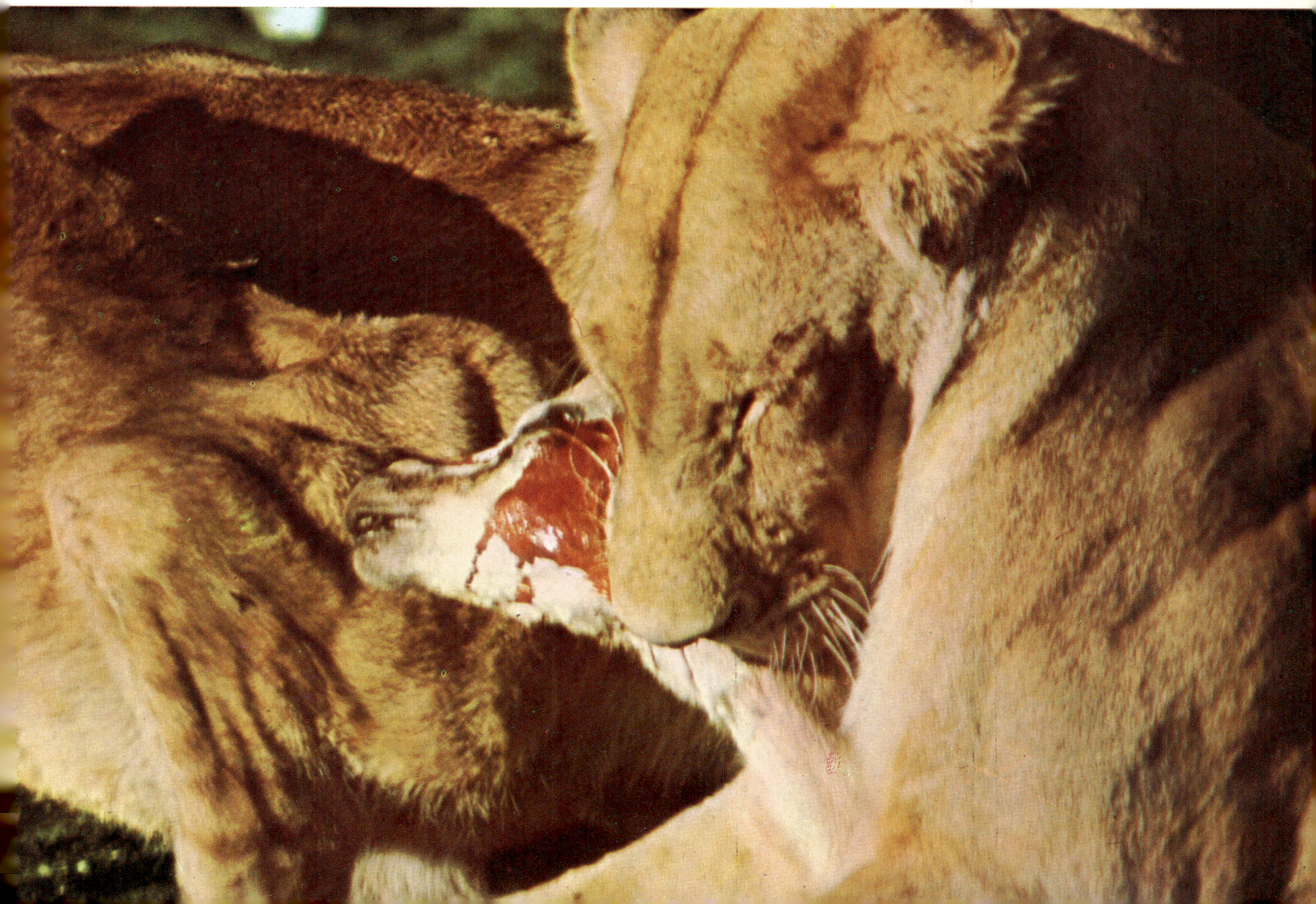

A lion and lioness take their ease in the tall grass of the savannah. The couple remain together during the mating season and the subsequent gestation period but separate as soon as the cubs are born.

moves from one clump to another through the tall grass, which is sometimes as high as a man, it satisfies its hunger by nibbling the leaves of dwarf acacias. This is its undoing. Simba, ever watchful, creeps through the grass and crouches down behind a bush, still as a rock. Twiga carries on, not suspecting anything, lowering its head to munch at the leaves. Then Simba pounces, swift as an arrow, biting Twiga on the muzzle, paws crushing the long neck and bringing the unbalanced animal crashing heavily to the ground. Simba does not relax its grip until Twiga has been strangled, though sometimes other lions which have been lying in wait close by will help to finish the work.'

Heavyweight contest

Dr Schaller's surveys in the Serengeti have shown that during the rainy season about 90 per cent of a lion's food supply consists of gnus and zebras, half a million of which graze on the high plains. When the dry season returns they switch their attention to Thomson's gazelles, hartebeeste, topis, warthogs and buffaloes.

In the Nairobi National Park there are about 25 young and adult lions and roughly 4,000 ungulates of various species. According to a detailed survey covering the years 1961-1966, taken by J. B. Foster of Nairobi University and published in the *East African Wildlife Journal*, these lions show a marked preference for gnus. When the latter were plentiful in the reserve twice as many of them were killed by lions as all the other ungulates put together. Naturally this reduced their numbers in drastic fashion, while at the same time the efforts of their predators were redoubled. It came to the point where there were serious fears that the entire gnu population in the reserve might be wiped out.

This survey of hunting patterns in a confined and closely controlled area proves clearly that given the choice between a large animal and a small one the lions will always select the former. This is why they will attack a 400-lb gnu rather than a 300-lb hartebeest or even lighter-weight gazelle. But it does not explain why they do not show more inclination for zebras, which may weigh over 500 lb, or elands, which can tip the scales at 1500 lb. Obviously gnus, for reasons already given, are far easier to catch. Dr Kruuk has confirmed that the hyenas of Ngorongoro Crater also kill almost as many gnus as zebras, despite the fact that they chase the latter more frequently, but apparently not as successfully.

We must also remember that the lion's hunting tradition is inherited. It is very possible that their partiality for gnus may be in response to a simple acquired taste passed on from one generation to another. Whatever the reason, it is an indisputable fact that the animals making up the basic diet of lions—gnus, zebras and large antelopes—are all about the same size, and that lions will much less frequently attack smaller animals such as gazelles and impalas, or larger creatures such as giraffes and buffaloes.

Lions will not often take on an elephant, and should they find themselves confronted by a herd they will usually yield them right of way without argument. There have been reports of elephants being eaten by lions, and at least two confirmed cases of lions killed by elephants in defence of their young. But such instances are rare for the baby elephants do not normally stray far from their mothers.

Several naturalists have given vivid descriptions of fights between lions and rhinoceroses. The Canadian zoologist John Goddard saw a lion in Ngorongoro attack a black rhinoceros which had wandered away from its mother, leaping on it from ambush. But the baby rhinoceros had enough time to utter a cry for help and its mother came charging up with astonishing alacrity. Before the lion knew what was happening it had been bowled over by the female rhino's horn. A second charge promptly ripped open the carnivore's throat and the unfortunate creature was dead within minutes.

Mervyn Cowie tells of a very different incident in Tsavo National Park when an adult rhinoceros sought human refuge and protection after being seriously wounded by lions. The dying animal limped into an encampment and even tried to enter one of the tents. It was in such a terrible state that the naturalists had no alternative but to put an end to its sufferings by shooting it.

There was another case of a rhinoceros in the Amboseli reserve which was devoured one night by a couple of lions opposite the park's safari lodge. This tragic incident had its origins five years earlier when a female rhinoceros with her two babies had taken up residence on the banks of a stream not far from one of the park buildings. The two youngsters grew up to the familiar sounds of the encampment and the sight of crowds of visitors. The time then came for the family group to break up. The mother took one of her offspring to seek a new home elsewhere while the other one remained alone in the park. The solitary animal was now quite acclimatised, treating the whole area round the stream as its own territory.

A group of lions were also in the habit of coming down to the stream to drink and to rest in the shade of the neighbouring trees, and for five years lions and rhino lived peacefully together. The rhinoceros now possessed good-sized horns and weighed about a ton. One night the park director was wakened by a series of shrill cries. At first he thought it was a pack of hyenas but when the noise persisted and became even more ear-piercing he decided to go out and investigate. He took his car and drove in the direction of the sounds, until his headlights illuminated a terrible scene. Two lions had bowled over the rhinoceros and while one of them grasped its head the other was ripping its throat open with its teeth. The unfortunate creature was struggling hard but unable either to get up or shake its tormentors loose.

The warden did his best to frighten off the lions by blowing the horn, flashing the headlights, slamming the car doors and shouting at the top of his voice. Finally the lions relinquished their prey and slunk off into the darkness. The badly wounded

Lionesses rely on the strength of their teeth for hunting but these deteriorate with age. The older lioness (above) is no longer an efficient huntress but will share in the prey taken by the younger lionesses of the troop.

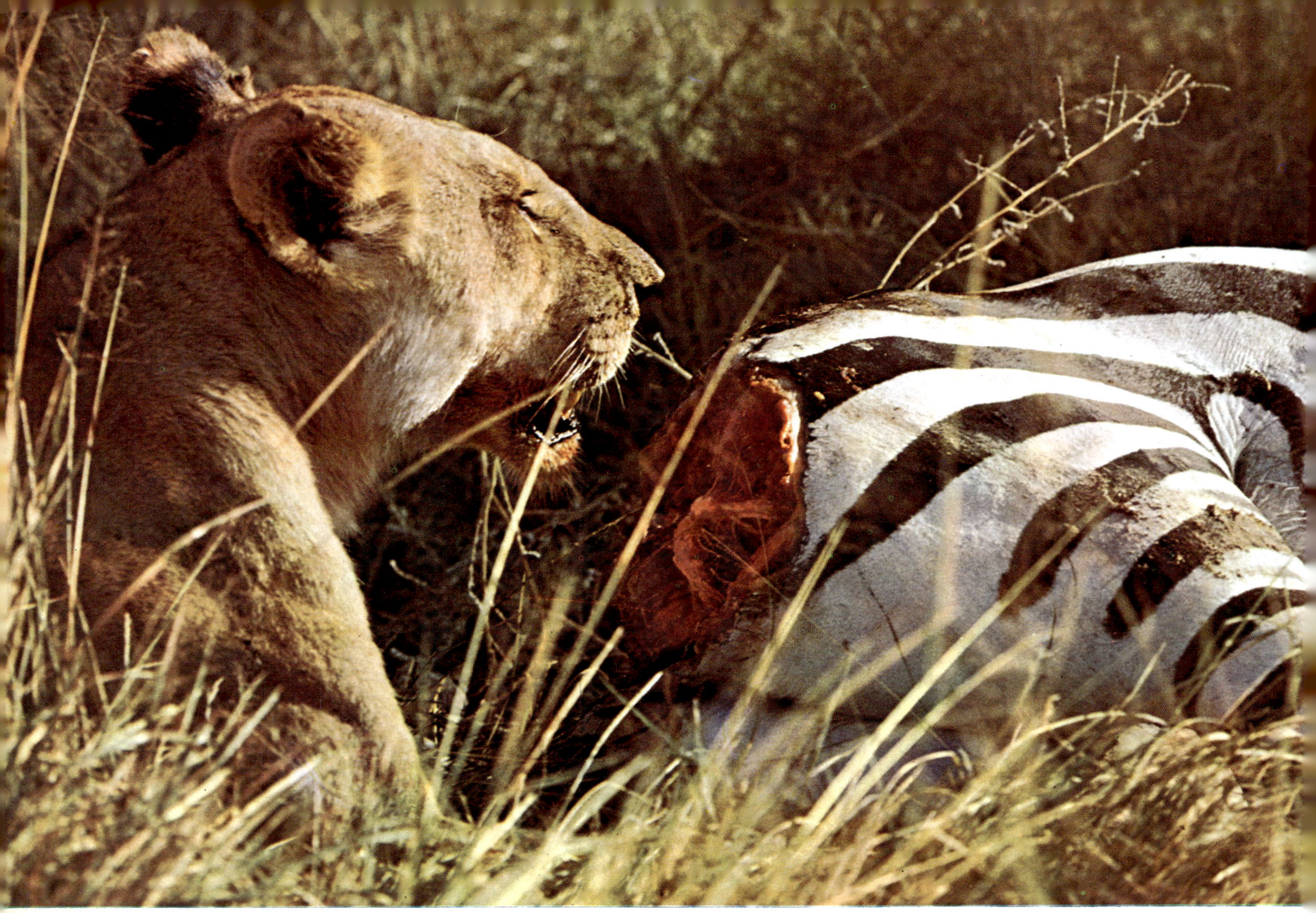

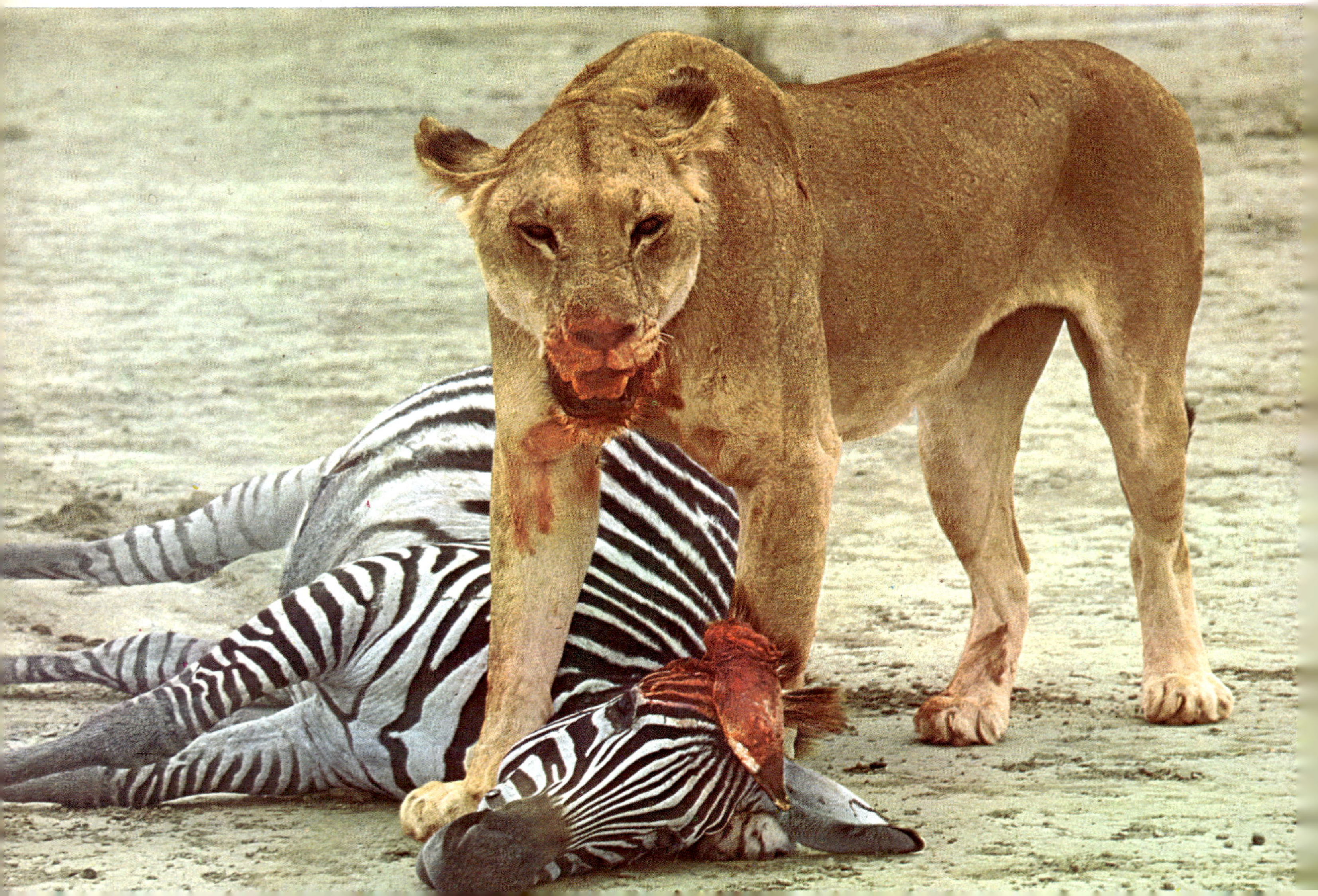

rhinoceros struggled uncertainly to its feet and appeared to be able to walk. But as soon as the warden had driven off, one of the lions returned to the fray and hurled the rhino once more to the ground. Its companion then delivered a flank attack and broke one of the rhino's front legs. By the time the warden arrived it was too late to do anything but put the animal out of its misery. Next day the lions were seen quietly feeding on the remains of the huge beast which had once been their constant companion.

Carrion eater and cannibal

Carnivores, like herbivores, have varied eating habits. There are those that are set in their ways and confine themselves to a fixed diet, while others are more like opportunists and take advantage of anything fortune lays in their path. Among the former category are the cheetah, which feeds exclusively on prey that it has killed by its own efforts, and the peregrine falcon, which catches other birds on the wing and has no relish for any other form of food. As for the vegetarians, many species sustain themselves with highly specialised kinds of food. Two striking examples are the koala bear, which eats only eucalyptus leaves, and the giant panda, whose every meal consists of bamboo shoots.

The lions – particularly those that have been studied by naturalists in Ngorongoro – come into the second category. They are the great opportunists of the carnivore world. If driven to it by dire circumstances they will satisfy themselves with any type of game, no matter what its origin or who has killed it. When it comes to finding food all methods are equally valid. By night they lie patiently listening for the slightest sound that may betray the presence of a pack of scavenging hyenas or a solitary prowling leopard. Either way there is a good chance of prey to be picked up without undue risk or effort. During the daytime too they are constantly on the alert for a flock of vultures circling in the sky over the spot where an animal may have fallen prey to a predator – another opportunity for a free meal.

In fact the lion is basically a lazy creature, only resorting to the hunt when absolutely forced to do so under the stress of extreme hunger. As long as it still has the smallest morsel belonging to the last prey taken it is content to leave matters as they are. No creature provides a better example of that fundamental law of nature which operates throughout the wild – the conservation of energy.

Although their sense of smell is not exceptionally refined it does sometimes serve them well enough to flush out young gazelles and other baby animals that may have been left hidden by their mothers in the comparative safety of the bush. Lions may often be seen feeding on these tiny creatures but it is difficult to make an accurate analysis because they are devoured much more rapidly than larger prey and leave behind hardly any traces. At other times the carnivores will turn to even more unconventional kinds of food, including tortoises and

The black rhinoceros, despite its size and weight, may come off the loser in a fight against one or more lions. In general lions avoid tackling such large creatures, having respect for their fighting powers, particularly when in defence of their young.

Facing page : Lions prefer medium-sized herbivores such as gnus and zebras, with a preference for the former. The reason may be that because of their speed, strength and sensory perceptions zebras are that much more difficult to catch.

The male lion has acquired his mane by
the time he is five years old. Its colour
may vary from yellow to brown according
to the region, as will its length and texture.
Mountain lions tend to have darker and
thicker manes than savannah lions.

reptiles. Two lions were even seen on one occasion eating a large crocodile.

In certain circumstances lions have been known to turn to cannibalism. George Schaller and Mervyn Cowie have both given instances of lions or lionesses eating their cubs, for reasons already suggested. Much still remains to be explained concerning this phenomenon. If it were really widespread it would obviously threaten the security and future of the species, and it would appear that there are inbuilt mechanisms—a complex structure of stimuli and responses—which inhibit these cannibalistic instincts. This applies to many species of carnivore, particularly the gregarious ones. Among lions it is possible that cannibalism only occurs between individuals of different troops or in the case of an accidental death—for example, by rifle fire.

The lion is also a typical carnivore in that it does not object to eating grass if the need arises. It may either feed on it directly from the savannah or ingest it together with the other stomach contents of its victims. Nor can it go long without drinking and it may take as much as a quarter of an hour without pause to quench its thirst. Although it prefers to drink from the running water of streams and rivers it will often lap up stagnant water quite contentedly. When serious drought conditions prevail it may resort to drinking the blood of a victim by slashing open its belly.

We have seen that the lion's favourite prey consists of large and medium-sized ungulates. It has been calculated that each individual kills about twenty such herbivores in the course of a year. A small troop comprising six or seven lions of varying ages will go hunting about twice a week, according to a survey made in the Nairobi National Park. This confirms the findings of George Schaller with his transistorised lion. This nomadic male was observed to feed seven times in three weeks, at more or less regular intervals.

Lions are able to go without food for more than ten days but when they do find an abundance of prey they glut themselves until they are almost incapable of moving. Five adult lions can manage to dispose of a zebra weighing 550 lb in a single day and Dr Schaller once saw a hungry lioness eat more than 70 lb of meat in one night—a fifth of her own body weight! But this is unusual. In general these huge cats are more moderate in their feeding habits. Two males and three females in the Serengeti killed four gnus at a time when such prey was plentiful. This provided each of them with about 15 lb of meat per day and appeared to be perfectly adequate.

The lions' marriage rites

Lionesses on heat give out a distinctive odour which announces their condition to the male, at the same time marking out a portion of territory with their urine mixed with secretions from their anal glands. Some authors claim that they also roar in a special manner. The female is at once joined by one of the dominant males and they retire together to a secluded

Facing page : Young lions are taught to hunt by the lionesses and at this age there is no clear division of labour according to sex. This young male has killed a reedbuck by leaping on its back and biting the muzzle until the animal suffocated. The victim probably lowered its head in order to protect its throat.

part of the communal territory, usually close to a watercourse. For an entire week the couple do not go hunting and eat very little, and it is noticeable that the male keeps very close to his mate at all times, rarely more than 15-20 feet away. During this period he shows many signs of tender concern and affection, continually licking her back and neck and frolicking playfully with her.

The sexual act itself is brief but is repeated several times a day. In the course of it the male may bite his partner's neck, while she utters loud roars, stretches herself on her back and contorts her body until he withdraws. At this mating season male lions from neighbouring territories, attracted by the female scent, may sometimes invade the private nuptial area, so that fierce fighting may result.

Not enough is known about the social hierarchy within a pride which leads to one male rather than another mating with a particular female; but once the choice is made the couple will be undisturbed and virtually ignored by all the other members of the pride.

As soon as the brief nuptial period is over both male and female rejoin the group, remaining together during the three and a half months of gestation. Immediately the cubs are born the couple separate. The lioness goes off alone into the bush to give birth while the male, apparently quite indifferent to her welfare, will promptly couple with other females and resume his normal daily activities.

The lions' nurseries

A lioness's litter may consist of between one and six cubs, three or four being the average. The newborn cubs are very weak and completely blind for the first 24 hours, but their eyes are fully open by the time they are three days old. The mother, after eating the membranes and the placenta, licks the cubs clean and then suckles them. Although she shows the closest maternal attention she has no choice but to leave them at times, either for short periods to quench her thirst at a nearby pool or stream or for longer periods in order to hunt and find food. The cubs are usually left resting in a well-hidden place but she is faced with a problem when setting out on a hunting trip that may last several days. Clearly the cubs cannot be left alone for that length of time.

There are conflicting theories concerning the way lionesses cope with this situation. Some authorities suggest that they retain the services of another lioness—sister, aunt, mother or close companion—to look after the cubs in her absence. Thus in troops of reasonable size a 'nursery' group may be formed, consisting of between six and ten females and perhaps upward of fifteen cubs.

In the Masai Mara reserve one such nursery group, in which seventeen cubs of varying ages were tended by six lionesses, was kept under observation. The day was spent resting but when at sunset the lionesses began to stir, all the cubs began to romp about them in the most playful fashion, apparently making

no distinction between their own mothers and the other females, lavishing licks of the tongue and other forms of caresses on all and sundry. Some of the lionesses set out on a hunting expedition and in due course their roars announced that a prey had been killed. This precipitated a general exodus so that all could share in the feast.

The close association of several lionesses within a pride probably originates in this double responsibility of hunting for food while at the same time keeping a watch on the cubs. The males, as we have seen, dissociate themselves entirely from such humdrum activities, their appointed task being to defend their common territory and nothing else.

This tendency of the females to join together during the critical period of rearing and educating the young plays' a vital part in the survival of the species. While some go out hunting on behalf of the entire community others stay behind to guard, clean and suckle the cubs. This kind of co-operative behaviour is by no means unique to lions. Penguins too have their nursery groups, arising from the same necessity of having to wander far from the nest in search of food and needing to provide protection for the chicks while they are away from the nest.

It would appear that this kind of communal endeavour occurs most frequently among lionesses from the same litter. The story of Blondie and Brunette, two lionesses from the Nairobi National Park, is now something of a classic—and a good example of such co-operation. It was clear from the beginning

A lion and lioness coupling. In the mating season the pair seclude themselves from their companions. The sexual act may be repeated several times, the male often holding the female down while gently biting the back of her neck.

Facing page : During the 'honeymoon' period, which may last a week, the male remains in close attendance at all times, licking the back and neck of the female and playing affectionate games with her.

that the two animals maintained close links, helping each other and sharing all the everyday tasks. While one was out hunting the other stood guard over the cubs and fed them. Since there was a town nearby it was possible to keep the two sisters under close and continual observation. Wardens, naturalists from the university and even tourists joined together in the survey.

On one occasion, when the two lionesses were tending eight cubs, Brunette was seriously injured in the course of her night's hunting. She was seen returning in the morning bleeding profusely from a deep gash in the stomach. Although her life was apparently not endangered the wound prevented her from going about her normal activities for some time.

Blondie immediately took over complete responsibility for her sister and the eight cubs. For two weeks the single lioness bravely carried the additional burden, carefully watched at a tactful distance by the wardens, who were ready to rush in with help should it be needed. But Blondie was well able to cope until Brunette was once more on her feet. The two lionesses with their cubs were for many years the star attractions of the park.

George Schaller disputes the 'nursery' theory, asserting that he has never seen true co-operative activity among females during the period of rearing the young in the Serengeti, and that if a lioness is ever found in the company of another female one of them is simply a looker-on, taking no part in the hunting and even less in caring for the cubs. The American naturalist says moreover that roughly half of all lions born in the Serengeti die during the first few critical days and weeks—either from hunger or by being devoured by hunting dogs, hyenas and other lions—and that this is simply because they have been abandoned, the mothers finding it impossible to cope with the double responsibility.

Dr Schaller tells how he himself witnessed a sequence of events in which the lack of adult co-operation had direct and tragic results for several cubs. A lioness had left her two cubs hidden in a clump of tall grass and was out on a hunting expedition. She managed to kill a gnu but since it was too heavy for her to drag off alone she left it and went back to find her cubs. Collecting one of them and returning to the spot where she had left her prey she found that a leopard had stolen it. She immediately charged and sent it scrambling up an acacia. But now she was faced with a perplexing problem. Should she stay with her cub until it had eaten its fill or should she leave it in order to fetch the other one? She chose the second course and set off again into the bush. The leopard waited until she was a good 200 yards away, then pounced on the defenceless cub and killed it. The victim just had time to utter a piercing cry before the leopard's claws sank into its body. The lioness rushed back to the rescue but she was too late. The cub was already dead and the leopard back once more in the safety of the branches. The lioness turned and loped off into the thickets, Schaller following at a distance. But when she reached her destination she found her second cub also dead—apparently killed by another lion.

Facing page : Lions will not often attack giraffes, which because of their height can easily spot a predator on open ground. The lions get their chance when a giraffe is forced to bend its head to browse on the leaves of small trees. The neck is particularly vulnerable and once caught off balance and hurled to the ground its plight is hopeless. In these pictures lionesses and cubs feed on a freshly killed giraffe. They will leave nothing but hooves and bones.

What conclusions can be drawn from these conflicting views and reports? Is the maternal behaviour of the lions of the Serengeti markedly different from that of the lions of the Nairobi National Park or the Masai Mara reserve? This seems highly unlikely–but there may be a simple explanation. All or almost all the females isolate themselves with their newborn cubs during the first few crucial weeks. They have to watch them, suckle them and find them food–and they have to do it all on their own. It is inevitable that they should have to leave the cubs unprotected at such a time and that the latter should thus be exposed to all kinds of hazards; but when they are able to move about freely the lioness conducts them back to the relative safety of the troop and teams up with other females. It is then that they establish their communal nursery, and from that moment the baby lions are provided with everything they could want–plenty of milk, constant supervision, the company of other cubs and an introduction to the techniques of the hunt–all of these duties devolving on the adult females in common.

The joys of youth

Anyone who has watched a group of young lions in the company of their mother will have remarked on the mutual demonstrations of tender affection. But do such exhibitions have any special significance? Adult lions are born hunters and killers and it is essential that they do not extend these activities indiscriminately to include their own young as well as their traditional prey. So all these carefree and lighthearted actions of the cubs–especially the lavish attentions they pay to the females in charge of them, such as licking the lionesses' lips, lashing out clumsily with their paws and rolling on their backs to be licked in their turn–have a particular purpose. They are designed to repress the adults' aggressive instincts towards them.

At this stage in their lives the cubs are quite uninhibited, tumbling all over the lionesses and expressing their pleasure at the appearance of a male by urinating on the spot. As the lions grow older such games and ritual postures take on even more importance, for they are defence mechanisms intended to ward off the rebukes and even attacks of their elders. Soon the young males, already showing signs of a mane, are big enough to be regarded as hated rivals by the leaders of their own pride. No longer cared for by attentive females, they are now left to their own devices and instinctively revert to the behaviour of their infancy in an effort to shield themselves from the teeth and claws of the adult males–rolling over on their backs and making use of tongue and paws to signify submission and obedience. But eventually–at about three years old–they will be forced to leave the pride and find their livelihood elsewhere. They may be able to establish their own territory in some remote, inhospitable part of the savannah, or they may be condemned to wander aimlessly for the rest of their lives, accompanied by other equally hapless males.

Facing page (above) : Lionesses often lick each other, the process having a dual purpose–cleaning the fur and getting rid of parasites. *(Below)* the lioness removes her cub to safety by the simple method of picking it up by the neck.

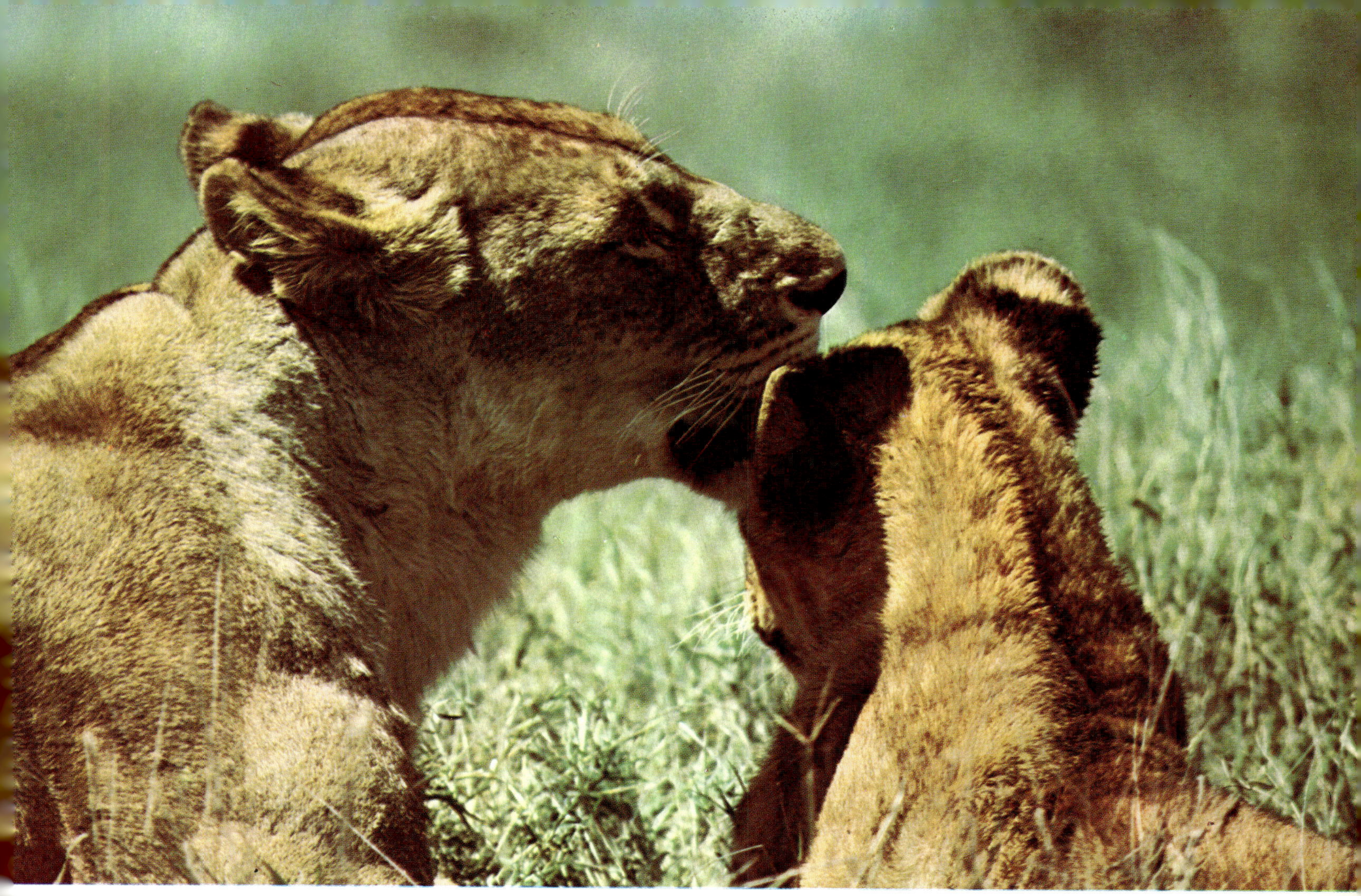
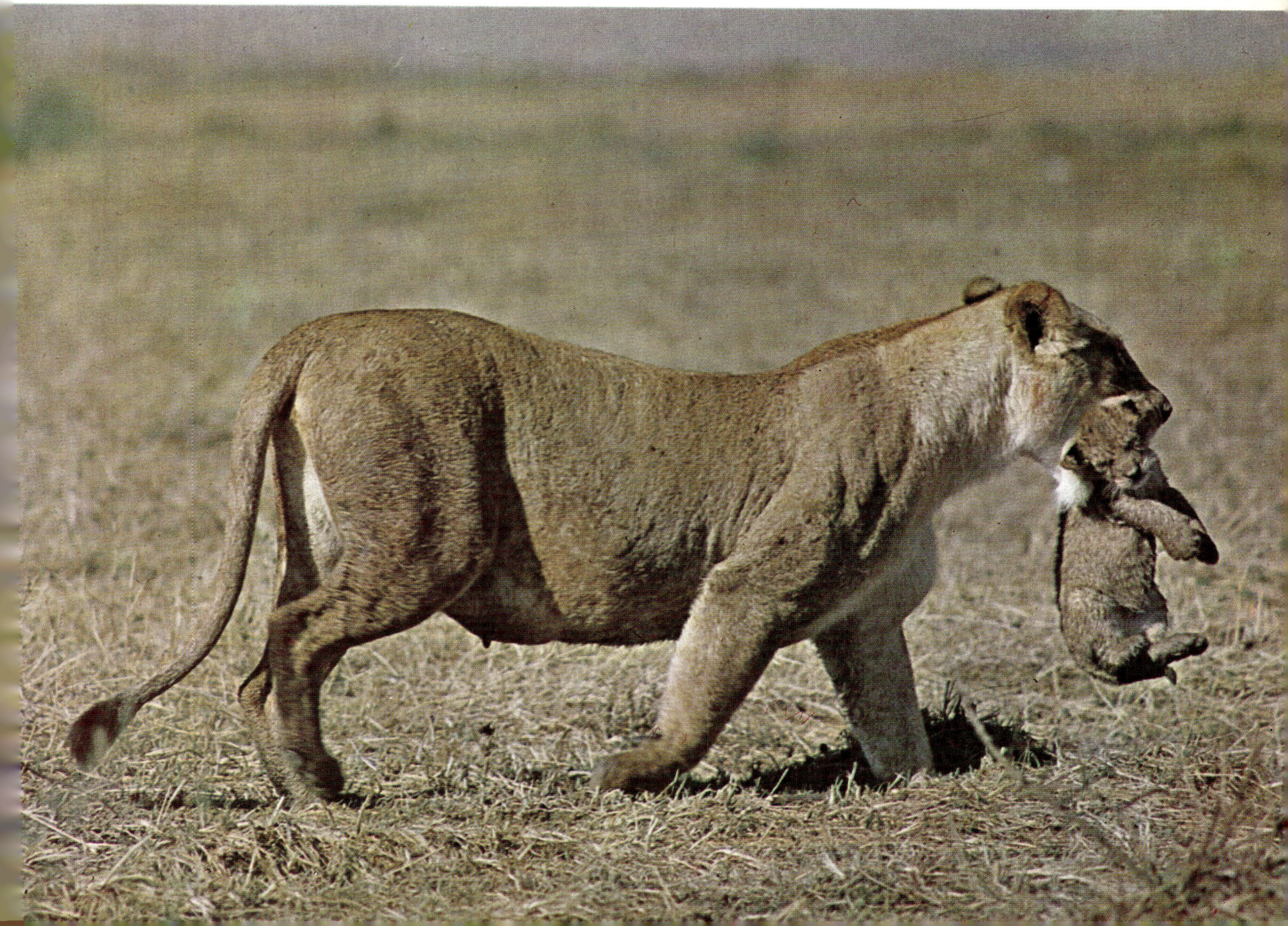

One way of determining the sex of a lion cub is by the presence or absence of black marks behind the ears. Only the female has these marks.

Is it possible that lions, even in later life, adopt the postures and movements remembered from infancy in order to avoid being hurt by a stronger enemy? Can they ward off a fatal bite simply by lying on their backs and offering their unprotected neck and throat to an adversary, whether of their own species or another? There is some evidence to show that this may be the case, but there is no conclusive proof. It is known that wolves—also gregarious hunters—adopt such ritual attitudes throughout their lives, specifically for this purpose. The same may be true of lions.

The call of the wild

The film *Born Free*, which was made several years ago, described the adventures of a lioness named Elsa which was captured when very young and brought up by the writer and painter Joy Adamson with her husband George, a Kenya game warden. Elsa grew up like a carefree domestic cat but when the Adamsons had to return to England they decided to allow her to revert to the wild rather than have her end her days in a zoo.

The attempts made by Elsa to readapt to a scarcely remembered wild state almost cost her her life. She had never learned to hunt, so she would trot happily behind wild animals, sometimes bowling them over with her paw but quite unable to apply the necessary skill and pressure to immobilise and kill them. During the time that she was learning from experience how to fend for herself she was fed by the Adamsons, otherwise she would have died of starvation. A few months after going back to the wild she returned to her foster-parents accompanied by three adorable cubs. The story of Elsa, first appearing in book form and then in the cinema, captured the imagination of a worldwide public. And it proved very clearly that a carnivore must have learned how to hunt in order to survive in the wild.

At the age of three months the cubs begin to follow the lionesses around, at first remaining hidden in the grass and watching their elders' activities from a distance. But as soon as prey is taken and divided up all their special prerogatives disappear (contrary to the practice of members of the dog family). The males send them flying with a blow of the paw and even the females will not allow them to get near until they themselves have eaten their fill. This may be the reason why the lionesses exclude their offspring from hunting groups as soon as they are old enough to fight the adults for their share of food. Males are even capable of killing young lions to defend what they consider their due. George Schaller suggests that such behaviour indicates that the lions' social evolution is still incomplete.

The young lions are able to take an active part in hunting as soon as they have their permanent teeth, at about a year. At fifteen months they may have captured their first prey and by the time they are two years old they are already experienced hunters. At three years the males leave the pride to make their own way in life while the females, now sexually mature, form an integral part of it.

First come, first served

Lions are often seen, and depicted in photographs and films, showering one another with little marks of affection—tongue-lickings, gentle bites, friendly taps, mutual grooming and the like—all making up a very reassuring and friendly image. Unfortunately this picture of family harmony does not altogether square with the facts—though the myth is hard to shatter. Consider, for example, how the hungry adults behave when they have killed their prey. In the Serengeti we watched several scenes in which the attitude of these much-praised creatures was, to put it bluntly, quite despicable.

At the height of the dry season, on the banks of the Seronera, fire had played havoc with the savannah and the entire region looked like a blackened wasteland. Under a large acacia two lionesses and five cubs—between four and six months old—were dozing. The sleek and healthy appearance of the adults contrasted with that of the cubs. They were unbelievably thin, their coats sparse and dirty—a sorry picture of neglect. About 300 yards away two huge male lions, with superb manes, were comfortably resting. Suddenly one of the females sat up and, without paying the slightest attention to one of the cubs which staggered towards her, fixed her unflinching gaze on a spot in the far distance where several hundred Thomson's gazelles were grazing. From where we stood we could clearly see the white bellies and black flank markings of the 'Tommies' as they wandered in groups down to the river.

After a close inspection of their movements the lioness rose to her feet and began pacing the river bank, followed by the five cubs. Now that they were standing we could see that they really were in miserable condition, their bodies beneath the dust-flecked coats so thin that we could count their ribs. With drooping heads and flattened ears they followed the lioness in Indian file, falling behind with every stride. The other female also bestirred herself and brought up the rear.

As the little group came level with the males the first lioness lifted her tail in a brief salute, then quickened her step and made for the open plain. We could see, through our binoculars, four gazelles nibbling at the stunted grass. Now creeping through the odd clumps of bushes spared by the fire, the first lioness drew steadily nearer to her prey. It was not until she was 40 yards away

Lion cubs have a spotted coat and lightly-ringed tail. This marking is an effective form of camouflage during the dangerous early days and weeks when the lioness may have to hide them among the bushes. Some authorities believe that in prehistoric times all lions may have been similarly marked for protective purposes.

The head of an adult male lion, showing the large canine teeth and the distinctive mane. It is the male, defender of territory who reserves for himself the largest part of any prey killed by the female.

that the antelopes spotted her. Immediately they took to their heels and seemed to be at a safe distance when one of them, for no reason at all, stopped, looked at the pursuing carnivore and resumed running. The pause proved fatal for the lioness had made up vital yards and now cut off the gazelle's escape route. Both animals fell to the ground in a tangled heap and within seconds the victim had been despatched.

We came closer to see what would happen next. The lioness was grasping the gazelle—a young male—by the throat. Her eyes were closed, her jaws were twitching and she was breathing heavily. Now the other lioness leapt forward and plunged her fangs into the dead beast, applying all her weight in an effort to snatch the carcase away from her companion. Imitating her, the cubs too began to fling themselves ineffectually against the flanks of the dead animal. It was a scene of complete chaos with everyone fighting one another, to the accompaniment of confused snarls and angry roars. In the end the prey was torn in two. One of the lionesses made off with the rear portion of the dead gazelle in her jaws, dragging a cub along in her wake. Halting a short distance away, she held the bleeding trophy down with her paw, trapping the unfortunate cub at the same time. For five minutes, and showing no sign of emotion, she held the youngster prisoner. Then she got up and released it from what had looked like certain death.

Meanwhile the other female, holding on to her half, was using her claws to fight off the other starving cubs. All that was left to them were a few scattered scraps of flesh and the odd pool of blood that had spilled into the dust. Forty minutes after its capture there was nothing to be seen of the gazelle. Not only had the young lions had nothing to eat but they had been brutally bruised in the fracas. Imagine our surprise when they staggered to their feet, covered with dust, and proceeded to lick both females, apparently with affection, on paws and muzzle! In fact, their probable motive for such behaviour was that they were attracted by the traces of blood on the adults' bodies.

After about five minutes' rest the group set off once more for the river. As soon as they had reached the area of high grass along the bank one of the lionesses bounded into the thickets. It was another five minutes before we traced her. When we caught up with her she had caught another male Thomson's gazelle, apparently on the spur of the moment. She was sprawled across its body, ready to defend it against the cubs. But this time they were too scared even to come near. As it happened, she soon had a more serious problem to deal with, for one of the male lions suddenly came hurtling out of the bushes. He paused for an instant, brushed aside one of the cubs which was rash enough to stand in his way, then pounced with a loud roar and snatched the gazelle from the lioness's jaws. Once again the prey was ripped into two parts. The cubs managed to trap a few pieces of flying flesh and one of them boldly took hold of the morsel which the lion was dragging through the grass. History then repeated itself as the cub found itself crushed under the lion's massive body. Eventually the huge beast got up, shook his head vigorously, prised the youngster loose and sent it crashing into the trunk of an acacia. He then settled down to enjoy his meal, but every time the cub, now quite famished, tried to come near he cuffed it away with his heavy paw.

By nightfall there was nothing left of the second gazelle and the unfortunate cubs had still had nothing to eat. They curled themselves up into brown balls and slept alongside the replete females while the male rejoined his companion.

A week before these events took place there had been nine cubs. Presumably the missing four had already died of hunger and been devoured by hyenas or even by their own starving brothers and sisters. In fact, during the annual dry season, when there is no prey available except Thomson's gazelles, some 90 per cent of the local population of lion cubs are condemned to die for lack of food. This is between 10 and 30 per cent higher than is usual in more favourable circumstances.

What can be the reason for this callous assertion of priority in which the strongest make off with the tastiest morsels and the feeblest go hungry? Despite the apparent cruelty involved, a case can be argued on the grounds of sheer necessity and very survival. If the adult males did not appropriate the largest share of prey they would be seriously weakened and this would threaten the security of the territory and the group. Furthermore, if the lionesses were to give away their part to their offspring they would have insufficient strength left to kill prey and both young and old would die. The painful spectacle on the banks of the Seronera shows exactly what role is expected of every individual in order to guarantee the continuation of the species. It is a rigidly ordered hierarchy. First and foremost are the males, defenders of the communal territory against intruders; then come the females, hunters and universal food providers; last of all, and expendable, are the cubs, not yet old enough to fight or hunt. As long as there is a piece of territory and enough females to find food and to breed when the time arrives, there will be cubs born— and enough of them will survive the rainy season and grow to maturity so as to ensure the species' survival.

The young lion of the Serengeti which completes its weaning by May is faced with a grim prospect between June and September. This is the season when hunger pangs will subject it to the implacable laws of natural selection that operate throughout the animal kingdom. The few survivors will be those cunning and strong enough to snatch their food from the claws of their parents and deny it to their brothers and sisters—and these will be the future leaders. Nature is much more concerned with the interests of the community as a whole than in the fate of any single individual.

To end on a slightly more cheerful note, before leaving the Serengeti we did have the consolation of seeing the family reunited around the carcass of a freshly-killed buffalo. This time both adults and young seemed to have eaten plentifully and there was still enough meat for three or four days.

In October, when the rains return, the blackened soil is once more carpeted with green grass. Now the gnus and zebras journey back in their thousands and the Serengeti lions have more food than they need—drought and hunger just a memory.

A hyena emerges from a mud bath. The front part of the body is far stronger and more developed than the rear part, enabling this killer and carrion-eater to drag its dead prey away in its jaws.

Who dares challenge me?

In theory a lion may kill any animal which shares its habitat—apart from adult pachyderms—without risk or fear of reprisal. At the apex of that imaginary pyramid which ranges the creatures of the African savannah according to strength and appetite, the powerful lion, the super-predator, stands alone.

It could of course be argued that man, with his poisoned arrow and rifle, is the absolute super-predator, here as elsewhere. He is certainly not an enemy to be trifled with—whether he comes in the guise of a Masai, spear in hand, eager to merit the title of warrior, a livestock breeder shooting to kill in order to protect his herds and his land, or a professional hunter craving the honour of being photographed with one foot proudly placed on his slaughtered victim's head. But man is not the lion's *natural* enemy, so we must leave him out of our present reckonings. We are concerned only with those animals which, if they do not attack the lion directly, tease and torment it, threaten its cubs and compete with it for food and available living space.

The boldest of the lion's rivals is the hyena, and the outcome of a confrontation between these two carnivores depends on how many are involved. A dozen hyenas can relieve a lioness of her prey but if two lions reinforce her the hyenas will make themselves scarce. Protecting prey is one of the male's responsibilities, though not one that is always faithfully discharged. But a solitary lion may be attacked by hyenas under other circumstances. Dr Schaller saw an exhausted lion fall victim to a pack of hyenas after a chase of more than a mile. Aged lions are also likely to end their days between the jaws of hyenas, as are cubs abandoned by their mothers.

It is said that hunting dogs also attack elderly lions and cubs, though there is no eye-witness account of such incidents. In any event—possibly because they are fewer and their hunting methods quite different—these dogs do not represent a serious challenge to healthy lions.

Lion cubs in the Serengeti are sometimes killed by leopards, which is consistent with these felines' habit of attacking animals smaller than themselves. But a leopard is no match for an adult lion and a warden once surprised a band of lions tormenting a leopard which had taken refuge in an acacia. One of the lionesses made a mighty leap and pulled it down from the branches and the others devoured it on the spot.

The ratel (*Mellivora capensis*) is a small, thickset and extremely aggressive African carnivore which feeds on rodents and small mammals as well as on honey and eggs. It appears to have scant respect for any animal—neither the strange beast known as the Landrover, whose wheels it has been seen to attack, nor the mighty lion itself. In the Nairobi National Park a group of young lions were forced to yield up a gnu which they were consuming as four ratels advanced towards them with an air of determined aggression.

The herbivores which defend themselves against the attacks of lions cannot be regarded as real enemies; but their power to inflict wounds should not be under-estimated—whether sturdily-

Facing page : Lionesses, who are usually more active than the males, are responsible both for feeding the family and rearing the young. It is usual for females of the same troop to remain together in the interests of common security and to provide the growing cubs with everything they need, in preparation for the day when they assume their adult role in the pride.

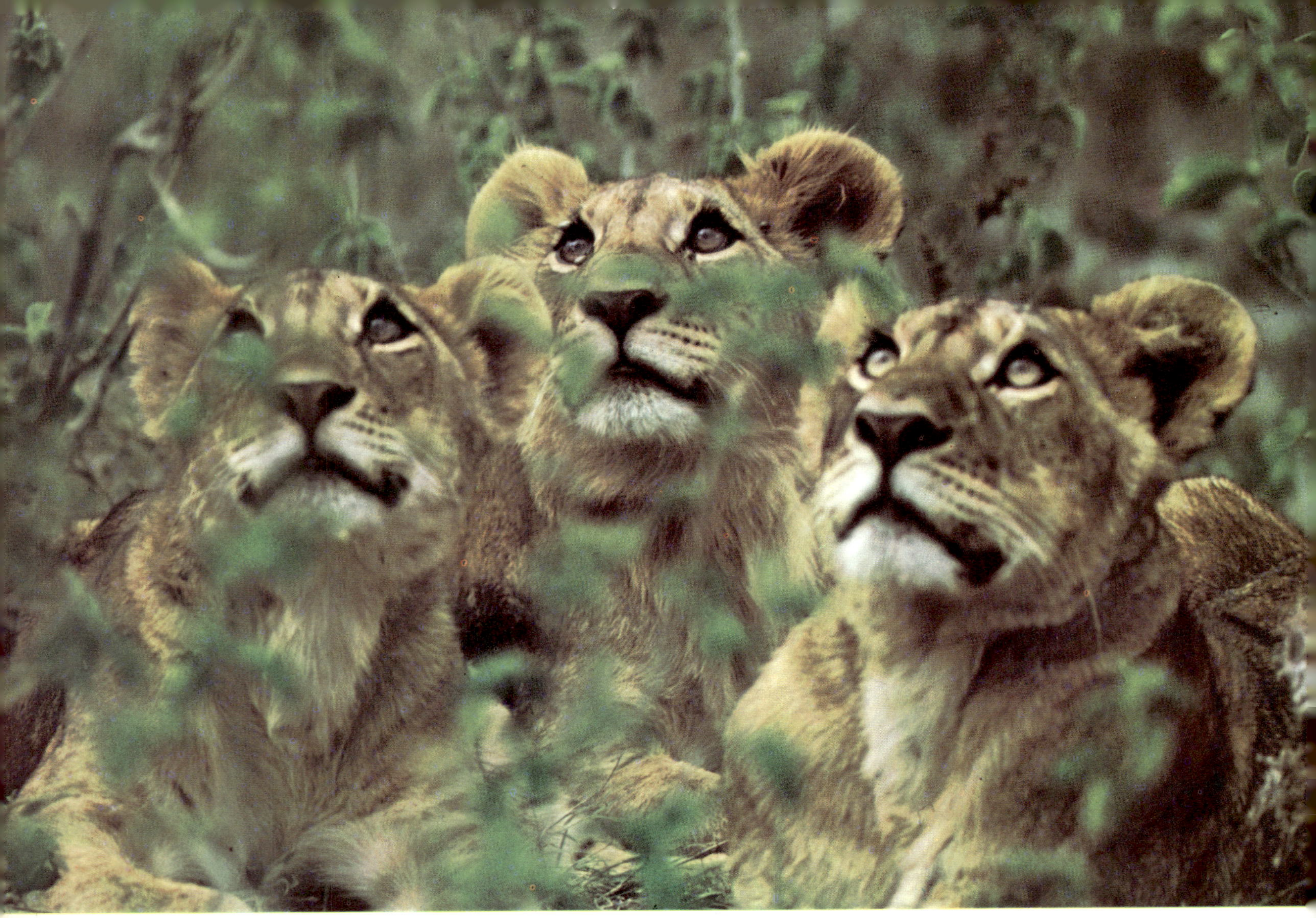

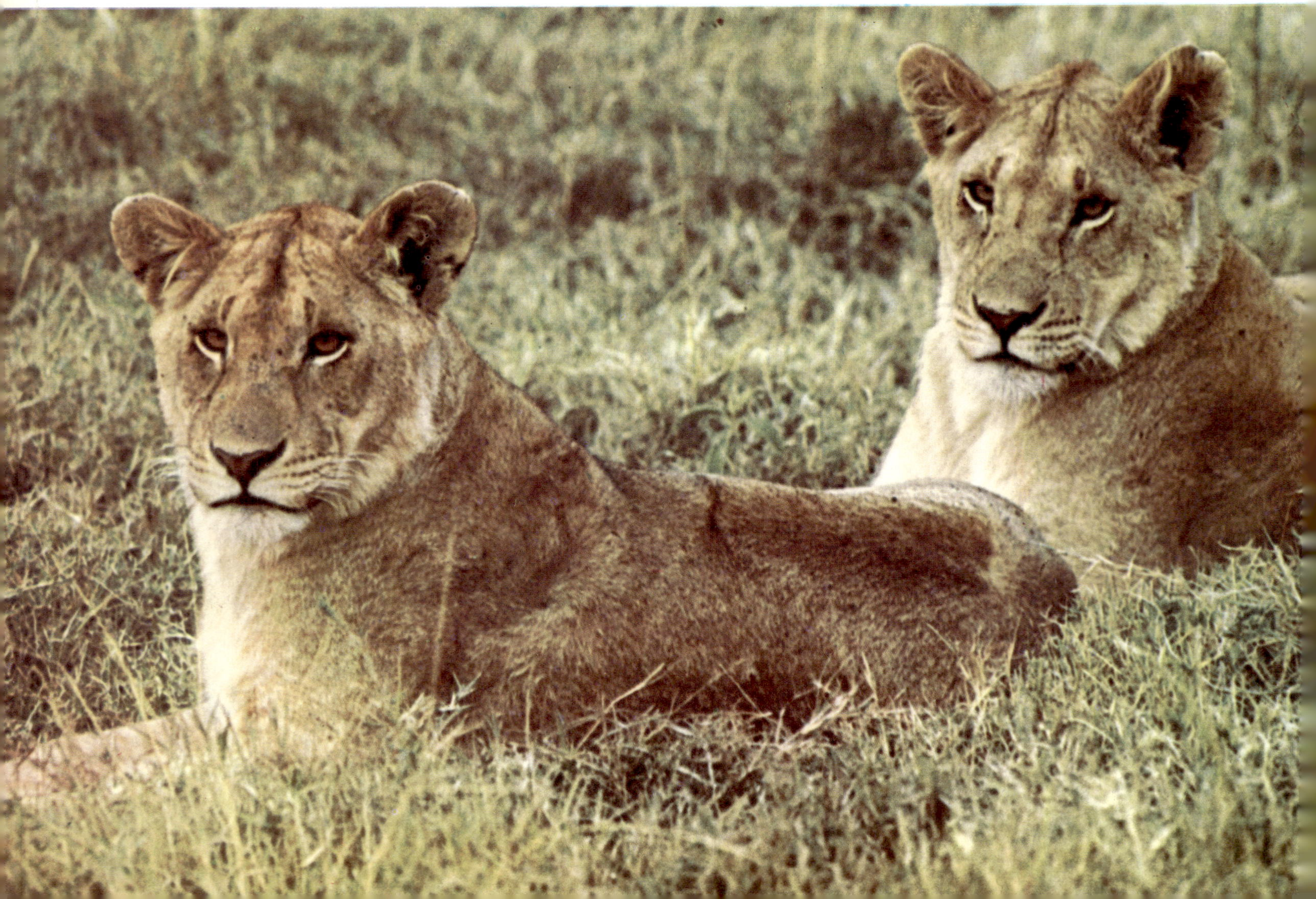

horned animals such as buffaloes, gnus and large antelopes, or creatures packing powerful kicks, such as zebras and giraffes.

Although they share the same habitats as a number of species of snake – including such dangerous kinds as mambas and vipers – lions do not seem to be harmed by them. The explanation may be that they know the reptiles' habits well enough to mistrust and keep well away from them. It is, however, possible that the snakes themselves avoid the presence of the lion, for in spite of their known aggressiveness when provoked, they tend to avoid danger whenever possible.

The real enemies of the lion – and the only creatures that are capable of making its life sheer misery – are insects. The dreaded army ants can devour a live cub unless the lioness is sharp enough to scent their presence and snatch up her offspring in her jaws. Ticks adhere to ears and eyelids no matter how assiduous or frequent the coat-cleaning procedures. Worse than either of these are the flies, which are quite merciless, particularly if the lion is covered with the blood of its victim. In Ngorongoro Crater we once saw an injured lioness whose wounds were barely visible under the thick black crust formed by these insects. The unhappy animal was worn out with efforts to get rid of them and had seemingly given up all hope of evasion or escape.

In 1961, also in Ngorongoro, the November and December rains lasted until the following March and the abnormal conditions led to a disturbing invasion of hordes of blood-sucking flies (*Stomoxys calcitrans*) of the stable fly family. Zebras and gnus did not seem unduly troubled but the lions were at their wits' end to avoid the pests, even climbing trees and seeking refuge in hyenas' lairs. Furthermore they were forced to modify their normal behaviour, wandering into the nearby Masai villages and attacking domestic animals. One lion was killed by a shepherd with a spear and two others were shot by the park personnel. Out of 60 lions that had been living in the crater when the scourge began only 15 were left when the dry season ended. The others were either dead or had abandoned the region to escape the insect plague.

Conditions of intense heat do not suit lions. Even in temperatures that humans can endure lions may be seen panting in the shade of acacias, especially when their stomachs are full of food. Yet at a height of almost 8,000 feet – as in Ngorongoro Crater – where the mornings are really cold, lions seem to be completely at ease.

To sum up – discounting the effect of man's own depredations – there are a number of factors which tend to restrict the growth of the lion population in the wild. The first and most significant is the high death rate among the cubs. Next in importance is accidental death – resulting from struggles within the troop for domination and for possession of prey, or in the course of hunting. Finally there are deaths caused by internal or external parasites and infectious disease. Between the ages of three and twelve the mortality rate is very low. At the age of fifteen a lion is already old and very few of them reach the age of twenty. They are devoured by the hyenas – the very animals against which they have fought during their lifetime – giving the scavengers the last word.

Facing page (above) : These lion cubs, wonderfully well concealed among the bushes, already have the keen, steady gaze of the adult predator. *(Below)* two lionesses rest in the grass of the savannah, ever alert to the faintest movement which may spell food or danger.

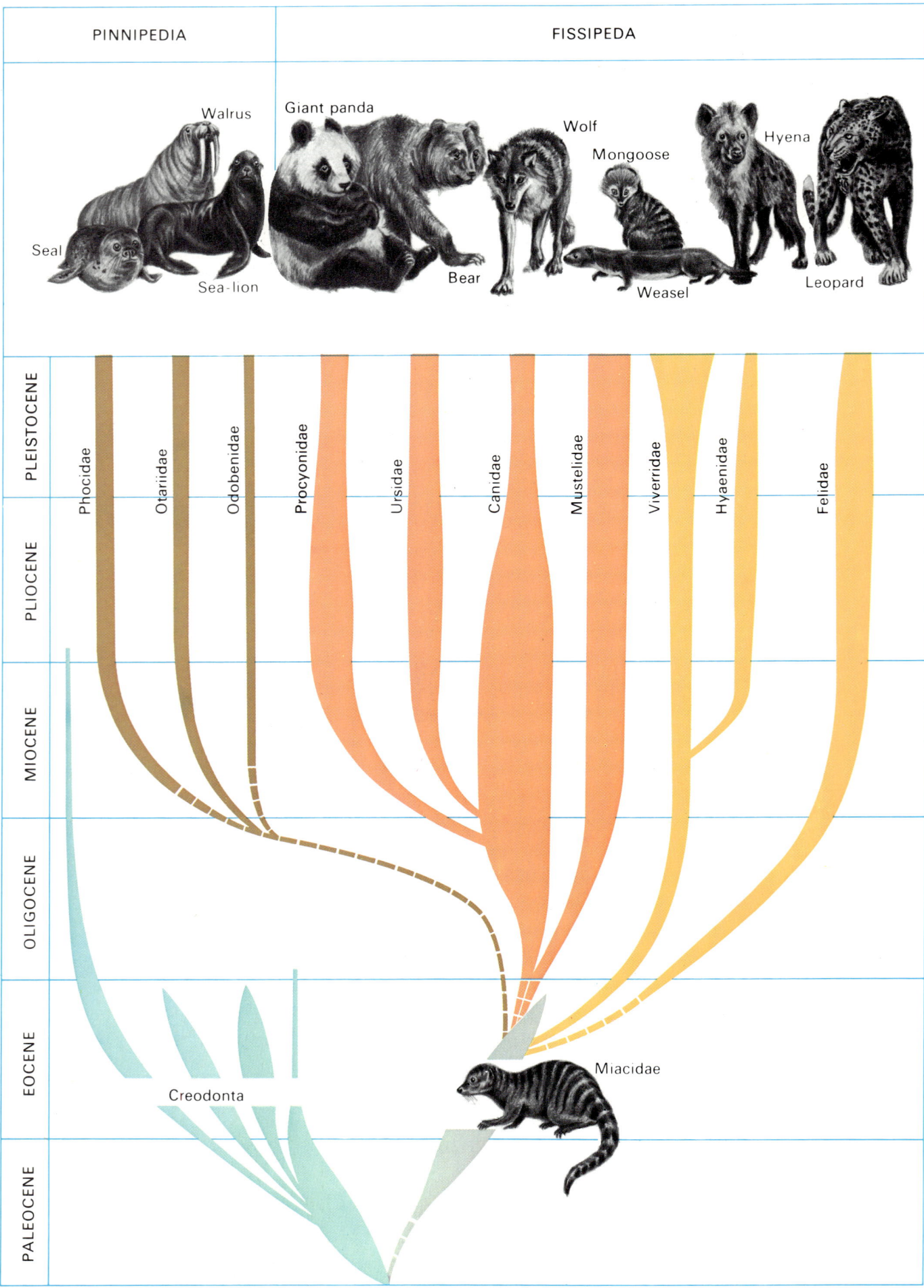

PINNIPEDIA

FISSIPEDA

Walrus

Giant panda

Wolf

Mongoose

Hyena

Seal

Bear

Weasel

Leopard

Sea-lion

PLEISTOCENE

PLIOCENE

MIOCENE

OLIGOCENE

EOCENE

PALEOCENE

Phocidae

Otariidae

Odobenidae

Procyonidae

Ursidae

Canidae

Mustelidae

Viverridae

Hyaenidae

Felidae

Creodonta

Miacidae

ORDER: Carnivora

The carnivores, literally 'flesh-eaters', are mammals whose diet consists primarily of meat. These mammals constitute an extremely important and highly diversified order of the animal kingdom, including creatures as contrasted in size as the weasel and the lion. Some of the carnivores are diurnal and others nocturnal in habit and they are found all over the world, both on land and in the sea. Outstanding among marine carnivores are the seals, numerous throughout the world, especially the polar regions.

Not all the carnivores have the identical number or distribution of teeth but their dental formula is based on that of their primitive ancestors, as revealed through fossil remains. The prehistoric carnivores had 44 teeth, arranged as follows: six incisors, two canines, eight premolars and six molars (in both the upper and lower jaw). The full complement of 44 teeth is rarely found in modern carnivores, though the Canidae (dogs) have lost only two. The Felidae (cats), on the other hand, have, during the course of evolution, been reduced to only 30 teeth.

The incisors are small and sharp, the canines strong and cone-shaped, developed into pointed fangs. The premolars and molars are the cheek teeth. The fourth premolar of the upper jaw and the first molar of the lower jaw are known respectively as the upper and lower carnassial teeth, a speciality of the carnivore order. These are larger than the other teeth and have sharp cutting edges—particularly effective for slicing flesh. In some species, such as the predatory cats, hyenas and weasels, the carnassial teeth are highly developed.

The ancestors of the modern carnivores lived at the beginning of the Eocene period over 60 million years ago and belonged to the suborder Creodonta. They were varied in size but most of them were heavy, low-slung animals, semi-plantigrade, with true claws. Their head was large in relation to the rest of the body but the cranial cavity and hence the brain were exceptionally small. They possessed the full number of 44 teeth, with well-developed canines. All the evidence shows that the creodonts were able to cope with many diets in different habitats. Some were omnivores, others frugivores (fruit-eaters). Most of them, notably the Hyaenodontidae and Arctocyonidae, disappeared at the start or during the course of the Oligocene period. But the family Miacidae, whose representatives had carnassials similar to those of modern species and markedly larger brains than other primitive carnivores, were the true ancestors of the majority of today's flesh-eaters. They too became extinct during the Oligocene, about 40 million years ago.

In the course of evolution the order Carnivora became divided into two branches or suborders—the Fissipeda, terrestrial hunters with paw-like feet, and the Pinnipedia, aquatic mammals with paddle-like feet. The former are further divided by certain authorities into the two super-families of Canoidea and Feloidea, on the basis of one distinctive feature of their internal anatomy—the Canoidea possessing an undivided tympanic (ear-drum) cavity and the Feloidea a partitioned cavity. It is a scientific rather than a visibly useful distinction but one that conveniently divides the cat tribe from the dog tribe.

The Canoidea include the families Canidae (dogs), Ursidae (bears), Procyonidae (raccoons) and Mustelidae (weasels); the Feloidea the families Felidae (cats), Viverridae (civets) and Hyaenidae (hyenas).

The Pinnipedia are aquatic, mostly marine, mammals whose limbs have been transformed into swimming paddles or flippers. Their teeth, lacking carnassials, are especially adapted for catching fish. The Pinnipedia are made up of three families, Phocidae (true seals), Otariidae (eared seals) and Odobenidae (walruses).

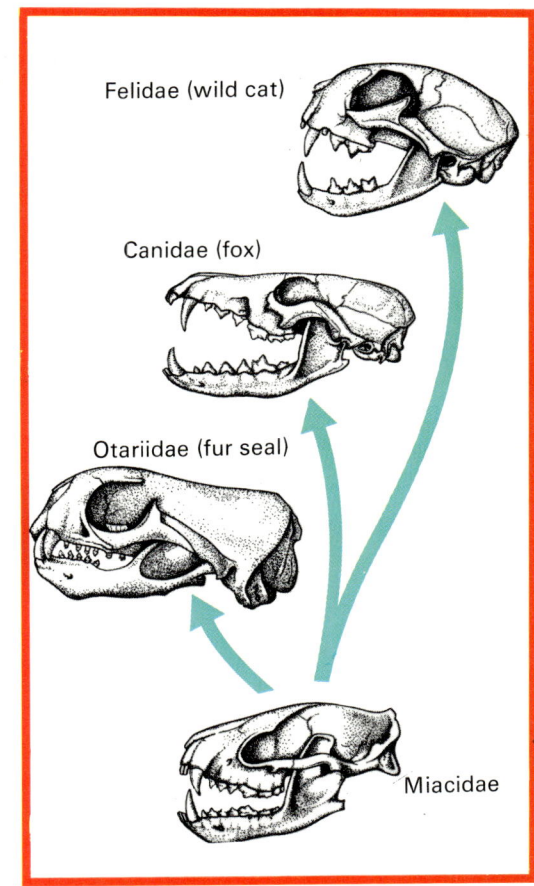

The teeth structure of the carnivores has not evolved as dramatically as that of the herbivores. The Miacidae, prehistoric carnivores living 60 million years ago, had the full complement of 44 teeth—including incisors, canines, premolars and molars. The Canidae have lost only two upper molars whereas the Felidae have lost six cheek teeth from the upper jaw and eight from the lower. The Otariidae, one of the families belonging to the Pinnipedia, have more or less undifferentiated teeth—apart from canines—suited to an aquatic life. The conical crowns help them to keep hold of slippery fishes.

Facing page : Evolutionary family tree of the order Carnivora.

FAMILY: Felidae

The Felidae have become distributed all round the world, with the exception of Madagascar, Australasia and Oceania. There are representatives of the family in steppes, grassy savannahs, forests, bush country, rocky hills and mountains. Apart from this capacity for adaptation to widely varying surroundings the Felidae show much diversity in size—from the African wild cat, which is smaller than·a domestic cat, to the huge tiger of the Amur river in eastern Asia. Their prey, in consequence, ranges from the diminutive mouse to the Cape buffalo.

The Felidae are the most voracious of flesh-eaters and the most adept hunters. With their powerful muscular system they are capable of exerting tremendous strength but because of the amount of energy thus expended they cannot maintain it for long. Quite simply, they lack endurance. Since their hunting technique does not involve pursuing and exhausting a victim they are experts in the art of ambush and the stealthy, concealed approach. Hiding behind any available natural obstacle they stalk their prey silently in order to surprise it and kill it as quickly as possible. Each stage of this hunting procedure demands the exercise of special skills and aptitudes. For the identification and localisation of prey the Felidae are endowed with remarkable eyesight, for both day and night use. The face is flat and the eyes are directed forward. In bright light the pupils are not round but vertical slits, and it is their far-ranging binocular vision which enables them to pick out contours easily and calculate distances with accuracy. The sense of hearing is also well developed—a valuable asset in detecting the presence of animals that cannot be seen—but their sense of smell does not appear to be particularly keen.

The Felidae must therefore remain concealed so that their potential victims do not spot them. Here too nature has provided them with the means of disguising themselves by blending with their immediate surroundings. Their coat may consist of a mixture of subtly coloured spots—as with the cheetahs, servals or leopards—or stripes—as in the case of tigers. Alternatively it may be of a uniform colour, though varied in tone, as with adult lions.

Another indispensable condition for this form of hunting is the absolutely soundless approach. For this purpose the Felidae possess soft pads at the bases of the toes (five on the front paws, four on the back paws), which deaden the impact of foot against ground when the animal leaps to the attack and enable it to retract its claws when not in use. It is this last specialised mechanism which prevents the felines maintaining a high speed over a long distance, since to achieve this an animal needs to be able to grip the ground with non-retractile claws, which serve the same purpose as the cleats or studs of an athlete's shoe. But since the Felidae are not long-distance runners this is no particular handicap. The only representative of the family which is renowned for its speed and thus prefers to catch its prey by chasing it is the cheetah, which has rounded, semi-retractile claws.

Apart from this last exceptional case the claws of the Big Cats do not get worn down nor do they break since they are only brought into action at the last moment; so they are especially effective as sharp weapons in any kind of fighting. Furthermore the great toe, which is relatively independent of the others, is very useful in capturing small animals. This is how the lynx manages to catch a bird on the wing and how a cat retains its hold on a mouse.

Because the facial region is comparatively compressed, the muscles of the rounded feline head ensure an efficient distribution of forces and convey unusual strength to the jaws. The joint of the mandible is so structured that only vertical movements of the jaw are possible, making the entire mechanism extremely firm and strong. This efficient apparatus is completed by differentiated teeth consisting of six incisors and two strong, pointed canines in each jaw, and fourteen cheek teeth (the premolars and molars), six to the lower jaw and eight to the upper jaw. The teeth have a particularly important function during the final stages of the hunt. After the feet and claws have achieved their purpose of knocking the animal to the ground and immobilising it, the fangs are sunk into the victim's throat and remain embedded there until death from strangulation results.

The principal genera of the family Felidae are *Felis*, *Lynx*, *Acinonyx* and *Panthera*.

In the genus *Felis* are the wild cats of Europe, Africa and Asia–including the European wild cat, the African wild cat and the jungle cat.

There are five species of *Lynx*, including the common lynx and the caracal of Africa and Asia.

The genus *Acinonyx* is represented by only one species–the cheetah (*Acinonyx jubatus*)–distinguished from the other felines by its aptitude for running and its semi-retractile claws.

The so-called Big Cats are grouped in the genus *Panthera*. One of their anatomical features is the voice-producing ligament of the larynx, which is extensible and magnifies a growl into a roar. Other Felidae are unable to do this because the ligament is part of the bone and thus restricts the movements of the larynx. Among the most important species are the lion, the tiger, the leopard, the ounce and the jaguar.

A caracal or desert lynx guards a freshly-killed hare. Like other Felidae the caracal has sharp retractile claws which enable it to seize and keep hold of its prey so that the fangs can add the final touch.

CHAPTER 8

The cheetah, world speed champion

In the middle of the parched savannah a lonely acacia is now the focus of our attention and interest. It is here that a female cheetah and her cubs have set up home—a sensible choice in view of the fact that it is about the only place in sight that offers any shade. Our aim is to keep this mother and her two three-month-old cubs under close observation for three days and nights so that we can learn something of their behaviour. There are plenty of gazelles—their favourite prey—in the vicinity and as long as they do not move away the chances are that the cheetahs will be happy to stay here.

It is about 7 o'clock in the morning when we decide to move in on the little family. The cubs are frolicking around their mother like graceful kittens, full of energy and the enjoyment of life. The light colour of their coats is nearer to grey than to the adult hue of golden-brown and they have an incipient mane of somewhat longer hair in the neck region. The adult female, stretched out on her back, lets her lively offspring tease and nibble her, calling them to order now and then with a gentle cuff of the paw, then tenderly licking their faces and ears. It is a peaceful group which appears to be paying no attention whatever to the equally tranquil herds of Thomson's gazelles grazing all round them on the empty plain.

After about an hour and a half the cubs, worn out from a non-stop round of games, settle down beside their mother. In the meantime five 'Tommies' slowly drift off towards the river bank some 300 yards away. Duma (to give the cheetah her local Swahili name) has not so far moved but now she sits up, takes a long look at the antelopes and stretches herself langourously. It is evident that she is a beautiful creature, in some ways not unlike a greyhound—the same streamlined body, long and

The black lines running down on either side of the cheetah's face give it a rather sad expression. Although the animal has many dog-like features, its eyesight is as remarkable as those of other Big Cats.

narrow chest, slender limbs, curving neck, supple back and flanks, and easy, casual gait. But in other ways she is a typical cat, with her round, foreshortened head, golden and black-spotted coat and powerful, ringed tail.

The cheetah therefore has some claim to be regarded as the perfect go-between of the dog and cat families. It is significant that its style of hunting is completely different from that of other felines. Instead of stalking its prey so that it can get as close as possible to deliver a lightning attack with flailing claws–the methods employed so effectively by lion, leopard, lynx and jaguar–the cheetah goes in for a chase over the open plains, pursing its prey as a greyhound does a hare. It can justly claim to be the fastest land animal in the world (accelerating to up to 70 miles per hour in a few seconds) although it is a comparatively poor performer over a distance. In fact it can only keep up its high speed for some 500 yards because of the high expenditure of energy involved. It then needs a long rest to gather its strength again.

If, like the other felines, the cheetah possessed curving, sharp, retractile claws it would not be able to make use of them–except at risk of damaging them–in chasing its prey. It is because these claws are rounded and only partially retractile, as in the dog, that they are suitable for running over hard surfaces.

While the cubs are yawning and their mother still lazily stretched out under the tree (though keeping a watchful gaze on the gazelles feeding on the grass near the river) we can get a good view through the binoculars of her large orange-yellow eyes and the two black lines that run like tear-streaks down the nose to the point where the lips meet. These marks give her a strangely sad, melancholy expression. But the eyes, so fixedly and unwaveringly intent, flash a warning of hidden reserves of pent-up power. They are perfect instruments of vision, but their sharpness seems to extend beyond this so as to endow the cheetah with a kind of 'second sight', enabling it to detect unerringly the slightest signs of weakness in a potential victim. How else can one explain the way in which a solitary cheetah always seems to pick on an undersized or otherwise handicapped animal for its prey?

Our cheetah is now on her feet and moving slowly forward, head lowered and limbs lightly flexed. She pays no attention to the direction of the wind, which, as it happens, is not in her favour. Step by step, with muscles rippling and body tensed, she draws closer to the grazing antelopes. Suddenly one of the gazelles lifts its head suspiciously. The cheetah immediately 'freezes' on the spot, muzzle thrust forward and one paw bent like that of a pointing dog, until the gazelle resumes its quiet cropping. Then she inches forward once more, taking cover behind the rare tufts of grass spared by the savannah fires, gradually getting nearer and nearer to the little group of gazelles. This procedure is repeated every time one of the herbivores raises its head. At this rate it takes her a full quarter of an hour to travel 25 yards.

Eventually the wind gives her away and brings all her plans to an abrupt conclusion. The gazelles scent her and after a few moments of hesitation–although apparently not greatly per-

Facing page : Of all the animals of the African savannah, the cheetah is without doubt the speed champion, capable of attaining 70 miles per hour over short distances. Less speedy carnivores, such as the hunting dog, possess greater endurance–essential for their style of hunting. But carnivores and herbivores alike rely on speed whether pursuing or being pursued. The figures in the chart indicate maximum, not average, speeds.

Elephant
25 mph

Giraffe
30 mph

White rhinoceros
25 mph

Hunting dog
30 mph

25 mph

Zebra
40 mph

Warthog
30 mph

Buffalo
35 mph

Spotted hyena
40 mph

40 mph

Gnu
50 mph

Ostrich
50 mph

Gazelle
50 mph

Lion
50 mph

50 mph

Springbok
60 mph

Cheetah
70 mph

65 mph

turbed by her presence only 200 yards away—they quietly move farther away across the blackened plain to a spot where they can resume their interrupted grazing. The cheetah does not make the slightest attempt to follow them but stretches, rolls over on her back, gets up and rejoins her cubs with complete composure and lack of concern.

Half an hour later she is once more on the look-out as a new group of Thomson's gazelles wanders down to the river, this time in the face of the wind. One of them appears to be suffering from an injured front hoof. In due course the cheetah abandons her nonchalant attitude, rises to her feet, fastens her gaze on the limping antelope and saunters about 25 yards in the direction of the river, without taking any of her previous precautions. She crouches down in the short, dry grass and lets the gazelles come to her, for on this occasion they are unable to scent her. When they are about 250 yards from her she breaks into a trot, gradually increasing her pace, though not yet using her maximum amount of energy. When the gazelles eventually notice her pounding towards them they do not stand on ceremony but turn tail and flee. The female which had appeared to be maimed now shows no sign of disability and makes off with the others, in front of a male. But the cheetah, now in full cry, has selected her as a victim and rounds on her, putting on an amazing burst of speed. The terrified gazelle swerves first to the left, then to the right, but these evasive tactics are of no avail. One blow from the cheetah's paw bowls her over. She turns a somersault, briefly recovers her feet but is once more hurled to the ground and gripped by the implacable predator.

We move cautiously towards the scene of the killing to find the cheetah lying across the body of her victim—still moving feebly—fangs sunk into its throat. For some seven or eight minutes she stays there, unmoving, breathing noisily through nostrils and mouth. The cubs, which were following their mother at a discreet distance, are scuttling around her, sniffing at the now-dead gazelle, but showing signs of playfulness rather than greed. One of them nibbles at the gazelle's neck, at the exact spot where the mother has fastened her teeth, shakes it and then lets go. The female cheetah, still breathing fast, her chest heaving, finally gets to her feet. She is visibly exhausted but after a few minutes picks up her prey by the throat and starts to drag it wearily off towards a patch of yellowing grass near the acacia. While she is pulling the dead animal along the cubs tumble joyfully at her heels, one of them grabbing hold of the trailing carcase and getting a free ride. The mother is forced to drop her load three times to recover her breath. When she eventually reaches her destination she hands over the prey to the cubs, which start nibbling away at the belly, and goes off to have a well-earned rest.

In a quarter of an hour she is back again and taking part in the feast. While she tucks in she keeps a careful watch along the river bank in case any lions are lurking. Stretched out beside one another the little family rip up the dead animal's belly muscles and pick out the particularly tender pieces. They do not touch the stomach or the intestines, but eagerly devour the liver and

CHEETAH
(*Acinonyx jubatus*)

Class: Mammalia
Order: Carnivora
Family: Felidae

Length of head and body: 84 inches (210 cm)
Length of tail: 30 inches (75 cm)
Height to shoulder: 28-32 inches (70-80 cm)
Weight: 90-145 lb (42-65 kg)
Diet: Flesh—favourite prey gazelle
Gestation: 90-95 days
Number of young: 2-4
Longevity: 16 years

Adults
Small rounded head, short ears with black marks on back, large expressive eyes. A black line runs down from either side of the muzzle, past the eyes to the point where the lips meet. The yellow-ochre colour of the coat is broken up by circular black spots distributed over the body, limbs, tail (partially ringed by these marks) and a part of the head. The hair is short, but a small mane, more noticeable in the male, covers the neck where the dorsal region begins. The limbs are long and slender, like those of other Felidae.

Young
For the first two months the belly and lower part of the flanks are black, and the back and top of the neck covered by a mane which gives the body a greyish rather than golden hue. At two months the dark spots become visible and the cubs begin to take on the distinctive colour and appearance of the adults.

other internal organs. They lap up the blood collected in the hollows of the abdomen and from time to time change places or take a brief rest. The adult cheetah, face bloodstained, still breathes nosily.

The cubs soon lose interest in the gazelle. In fact they do not really appear to have been very hungry. A little later their mother rejoins them after checking the grass and the ground for any bits and pieces she might have overlooked. It seems to be more of a stereotyped, reflex action than a procedure of any practical value.

When the family has gone back into the shade of the acacia two jackals and a dozen vultures appear from nowhere to enjoy the remains of the meal. In fact there is plenty left for them, the cheetahs having eaten only about one-third of the animal, not touching the head, neck, skin, intestines and much of the carcase itself. But the scavengers make short work of it and in a few minutes everything has gone.

The sun is still high in the sky and its powerful rays are scorching the surface of the plain. The outlines of the antelopes are hazy in the rippling heat currents that rise from the hard soil. The only shade is provided by the acacias that line the banks of the Seronera – scattered oases frequented by lions – and it is in this direction that we now head to escape from the heat.

Chases between cheetahs and gazelles are common in the wild, and in the Middle Ages similar spectacles were staged as entertainments. Persian noblemen, for example, would spend large sums of money to capture and tame cheetahs, making them show off their racing and hunting prowess against the only animals that could begin to match them for speed – the gazelles. The issue must have been beyond doubt.

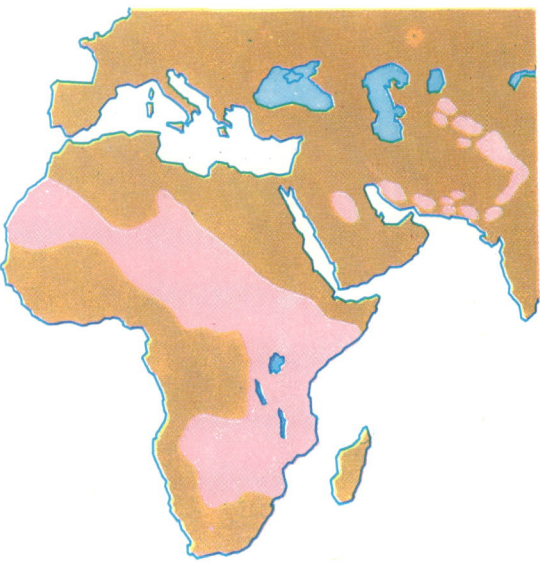

Geographical distribution of the cheetah.

Away from it all

Since cheetahs are especially partial to gazelles it is usual to come across them in many of the regions habitually frequented by these little antelopes. So we find the solitary hunters prowling all over the plains of Africa, whether savannah country such as exists in the Serengeti, open steppe of the type found in Amboseli, or even certain desert regions on the fringes of the Sahara. Moreover cheetahs are quite widely distributed through Asia. In spite of this extensive range, however, cheetahs are not very numerous and, as far as Asia is concerned, rank among those animals threatened with extinction. The fact that the species is comparatively rare is once again due to the machinations of man, who has pursued these splendid creatures, either with a view to taking them alive and training them to hunt or killing them for their pelts. In one region of Asia the last surviving pair of cheetahs were shot in order to prove that the species still existed!

In East Africa cheetahs live principally in the national parks and reserves. According to recent counts they seem to be most numerous in Nairobi National Park and in Amboseli. It was in the Nairobi park that the American biologist from Washington University, Randall Eaton, made a detailed survey of their

A young cheetah drinking from a pool. The hair of the beautifully spotted coat is fairly short but a modest mane of rather longer hair adorns the neck.

behaviour. It is his findings, coupled with those of George Schaller in the Serengeti and confirmed by the work of Dr Felix Rodriguez de la Fuente, that have provided the fullest information on the life and habits of cheetahs in the wild.

In the Nairobi National Park the cheetah can find one type of habitat which suits it admirably—savannah terrain where the grass is fairly short and where there are plenty of small trees of the *Acacia drepanalobium* species. So it is fairly easy to predict the limited regions where it will normally be found, both by day and night.

This preference, observed by Randall Eaton, is echoed in the other parts of Africa inhabited by the cheetah. Since the animal cannot maintain its high speed for much more than a quarter of a mile it still needs suitable vegetation cover in order to lie in wait for its prey and get near to it without being seen. The grass, however, must not be too high nor the bushes too numerous to interfere with its sudden bursts of sprinting; thus the savannah with short grass supplies almost ideal conditions—sufficient shade, a good selection of hiding places and extensive open stretches for the chase. Furthermore, wide plains interspersed with trees are the favourite haunts of the impala—another of the cheetah's coveted victims.

When cheetahs elect to hunt in groups rather than on their own the question of vegetation cover is less important. The fact that they may easily be detected by herbivores is of little concern since in such situations sheer weight of numbers is the decisive factor.

There are other reasons determining the choice of habitat. In the first place, climatic variations which affect vegetation growth also have an important influence on the movements of the herds. Secondly, there are enemies to be considered. During the period when the females are rearing their cubs there is the customary threat posed by hyenas and hunting dogs; but at all times there is one animal in particular which it prefers to avoid. For some reason the cheetah stands in dread of the lion, despite the fact that no special threat or danger is involved. This has been confirmed by surveys made in the Nairobi National Park where lions are plentiful. Both species appear to give each other a wide berth. It is true that lions have been seen devouring a cheetah but nobody has actually watched a lion killing one, so this may be a further example of lions indulging their habit of consuming carrion.

The cheetah, active by day and sleeping at night, has a special fear of nocturnal predators. Hyenas have sometimes been seen attacking cheetahs in the Serengeti while in Amboseli a leopard was observed to capture and kill a young cheetah.

In regions where there are large numbers of lions, hunting dogs, hyenas and leopards, it is therefore unusual for cheetahs to abound. They will keep well away from these centres of population and from the hunting precincts of rival predators. From their observation posts on the open plain they can keep a daytime watch both on the activities of likely prey and the approach of enemies. Whatever the circumstances they can use their remarkable speed either for pursuit or escape.

Naturalists in the Serengeti once spent several weeks watching the movements of two female cheetahs and their cubs, which had set up home on open ground about 500 yards away from the river bank and the shady spots used as a resting place by a troop of lions. Their only shelter was provided by a solitary acacia and from time to time they would turn their eyes towards the river where during the daytime the lions could be seen stretched out in indolent comfort.

One morning the observers' attention was drawn towards a cloud of dust which turned out to have been caused by one of the cheetahs in the process of killing a Thomson's gazelle. The panting huntress was straddling her victim while the cubs were gathered round, nibbling at the skin in a desultory fashion. Suddenly the cheetah straightened up, obviously uneasy, head turned in the direction of the river. She let out a single dry snarl and abruptly bounded off across the plain, followed by the cubs. The naturalists, positioned half-way between the river and the feeding cheetahs, could see nothing that might have precipitated the sudden flight, nor had they betrayed their own presence. But as they glanced towards the Seronera the answer was provided. A male lion, resplendent with flowing mane, was trotting resolutely in the direction of the dead gazelle. While the cheetah and her cubs found refuge near a clump of acacias the huge beast settled down to enjoy the meat which the cheetahs had barely had time to touch. When he had satisfied his hunger he sauntered back to the river bank, leaving the remains of the feast to the vultures that had been hovering

After killing her prey this female cheetah pauses to recover her breath before starting to feed. The cubs will also get their fair share but a watch must be kept for other predators, especially lions.

nearby from the moment the gazelle had been killed. Quite clearly it was the presence of the scavengers that had alerted the lion to the possibility of a free meal. The naturalists were unable to return on the following day so could not determine whether the cheetah had gone back to its lonely vantage point or whether the family had left the locality for good.

Why be exclusive?

Most of the carnivores are territorial hunters. Some, like the leopard, set up a private estate; others, such as the lion, establish a communal domain. All of them laboriously mark out the boundaries of their territory by means of scent posts (excreta) and defend it fiercely against outsiders.

There is enough general evidence to suggest that the cheetah is a territorial hunter like the other Big Cats. But judging by the studies carried out by Randall Eaton and other naturalists in the Nairobi National Park and other parts of Africa, there is some doubt as to whether the cheetah is really an exclusive owner of territory.

This feline is semi-nomadic in habit and although having lived for some time in one area will not bother to defend it against others of its kind, even if they are strangers. Thirteen cheetahs belonging to three different groups were seen sharing the same ground in the Nairobi National Park, indicating that their territories overlapped or coincided.

Female cheetahs with young seem to be more attached to a fixed area, extending perhaps five square miles, moving about very little when the cubs are small. Often the entire family group may spend two or three days in one place and then, following a kind of rotation system, may roam all over the region within the next week, always coming back to the same spot at night. The little groups return to these meeting points again and again, even if they are far from the hunting areas.

When adult cheetahs from different troops meet one another they do not show any signs of hostility but are content simply to exchange curious glances. Yet females accompanied by cubs are often uneasy in such situations. One evening some observers watched two males and one female settling down a few hundred yards from a mother and her cubs installed in their customary nightly resting place. The mother was clearly terrified by the intruders and would not take her eyes off them. In the end she gathered her cubs and ushered them off to another refuge, looking warily behind her as she went.

Cheetahs, like most other carnivores, signal their presence by urinating, though in their case it does not appear to be associated with territorial claims but connected with reproduction. This is the way the male goes about attracting the females or marking the route he has taken in order to reassemble his scattered companions. When another cheetah comes across one of these signals it spends some time sniffing the spot, then deposits its own message. The gesture is particularly significant in the case of the females, who never behave in this manner except when they are on heat.

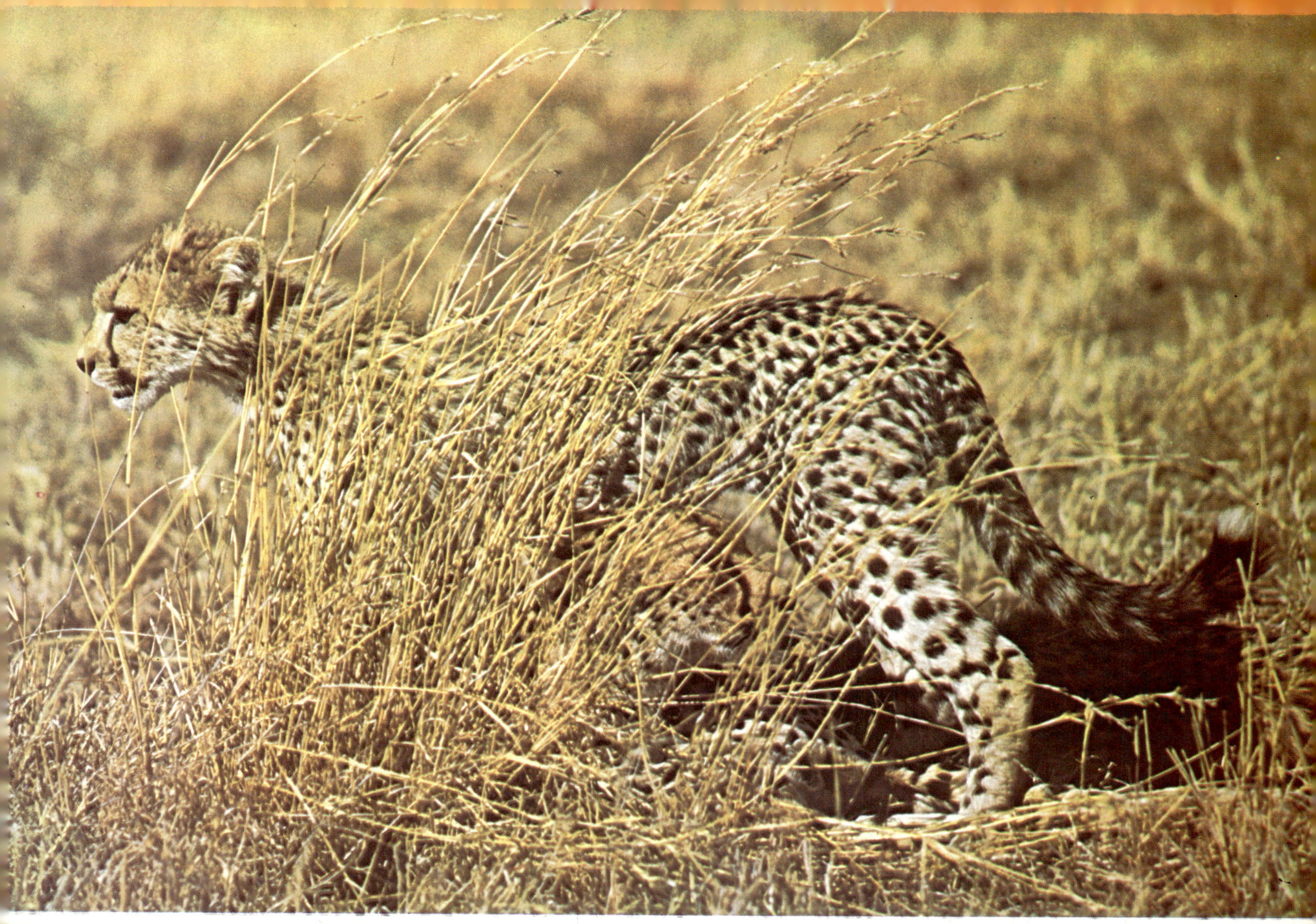
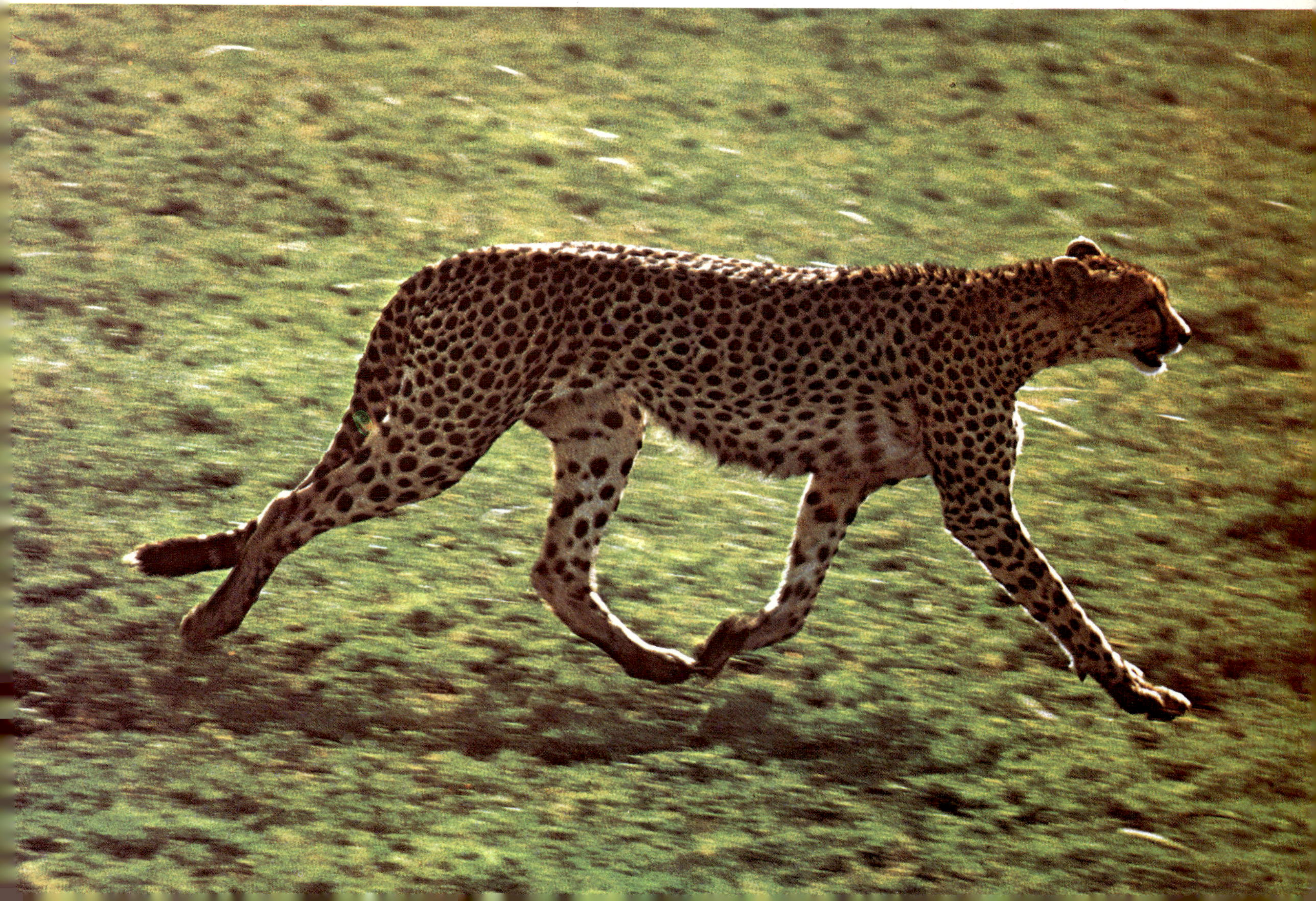

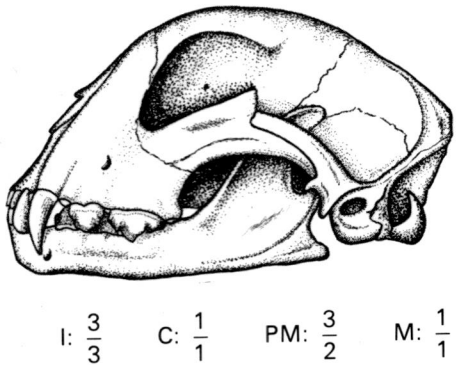

$$\text{I:}\ \frac{3}{3} \qquad \text{C:}\ \frac{1}{1} \qquad \text{PM:}\ \frac{3}{2} \qquad \text{M:}\ \frac{1}{1}$$

Skull and dental formula of cheetah.

Preceding page (above) : The spotted coat of the cheetah helps to conceal its presence in the dry grass of the savannah when it uses available vegetation cover to stalk its prey. (*Below*) the cheetah has spotted its prey and is now committed to the attack, beginning at a trot and quickly accelerating to a speed of between 60 and 70 miles per hour.

The easy-going cheetah

Almost all the Felidae are specialised hunters, active at all times except during the mating season or, in some instances, whilst rearing their young. The lion and the leopard may be taken as extreme cases. The former is part of a group comprising as many as forty individuals, including males, females and cubs of over three months of age. The latter is a confirmed solitary hunter, apart from the period when the female is bearing and suckling her young, when the male hunts on her behalf. The social behaviour of the cheetah lies somewhere between these two.

Like the lion, cheetahs live in troops, although these are of more modest size and quite differently organised. There are in fact two types—one consisting of adult males and females, both sharing in the hunting, the other made up of a female with cubs of different ages.

Social animals maintain order within a group by means of diverse ritual activities. Apart from their vocal and olfactory signal systems, familiar procedures in most animal communities include cleaning and getting rid of parasites. Monkeys groom each other conscientiously, members of the dog family nibble one another and cheetahs lick each other's faces. These are not just matters of simple hygiene but are gestures intended to establish and maintain peaceful relations among the various members of the group.

Cheetahs do not as a rule engage in ritual combat and in this sense live together harmoniously. It is therefore hard to determine how they go about choosing a leader; for such leaders certainly exist and studies have shown that there is a definite social hierarchy. One individual gives the commands and the others respect his authority. We saw an example of this whilst watching two adult males and a female cheetah lying side by side in the shade of a bito tree to avoid the searing midday heat. One of the males suddenly spotted a warthog with two young, about 220 yards from the tree, got up and slowly sauntered in their direction. He had gone some 50 yards when he turned to look back at his companions, neither of whom had moved. The other male still made no effort to join him but the female, evidently watching events closely, began running towards him. Then she too stopped and glanced back at the male still stretched out under the tree. Abandoning the chase, the first male trotted back to join her and then both returned to their shady retreat.

A few moments later the cheetah which had thus far displayed no interest in these hunting manoeuvres—now clearly revealed as the leader of the group—got up and wandered a little way from the tree, followed by his two companions. All three now moved purposefully forward, hidden by tall grass, surprising the warthogs, flushing them from cover and giving chase. We had other chances of watching this trio in action and noted that it was always the same male that directed operations.

All the evidence indicates that it is the males who assume the role of leader and guide. One never sees groups composed solely of females. The leader of the troop appears to have a number of duties, the most important of which is the protection of his

companion, so much so that when they are resting or asleep he remains in a seated guard position, keeping watch on every side. He seems to be particularly attentive to the needs of his male companions, and neither he nor they pay much regard to the females with small cubs, who lead an isolated existence and appear to be wholly self-dependent.

A fallen tree trunk provides this cheetah with an excellent vantage point from which to survey the surroundings. From here it can make a leisurely examination of the grazing herds of antelope and pick out the weakest and most vulnerable animals.

Life without father

When it comes to sexual matters the cheetah is much less concerned with formal ceremony and ritual than the lion and not nearly as particular and exclusive as the leopard. In Nairobi National Park one female was seen to couple with three males in succession without provoking any rivalry. There is no such thing as a 'married couple' in a cheetah community and when the cubs are born, after a gestation period of 90–95 days, the mother takes them off to rear them. This behaviour is unusual among felines for even leopards hunt on behalf of their mates and defend them while they are with their cubs.

In the Serengeti a team of naturalists spent several weeks near a female cheetah which had just given birth, studying the behaviour of this fatherless family. With no male present, the female had to be constantly on the alert, never straying far from her lair except to hunt—and even this she tried to do with one eye on the spot where she had deposited the cubs. This self-imposed limitation of the hunting precinct obviously lessened her chances of finding food. When she was not hunting she

The female cheetah is abandoned by her mate as soon as she gives birth and is solely responsible for feeding, rearing and protecting her cubs in their first weeks.

licked, suckled and played with the cubs, to the accompaniment of prolonged, low, cat-like purring. Whenever the cubs, distinctive with their long grey ruffs, strayed from the lair she followed their every step, summoning those that were going too far with a short, harsh, admonitory growl—a strange, deep sound, and quite unmistakable—to which the youngsters replied with sharp, shrill cries. If this warning signal proved unavailing the mother would pursue the roaming cub, pick it up by the scruff of its neck, carry it back and toss it down in the high sheltering grass.

The female cheetah is thus severely handicapped by having to bear the sole responsibility of caring for and bringing up her cubs. Lionesses, with their communal 'nursery' arrangements, and female leopards, attentively protected by their mates, find life at such times far easier. This neglect of the females may be the result of the cheetahs' specialised and unprotracted hunting operations, but from the evolutionary standpoint it is a retrograde step compared with the social behaviour of the other carnivores. Whatever the reason, the female cheetah watched by the naturalists never wandered out of sight of her cubs, though admittedly she had the advantage of keeping to the open plain where predators were unlikely to approach undetected.

At the age of six weeks the cubs were big enough to leave their place of refuge and to accompany their mother on her travels across hunting terrain. They concealed themselves in patches of tall grass where, thanks to their greyish manes, they were extremely difficult to distinguish.

When they are two months old cheetah cubs lose some of their infant characteristics, notably the black hair that covers belly and flanks and part of the long mane down neck and back. They become increasingly active and playful, but should the female see or begin to stalk a prey that might be alerted by the cubs' antics, she will give a short, low cry and they will immediately come to heel, gathering around her, staying quite motionless and watching her attentively.

Even when the female moves off to pursue and capture her prey the cubs' eyes never leave her. Observers in the Serengeti confirm that even when a female cheetah has brought down a gazelle at a distance of 500 yards or more, the cubs, which have obviously been watching the entire action, are round her in a matter of seconds. They may well be assisted in finding the exact spot by the black mark behind her ears which stands out sharply from the rest of her body among the shifting contours and colours of the plain; but they can also clearly hear at that distance the characteristic short cry that she emits to announce the success of her hunting mission.

When the cubs are four months old and try to come near to suckle, the mother pushes them away. At this age their claws are still sharp and more retractile than those of the adult, enabling them to climb trees with ease. But regular runs over the hard surface of the plains and savannahs soon blunt them, while the muscles and tendons controlling retraction gradually lose their elasticity.

By six months the young cheetahs are two-thirds of the way to becoming adults. They spend much of the time chasing one

At about five weeks the baby cheetah is a strange-looking creature. The long greyish mane covering its head, neck, back and tail provides natural protection against predators in the long undergrowth.

another, some playing the part of the hunter, others of the prey. They roll on the ground, limbs entangled, nibbling each other's necks. Such simulations of hunting ritual never lapse into serious fighting–they are simply part of the cubs' apprenticeship. As a result of these juvenile games, movements and actions that will later be matters of life and death–the sudden feint, the crippling blow of the paw, the deadly thrust of the fangs– all gain in precision and efficiency. Sometimes the skills are practised on jackals prowling around prey killed by the adults.

The real lessons begin when they are seven months old. Now the cubs are taught to watch, to stalk, to knock down and to kill. The females may lead their youngsters towards groups of gazelle so that they can kill them unaided. In the course of these training expeditions the mother does not attempt to catch the prey but cuts off their escape route so that they cross her pupils' path. In Nairobi National Park observers watched a cheetah and four young stalking a group of warthogs which were quietly grubbing for worms about 50 yards away. The female cheetah advanced to within 20 yards before launching her attack, at which point the warthogs decided to take flight. Although she appeared to be directing her assault against the young warthogs, she ignored them completely and went after the adult. The young cheetahs promptly joined in chasing the baby hogs, which tried to escape by running in wide circles and squealing desperately. Meanwhile the female cheetah stationed herself between the adult warthog and her offspring so as to prevent them joining up, then altered her tactics to come to the aid of her own cubs in their pursuit of one of the baby hogs. Although she could easily have caught it she was evidently more concerned to supervise their hunting techniques and skills. As it happened, the lesson ended happily for the intended prey, for the young warthog proved too cunning for its relatively inexperienced pursuers and managed to reach its lair.

During this training period the female cheetah continues to operate on her own account when it comes to the serious matter

As it grows, and becomes capable of fending for itself, the young cheetah takes on the appearance of the adult, the coat colour getting lighter and the mane now being confined to the neck and shoulders.

of finding food for the family, for the young apprentices are not yet strong enough to fend for themselves and are still a hindrance rather than a help. Until they are ten months old they are unable either to bowl over an animal with one of their front paws or to strangle it decisively in the true adult manner. If they do have the good fortune to bring down a gazelle their teeth go for the back of the neck and other less vulnerable regions of the body. Age alone brings experience and it is only after much frustration and many failures that they learn to use their teeth and jaws to best advantage, so that a quick bite in the throat can bring about the paralysis and speedy death of their victim.

Cheetahs reach their full physical development and sexual maturity between the ages of thirteen and sixteen months. They still retain part of the mane on neck and shoulders, but it is now considerably reduced both in length and extent. Very soon the adult female will be ready to take another mate or mates and the youngsters must be prepared to leave her to join another troop or to take up a solitary life. Their choice will be largely determined by the comparative numbers of natural enemies in a given area. Where lions, leopards and hyenas abound, as in the Serengeti, cheetahs tend to stay on their own, but where the predators are few and far between, as in the Nairobi National Park, larger groups of cheetahs are frequently found. In the latter case a high survival rate among the cubs appears to ensure that such troops remain strong and secure.

The hunter with a difference

Of all mammals the Felidae are the supreme hunters. Nature has endowed them with powerful muscles, efficient weapons in the shape of tooth and claw, remarkable sensory perception and rapid reflexes, and amazingly co-ordinated movements. All this bodily and mental equipment enables them to live and hunt in widely contrasting environments. The domestic cat can take a mouse or a bird in the undergrowth, the lion can bring down a young buffalo on the savannah, the snow leopard chases wild goats over the mountain crags and the jaguar of the American tropics will scoop a cayman from the river bed.

Open plains with short grass affording little opportunity for an ambush are not ideal hunting grounds for the majority of Felidae, though highly suitable for the members of the dog family. But such terrain has become the principal habitat of the cheetah, whose hunting technique is especially adapted to such surroundings. Yet although the cheetah's physical attributes and capabilities might seem to be more attuned to a hunting method favoured by dogs, when it comes to the final assault and kill it behaves like all other cats, large and small. Despite the fact that it can outrun the rest of the feline tribe, it is still not the equal of the hunting dog, to give only one example, when endurance is involved. True, it can outstrip the latter in acceleration and in short sprints—a top speed of between 60–70 miles per hour has been recorded—but a distance of about 500 yards seems to be its limit. In the ordinary way a chase of

At dawn this female cheetah stations herself on top of an anthill, keeping watch on the movements of the herds of Thomson's gazelle, one of which is likely to furnish the next meal for her family. The cubs are not yet ready to join her in the hunt but will soon receive their apprenticeship.

300–400 yards will leave it exhausted and forced to take a breather before returning to the fray.

The cheetah, when hunting, stalks its prey in the 'masked approach' style–looking for, selecting and then getting near enough to its victim to launch an unexpected attack. Discovering the prey is the easiest part for it will have settled in a region where there is an abundance of antelopes such as gazelles or impalas. Dawn is the time usually chosen for going out in quest of food. On the open plains the cheetah will often select a vantage point in order to survey the surrounding terrain. It may be the top of an anthill or branches near the ground; the cheetah does not normally climb into the upper parts of a tree. This reconnaissance stage is likely to be long and painstaking. If the herds are too far away and unlikely to come closer to the cheetah's observation post it will then advance, without taking undue precautions, until it is some few hundred yards away. At that point it may take advantage of any available cover–a mound, a bush, a dead tree trunk, an anthill or a clump of tall grass–such natural obstacles perhaps spelling the difference between success and failure. Belly to the ground, the cheetah then moves slowly towards the grazing animals. Should one of them lift its head, the predator stops at once. It can afford to be patient. If it is spotted, after an approach which may have lasted over half an hour, it will start all over again without any hint of being discouraged. Its aim is always to reduce the distance between itself and its prey to a minimum so as to launch an attack without arousing any suspicion. The farther it can go undetected the better its chance of covering the intervening ground flat-out when the antelopes finally awaken to the danger and take to their heels.

The victim will be chosen in advance and the target will not change at any stage of the ensuing chase. It appears probable that the preliminary period of reconnaissance enables the cheetah to evaluate the physical condition of the animals under observation, for it invariably fixes upon the weakest and hence the most vulnerable individual. At the moment it pounces its speed may be equal to that of the fleeing herd but when the selected victim is overtaken by its companions or becomes separated from them, the hunter suddenly accelerates. With a single blow of the paw it catches its prey off-balance and tumbles it to the ground. In a flash it has sunk its fangs into the animal's throat, applying all the pressure necessary to bring about death by strangulation.

This hunting method, involving the careful preselection of an injured or sick animal or the leisurely pursuit of a herd to ascertain which is the feeblest and therefore quickest to tire, has predictable consequences. The weak, the young, the aged and the pregnant females can all be overtaken without difficulty and thus the predator is able to conserve its energy to best advantage.

The cheetah is also able, if need be, to use the technique of ambush like other felines. One female in the Serengeti was seen to station herself in a strategic position, camouflaged in the yellow grass not far from an acacia where her cubs were resting.

Facing page : In this sequence of pictures a female cheetah is seen (*left, top to bottom*) chasing her prey–a Thomson's gazelle. Once she has got as near as she dare to her prey without being detected she launches herself forward at full speed. The gazelle zig-zags in a vain effort to shake off its pursuer but over this short distance superior speed and strength tells. On the right the cheetah straddles her dead victim, joined now by her cubs. She starts to drag the carcase away, dropping it from time to time to recover her breath.

Throughout the day she maintained an uninterrupted vigil, never taking her eyes off the herds of gazelle wandering over the dry savannah and down to the shady banks of the Seronera. When a small band strayed near her she sprang to the attack. Either she was in prime physical condition or her selection of prey was unerring – perhaps both – for two out of three such attempts were completely successful.

In the Nairobi National Park the cheetahs use artificial prominences such as road embankments to await the passage of their antelope prey. From even these modest heights the wind is less likely to betray their presence.

In regions with more vegetation cover, such as parts of Amboseli, the cheetah may sometimes come across prey unexpectedly. In such situations the element of surprise is fortunate for the hunter and disastrous for the victim. When a herd notices a predator in good time it has the chance of making off in tight formation, but when it is taken by surprise it is more likely to stampede, with individuals running in different directions. In the Nairobi National Park a cheetah was seen to move away from its hideout and to saunter casually towards a thicket. Suddenly it spotted two young warthogs nearby and immediately changed course. The hogs caught sight of the fast-moving cheetah when it was only a few yards away and scampered off. Their mother, browsing close by, also turned tail, but in the opposite direction. When she realised that her offspring were not following her she swung round and rushed back in futile pursuit of the cheetah. She was too late, for one of the young warthogs had already been caught. The mother made ready to charge but dared not get too close. Finally she elected to make her escape with the surviving youngster. On this occasion surprise and luck had been on the side of the predator, causing the victims to panic, separate and be picked off without difficulty.

Although cheetahs prefer to go hunting at daybreak they are quite prepared to undertake it at any time of day and have even been known to kill their prey by moonlight. Given the choice, dawn provides them with more time to observe and select their prey at leisure whereas with the approach of evening they may be rushed into a premature and possibly ineffectual attack. But if cheetahs are genuinely hungry they will not be particular either as to the nature of the prey or the time.

In areas where herbivores are to be found in plenty the cheetah, with no problem of finding food, will pass hours on end stretched out in the shade of an acacia, in the grass or on top of an ant-hill. One female in the Serengeti never bothered to stir more than a mile from her retreat – except to hunt – during the year and a half that she was under observation. In the morning she could always be relied upon to be exactly where she had been the previous day.

The cheetah's hunting forays are not invariably crowned with success. In the Nairobi National Park, Randall Eaton saw one cheetah make no less than fifteen futile attacks in two days. One reason for failure must be that the animal takes no account of wind direction, another that in spite of its speed

Caracal

Cheetah

Fox

Mongoose

lack of stamina forces it to abandon the chase. In such unsuccessful pursuits the cheetah becomes acutely short of breath, its rate of respiration so fast that it has to stop to regather its forces for at least half an hour before trying again. Yet another reason for failure may be that it miscalculates its victim's potential speed so that a mistake is made in the approach stage rather than during the actual pursuit.

A group hunt is usually likely to be more promising than a solitary foray, not only because a group of cheetahs can help one another by cutting off their victim's escape route but also because larger and heavier animals can be tackled. In the Nairobi National Park the largest victim taken by one cheetah on its own was an impala weighing about 130 lb, but a group of four male cheetahs managed to dispose of a number of zebras and gnus, each weighing 500–600 lb. One favourite prey of cheetahs hunting together is the kongoni–which may tip the scales at about 330 lb. In considering these figures it should be remembered that a female cheetah weighs only 90 lb and a male little more than 110 lb–or perhaps, in exceptional circumstances, 130 lb.

A hunting cheetah may cover 200 yards at a rapid, half-bent trotting pace and then accelerate to cover the next 200 yards at top speed. Should the attack be successful it will drag its victim towards a spot where it can be hidden from the eyes of an enemy. It then eats greedily but with continual interruptions to glance in every direction. Vultures and roving jackals will cause it little worry or disturbance but the presence of lions or hyenas may cause it to abandon its feast.

Facts and figures

In the process of evolution the cheetah has attained a high degree of specialisation, but it possesses neither the strong weapons nor the sheer body weight of other felines. Its claws are semi-retractile, the head is small and the jaws are comparatively short and brittle. Nevertheless its method of despatching its prey by slow strangulation corresponds with that employed by its larger relatives. It simply selects victims better suited to its own needs and capabilities.

In normal circumstances the cheetah chooses relatively weak, lightweight animals for its prey–rarely over 130 lb. It all depends on the amount of meat that it will provide. A solitary cheetah has very modest food requirements so that a single Thomson's gazelle–usually a female or a young animal–may be quite sufficient. A male of this species is unnecessarily large, as is a Grant's gazelle, so these are rarely hunted. Although it takes care to drag away the remains of its prey, and a case has been reported of a female feeding on an animal she had killed the previous day, the cheetah normally only eats fresh meat. If it is unable to finish what it has begun, the vultures, hyenas, jackals and lions are the beneficiaries.

The surveys of George Schaller in the Serengeti and of Randall Eaton in Nairobi National Park have furnished interesting facts and figures about the cheetah's hunting and feeding

Facing page: Mother cheetah and cubs have finished their meal but for her there is no time to relax her guard, especially if there are lions in the neighbourhood.

habits in two contrasted environments – the former in open savannah, the latter in a region more densely covered with vegetation.

Of 136 herbivores killed by the cheetahs of the Serengeti during a one and a half year period, 90 per cent were Thomson's gazelles, the balance consisting of gnus, Grant's gazelles, Cape hares, reedbucks, topis, kongonis and dik-diks. More than half the 'Tommies' killed were fawns. The gnus, kongonis, Grant's gazelles and topis were almost all under a month old. Only two Grant's gazelles, a dik-dik and a reedbuck were adults. The largest animal captured was a young topi weighing about 200 lb. This count shows that when cheetahs hunt alone they pick out small or young animals and rarely adults. The choice is dictated more by their own food requirements than by the availability of game.

One female cheetah with her three four-month-old cubs killed 24 Thomson's gazelles and one hare during a four-week period – and all within an area of a couple of square miles. This was an average figure over the entire period. Sometimes she would fail to catch anything and was once unsuccessful three days running. On another day she killed two gazelles. Considering the weight of her victims she was managing to bring in about 20 lb of meat every day; but on two occasions prowling lions relieved her of her prey and another time it was filched by a hyena. In fact the cheetah was eating only about 9 lb of meat herself while her cubs seemed contented with little more than one lb each. Bearing in mind that they hardly touched the bones, the skin, the intestines, the neck and the head, roughly two-fifths of every animal she killed went to waste – or more precisely to scavengers.

In the Nairobi National Park, as expected, the statistics told a different story, with impalas taking the place of the Thomson's gazelles as favourite prey. But the only reason for this preference seemed to be that these antelopes were easier to catch than others. Naturalists following a cheetah's tracks for several consecutive days found that although it had come across at least seven different species of ruminants it had killed only impalas. Warthogs were sometimes attacked but adults never fell victim to the cheetahs.

The Nairobi survey demonstrated that the average daily food intake of each cheetah in the park was approximately $2\frac{1}{2}$ lb of fresh meat for roughly every 20 lb of the animal's own weight. Calculating from this figure, it was reckoned that all the cheetahs then living in the park required collectively about 215 lb each day for their sustenance and survival. But since only some three-fifths of what they killed could be said to be edible, the total weight of animals killed every day would have to be in the neighbourhood of 300 lb – roughly the equivalent of 110,000 lb of meat annually. Over a period of a year this weight of ungulates would be found compressed into a fairly small area of a few square miles. So it is quite obvious that the cheetahs in the park would be able to survive happily on a very small percentage of the species living in their vicinity, with no temptation to kill more than they actually needed.

Facing page (above) : A male Grant's gazelle lies dead at the feet of a cheetah. Because of their larger size gazelles of this species are not captured nearly so often as the 'tommies'. (*Below*) cheetahs devouring a zebra. Gnus and zebras are sometimes attacked by cheetahs but only by three or four hunting in company – usually males.

Guinea fowl

Cape or brown hare

Thomson's gazelle

Warthog

Grant's gazelle

Impala

Gnu

LIST OF PREY KILLED BY CHEETAHS IN EAST AFRICA

Thomson's gazelle (*Gazella thomsoni*)	38	Topi (*Damaliscus lunatus topi*)	3
Grant's gazelle (*Gazella granti*)	37	Kob (*Kobus kob*)	3
Impala (*Aepyceros melampus*)	27	Warthog (*Phacochoerus aethiopicus*)	3
Oribi (*Ourebia ourebi*)	7	Guinea fowl (*Numida mitrata*)	3
Cape hare (*Lepus capensis*)	7	Lesser kudu (*Tragelaphus imberbis*)	2
Brindled gnu (*Connochaetes taurinus*)	6	Harnessed antelope (*Tragelaphus scriptus*)	2
Kongoni (*Alcelaphus buselaphus cokei*)	5	Kori bustard (*Ardeotis kori*)	2
Gerenuk (*Litocranius walleri*)	5	Grévy's zebra (*Equus grevyi*)	1
Common zebra (*Equus quagga*)	4	Common duiker (*Sylvicapra grimmia*)	1
Steinbok (*Raphicerus campestris*)	4	Black-backed jackal (*Canis mesomelas*)	1
Ostrich (*Struthio camelus*)	4	Mole-rat (*Tachyoryctes* sp.)	1
Oryx (*Oryx beisa*)	3	Sable antelope (*Hippotragus niger*)	1
Dik-dik (*Rhynchotragus kirki*)	3	Total (25 species)	173

The coursing cheetah

History records that the first people to capture and train cheetahs for hunting were the Egyptians, as far back as 1500 B.C. Although we know nothing of the methods used it is likely that they were similar to those employed much later in the Middle Ages in many parts of Asia. Certainly they would have confined their captures to young animals already hunting in the wild. It is notoriously difficult to teach hunting techniques to wild animals removed too soon from their family, nor do such individuals become as adept as those that have learned it in their natural surroundings. Conversely, an animal taken into captivity too late in life, having known a measure of independence, will never be bent to human will.

When coursing with cheetahs was popular in Arabia, Persia and India five or six centuries ago the animals were in plentiful supply. They were captured either by means of strong cords stretched across their lairs or by being pursued on horseback. When, after a short chase, the cheetah was near the point of collapse, it was a simple matter to catch it in a net. Its captors knew that all the animal's reactions were triggered off by visual stimuli, so their first step was to cover its eyes with a leather hood. Temporarily deprived of its sight, the animal quickly became docile. The hood was kept on for some days, during which time the captive cheetah was fed with warm blood. Its trainers also taught it to come to feed of its own accord by tapping out a regular rhythm on the side of the wooden feeding bowl whenever they put it in front of the blind-folded animal. In the course of a few days the cheetah was able to recognise the sound of the tapping on the bowl and would be impatient for its meal even before the trainer had put in an appearance. This simple technique of conditioned reflexes—in which an auditory signal functions as a stimulus to the digestive processes—anticipated the experiments with dogs by Pavlov centuries later!

It was now time to remove the hood from the cheetah's eyes. This was done gradually, starting in a darkened room and not otherwise changing the feeding methods. The animal grew slowly accustomed to the sight and sounds of people, vehicles and domestic animals. Now that it knew its trainers, recognised their voices and responded to auditory signals and commands it was left to pursue its natural activity—hunting. After a number of training runs it was ready to provide entertainment. Eyes covered, it was led to a point about 100 yards from its chosen prey. Then, freed both of its hood and its leash, it bounded away in pursuit of its victim. When it had brought off the kill it was rewarded with another drink of warm blood.

The training of cheetahs for the hunt lasted in Persia and Arabia until the 19th century but because the animals would not breed in captivity the sportsmen had to look farther afield. They descended on Africa, threatening to reduce their numbers as drastically there as they had in Asia. Only rigidly enforced protective measures have managed to save the survivors, and it is essential that such efforts should not be slackened.

A young cheetah takes its exercise on the open plain while its mother stands guard.

Facing page (above): Seven species commonly attacked and killed by the cheetah. As the thickness of the arrows indicate, Thomson's gazelles are the favourite prey, followed by Grant's gazelles and impalas. (*Below*) this statistical report, prepared by the East African Wildlife Society as part of its 'Operation Cheetah Survey', confirms that Thomson's and Grant's gazelles and impalas supply well over half the food requirements of cheetahs in the wild.

CHAPTER 9

A silent killer, the African hunting dog

Once again we are driving across a dry, empty stretch of the Serengeti National Park—a grey wilderness of dust and sun-scorched earth on every side. For mile upon mile the scenery has been stiflingly monotonous and quite devoid of any features of interest. Now and then the patches of burned yellow grass disappear from view so that we get the impression of travelling across a desert, and occasionally the flat and unvarying surface of the plain is broken by shallow depressions, parched and cracked by the heat. At this time of year not the slightest drop of water gathers in these hollows. Not surprisingly, there is no sign of animal life anywhere.

Although we have taken the precaution of sealing up all the window, roof and door openings of the Landrover with strips of soft rubber, the fine dust of the savannah still manages to filter through, covering the windscreen and the camera lenses, irritating skin and nostrils. But the uncomfortable journey is almost at an end. In the distance we can dimly distinguish a black speck which grows larger and clearer as our vehicle jolts its way across the plain. This is our destination—the lair of a pack of wild dogs. What is special about it is that this is the only hideout recently discovered in the Serengeti belonging to that much-feared predator of the savannah—the African hunting dog. It has taken a team of naturalists, scouring the entire reserve mile by mile in Landrovers, two months to find this remote site.

The scientific name of this unusual animal is *Lycaon pictus*, a combination of Greek and Latin words literally signifying 'ornamental wolf'. It is a very apt description both as regards behaviour and appearance—although many observers and authors consider that it is more like a hyena than a wolf, even to the extent of classifying it, quite wrongly, in the hyena family. Admittedly, its massive skull, erect ears, powerful jaws and short, coarse hair do combine to give the animal a superficial resemblance to the hyena, but the latter creature is distinguished by its low hindquarters, thick neck and stocky legs. More detailed examination shows that it is not all that much like a wolf either, which is considerably heavier and also has shorter legs.

All the movements, characteristic actions and postures of the animal – including its peculiar way of whining – link it much more closely with certain breeds of domestic dog. Although it is known by a variety of common names, that of African or Cape hunting dog is by far the most accurate and least confusing for this fascinating hunter of the savannahs and open plains. Yet until quite recently hardly anything has been known of the animal's true nature and behaviour apart from the obvious and visible fact that it is one of the deadliest and most savage killers of the African wild.

Our purpose in travelling to this spot is to study the behaviour of a typical pack of hunting dogs in natural surroundings. The pack consists of eleven dogs and the first thing we are likely to notice is that no two of them are identical in coat colour or pattern. Some are dark brown, others almost yellow – and all of them are irregularly covered with white, brown, rust-coloured or black patches. Only their tails are alike, each ending in a pure white tuft.

The pack has apparently just welcomed the return of one of their number, which trots up to them with ears pricked and tail wagging. Their delight at the wanderer's arrival is plain from the little growls of satisfaction and the way they frisk about nibbling its mouth – the nearest dogs can get to an actual embrace. Watching them in this playful, relaxed mood it is easy to see that they are accomplished runners. Although the head, because of the highly developed jaws, may seem on the heavy side, the body is well proportioned, the limbs being long, sinewy and slender, ideally suited to rapid movement. The hunting dog can reach not much more than 30 miles per hour, but it does not need to rely on speed alone. Much more important is its stamina, enabling it to keep up a steady pace for a considerable distance in pursuit of much heavier and stronger animals.

The fact that no two hunting dogs are exactly alike in outward appearance is a great boon to all zoologists who study their behaviour, since there is no need to resort to special marking procedures or other ingenious methods of identification. The celebrated writer and naturalist Baron Hugo van Lawick (noted, together with his wife Jane Goodall, for his surveys of chimpanzees) has for some time been making a special study of the behaviour of hunting dogs in the Serengeti. We had the pleasure of visiting him in his camp on the shores of Lake Lagarja, where he explained the objectives of his survey, compared notes with us and took us to see some of the lairs used by these animals. Much of the following information is based on first-hand reports by the baron of his recent findings, supplementing and in many cases correcting information supplied by previous authors on the subject. Clearly the 'killer' has a softer side to its nature.

Facing page (above) : African hunting dogs are noted for the harmony of their family life. Here a female suckles her litter, which may comprise up to ten puppies. (*Below*) adult hunting dogs provide food for the entire pack by regurgitating pieces of meat that have been predigested at the site of the kill.

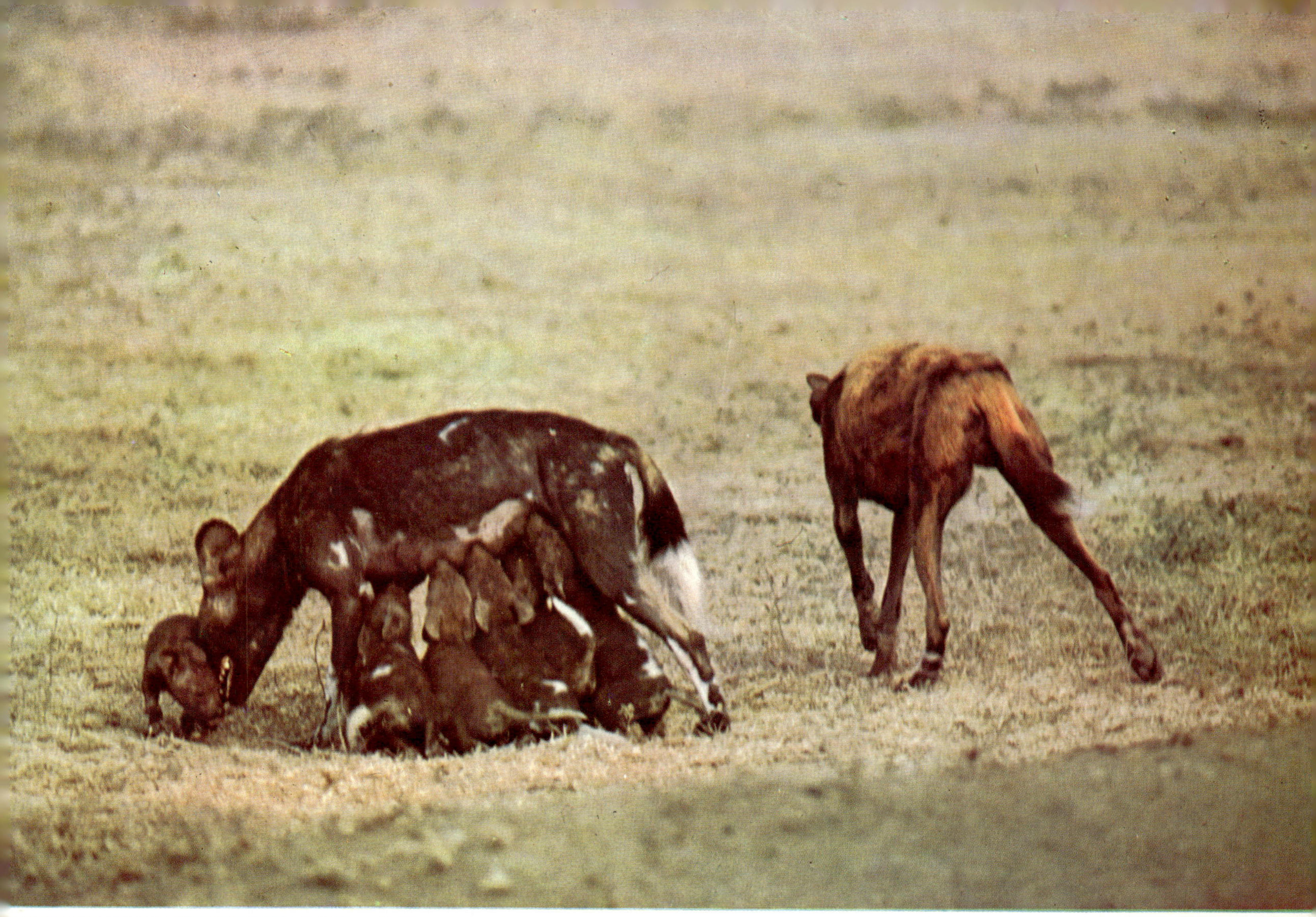
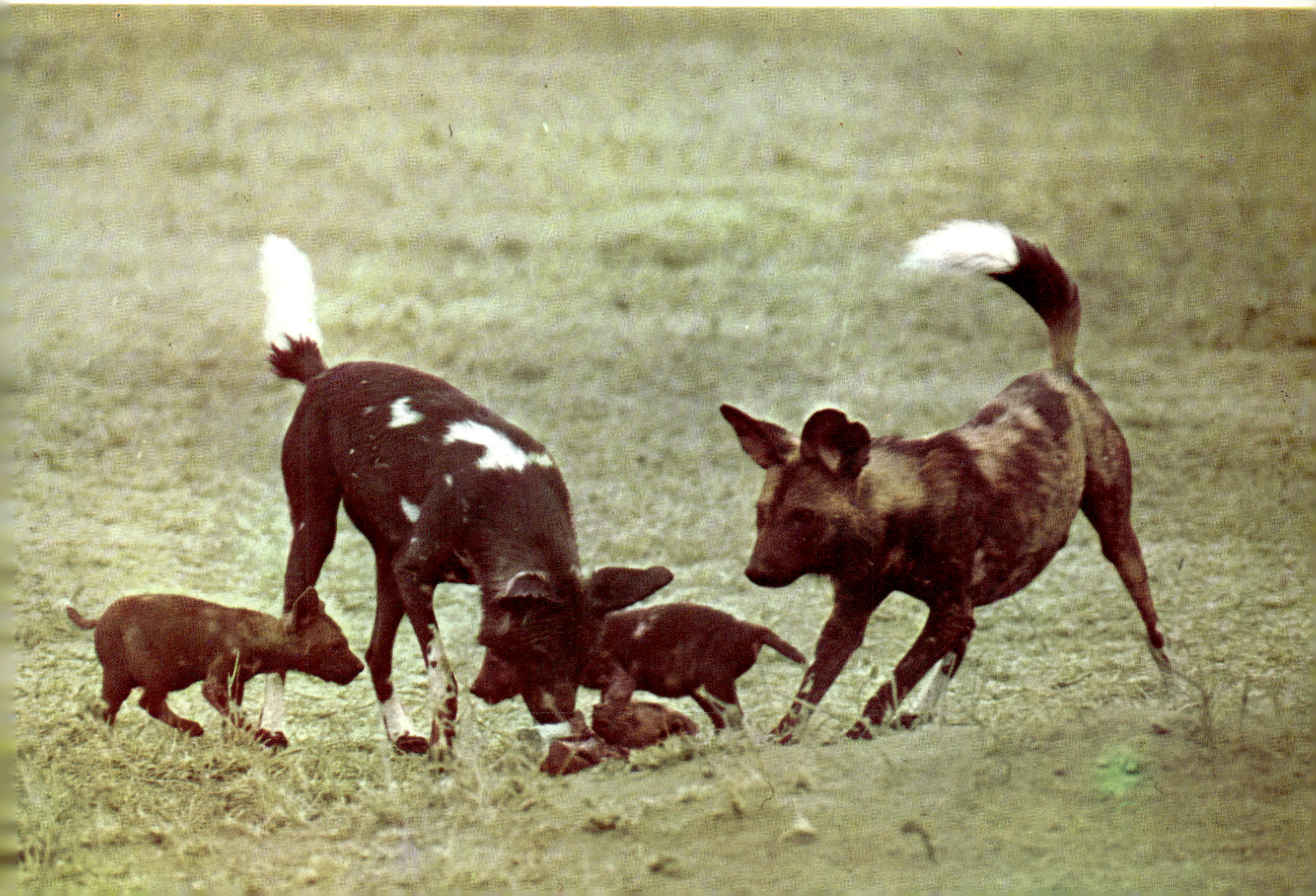

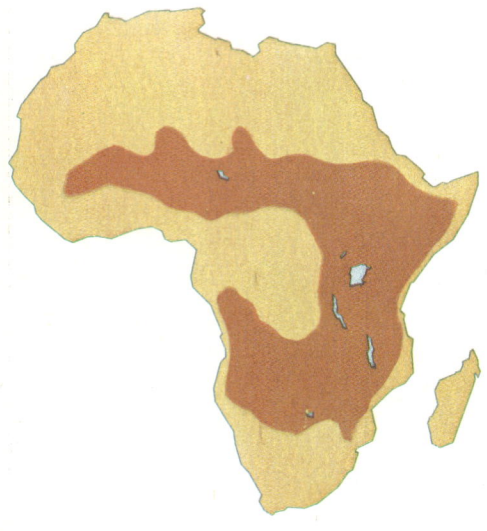

Geographical distribution of the African or Cape hunting dog.

AFRICAN (CAPE) HUNTING DOG
(*Lycaon pictus*)

Class: Mammalia
Order: Carnivora
Family: Canidae

Length of head and body: 40 inches (100 cm)
Length of tail: 14 inches (35 cm)
Height to shoulder: 24–30 inches (60–75 cm)
Weight: 60–80 lb (27–36 kg)
Diet: flesh
Gestation: 60–80 days
Number of young: 6–10

Adults
Large head with strong jaws, slightly pointed muzzle, large, upright and rounded ears. Muscular body, fairly long limbs. The coat, marked with brown, white, yellow and black blotches, varies considerably from one individual to another. All animals, however, have a black throat and muzzle, with a dark median line from ears to eye-level, and white-tipped tail.

Young
Squat, roundish bodies, pricked ears, uniformly dark brown coat colour.

Happy families

When the adult hunting dogs return from a foray to the lair they are normally greeted by a group of stubby, short-legged puppies scampering out to meet them. The adult dogs in their turn welcome their offspring with affectionate licks and gentle bites on the muzzle. It is a harmonious and appealing family scene, the puppies emitting sharp little cries and rolling about on their backs so that their parents can nuzzle them. But the sole aim of the adults' expedition was to find food and their pups are hungry. Now, in a spontaneous gesture characteristic of the species, they poke their tiny mouths between the jaws of their parents. One of the males responds to this recognised invitation by regurgitating several pieces of meat, and the puppies hurl themselves greedily at the food. A short distance away three other puppies adopt exactly the same procedure with a female. She too spits into the dust a soft pulpy mass of predigested meat and her youngsters make short work of it. When they have all eaten enough they complete their meal by turning to their mother, now stationed at the entrance to the lair, rubbing themselves against her side and then settling down to suck from her swollen teats. During the course of their suckling, two males, one strikingly larger than the other, wander over to lick the female's muzzle and flanks.

Having drunk their fill the puppies return to their games. At their present age—about two months—they are already very energetic but prudent enough not to stray too far from their home. The adults join in the fun, leaping around in a carefree manner and teasing the pups by knocking them over—a pastime evidently enjoyed by all concerned.

The sun sets and—as if in response to a given command—the young dogs trot back in single file into their burrow, while the adults settle down outside, huddling against one another for warmth. One of the females remains on guard at the entrance to the lair, keeping watch over the sleeping puppies. Now and then a male gets up and strolls over to lick her muzzle affectionately, then returns to his prostrate companions. Night has fallen. There is no sight or sound of any other animal on this huge, arid expanse of plain. Three Grant's gazelles, whose shapes were clearly silhouetted against the horizon as the sun dipped, have now vanished into the darkness. The hunting dogs—secure in their family circle—are sole masters here. The display of mutual affection and loyalty which we have witnessed is unexpected and strangely touching, considering that the animal has acquired such a terrible reputation for cold-hearted savagery—a reputation entirely undeserved.

African hunting dogs are prolific creatures. After a gestation period lasting from 60–80 days the bitch gives birth to a litter normally numbering six to ten puppies, though twelve is not uncommon and even sixteen has been reported.

The German naturalist Dr Kühme, from the Max Planck Institute in Seewiesen, near Munich, spent February to May 1964 studying a pack of hunting dogs, comprising six males and two females. One of the latter had four, the other eleven, puppies.

None of these was more than three weeks old when the survey began and all were still dependent upon the adults for everything.

African hunting dogs seek shelter for their puppies as soon as they are born and make sure that they stay there until they are old enough to look after themselves. The little community studied by Dr Kühme had its home in a burrow, but not one that the adult dogs had troubled to dig for their own needs. In fact hunting dogs always make use of burrows previously occupied and now abandoned by another animal, such as an aardvark, a warthog or a hyena. This group behaved like any other, the adults showing every sign of tenderness and tolerance towards their offspring, the latter reciprocating their affection. Sometimes the two females became quarrelsome but only to vie with each other in their attentions to the puppies. Neither appeared to make any distinction between her own progeny and that of her companion, and after a few weeks—as far as could be seen—the entire litter was being raised communally.

When the largest pups were about six weeks old and beginning to run about on their own, the burrow was discovered by a company of lions. The dogs reacted to this unwelcome intrusion by removing themselves to a new hideout, those puppies still unable to walk being picked up in the females' jaws. A few days later another emergency forced the families to make another move, and the procedure was repeated several times more before the youngsters were three months old. At that point the adults apparently felt secure enough to stay put and spend the night resting in the open, wherever they happened to be when dusk fell.

During the few weeks following the birth of the puppies the adult dogs were extremely resentful of the approaches of hyenas and jackals, attacking the marauders ferociously as soon as they ventured to within 500 yards of the lair, though never actually killing them. As the youngsters became older the adults relaxed their hostile attitude and by the time the puppies were three months old predators were roaming quite freely in the vicinity.

All the surveys thus far carried out indicate that males and females alike co-operate in bringing up the young; but it is a female that guards the litter during the day, standing like a sentinel at the entrance to the lair, while the rest of the pack is off hunting for prey. Sometimes, however, one or two males stay behind to keep her company. Both adults and young give a delighted welcome to the home-coming hunters and the latter, who have chewed and swallowed the remains of their prey in a cursory manner at the site of the kill, now regurgitate part of the food for the puppies and their guardians. The bits of meat are rechewed and digested by all present, the females carefully pulping it into minute lumps for the youngest puppies, which are ready to graduate to solid food at about the age of two weeks. In this way the food is evenly distributed, each puppy supplementing its meat diet with mother's milk, the process of weaning lasting up to ten or twelve weeks. The combination of meat and milk affords sufficient liquid to dispense with drinking water and surveys suggest that it is not until the puppies are ready to accompany their parents in their semi-nomadic existence— when they are about three months old—that they begin to look

The female suckles her puppies for ten to twelve weeks, during which time they have no need for fresh drinking water. Sometimes one bitch will care for another's litter, squabbles occurring between them for the privilege. The puppies are ready to accompany the adults on hunting expeditions when they are about three months old.

246

for water and to drink at regular intervals.

The zoologists Richard Estes and John M. Goddard, who made a two-year survey of the behaviour of a pack of African hunting dogs in Ngorongoro Crater, described one incident which, among many others, demonstrates the social instincts of the species while the young are being reared. Towards the end of one February, a female gave birth to ten puppies, nine of them males. For reasons never discovered, the mother died shortly afterwards inside the burrow. The rest of the pack, all males, promptly retrieved the corpse from the hole and assumed joint parental responsibility for the orphans, which were still too young to fend for themselves. Thus when the pack went out hunting one dog remained behind on sentry duty to guard the puppies.

In due course the youngsters were able to trot at the heels of

Hunting dogs make no effort to conceal their presence from gazelles and other prey. Here a group surveys the grazing animals from a distance prior to launching an attack.

their elders. But their mortality rate was unusually high. In spite of the attentiveness of their foster-fathers only four puppies managed to survive. This may either have been due to the continual attacks of predatory hyenas or as a result of starvation. It was obvious that they could not always keep up with the adults, sometimes lagging as much as a mile behind and thus giving the hyenas ample time to pounce on them at leisure. Alternatively, by the time they caught up and arrived at the scene of a kill, the carrion eaters might already be gorging themselves so that the puppies were left without food.

Lessons in hunting begin early in life–as among all baby carnivores–starting with half-serious games which help to accustom them to using their muscles, developing their senses and co-ordinating their different movements. When they first go out with the pack they take no active part in attacking or killing prey. They simply sit and watch their peers, then, as they grow older, take an increasingly positive share in the proceedings. But even though they are incapable of making a direct contribution, their elders make sure that they do not go hungry. Whether they are weak, late in developing or just unlucky, they all get their fair share of food, dutifully regurgitated for them by the adults. It is this unselfish behaviour which helps to ensure the health and strength of the pack and the species, so that all can later join in the communal activities of hunting, defending territory and providing food for the others, including a new generation. Some writers have described hunting dogs as possessing a common stomach and this is the literal truth. Indeed, when one compares their behaviour with that of lions, which often condemn their cubs to death by denying them any share in captured prey, one can only feel admiration for the solicitude and co-operative spirit shown by these wild dogs. In their case the survival of the group is assured, but not at the expense of any individual.

The large pricked ears of the hunting dog give it the appearance of being continually on the alert. Keen eyesight and scenting powers help to locate both prey and enemies.

Free as the wind

The English sport of foxhunting makes use of a special breed of hound which is trained to chase the fox to the point of the victim's exhaustion and possible death. The hunting dogs of Africa use similar methods–though they require no training for the purpose –but having said that, all resemblance to foxhounds ends there. This is no unruly, baying rabble, responding to human command and control, but a silent, superbly disciplined group, acting in its own interest and motivated by the simple need to survive in a competitive environment.

Having outpaced and overtaken their victim, hunting dogs do not, and are in fact unable to, achieve a clean killing. They attack the most vulnerable parts of the body–generally the flanks and the rump–with their teeth, literally ripping the flesh away. It is a highly unpleasant business and it has given the dogs the reputation of being the most callous and bloodthirsty of carnivores. In their defence it can be argued that they resort to such methods only from dire need, whereas trained dogs will often display comparable cruelty without such an excuse.

The area over which a pack of hunting dogs will roam in search of their prey depends entirely on the availability of suitable animals to kill. The pack kept under observation by Dr Kühme in the Serengeti needed a hunting precinct of only 15 square miles during February when herds of gnus and zebras were congregated on the plains; but by May, when the herbivores had departed to find fresher pastures, the wild dogs were forced to extend their area of operations to a district covering about 75 square miles. Four other packs roamed the fringes of their hunting territory but never actually crossed its borders.

Hunting dogs do not in fact usually come to blows over territorial rights. Hugo van Lawick was surprised to note that one pack which he had been watching on the previous day had been replaced by a different, larger one. There were no signs of a fight, nor did he ever come across the original pack again. These unexpected arrivals and departures are characteristic of the species. A pack may remain in the same spot for only a couple of days or for as long as several months. The birth of a litter is the only event that will guarantee their staying put for any predictable period. Once the young are reared and old enough to join the adults, the nomadic pattern returns, but pack movements are probably determined to some extent by the timing of the rainy season and its effect on the local herbivore population. It fails to explain, however, why a pack of hunting dogs may suddenly abandon such ideal terrain as can always be found in Ngorongoro for less suitable surroundings. Even the native-born wardens have no real answer. They say that 'mua muitu'—their local name for the animal—will come and go as it likes, as freely and unpredictably as the wind itself.

The chase begins

All the wild dogs studied in the Serengeti and in Ngorongoro Crater had fixed hours for hunting. The former group were in the habit of making their first sortie at dawn, between 6.30 and 7.30 am, and another at dusk, between 6.00 and 7.00 pm. It is clear, therefore, that these dogs are diurnal creatures which make use of their keen vision to locate and follow their prey, with the sense of smell serving as a secondary aid. They prefer to venture out during these cooler hours since long and tiring chases in the full heat of day are not to their liking. In exceptional circumstances they may sally out on a clear night or, during a cool, rainy period, towards mid-day. In Ngorongoro Crater, where there are fewer hours of sunshine, they normally advance their evening departure by about an hour.

When they are not encumbered with a litter hunting dogs will settle down anywhere on the savannah, sleeping huddled against one another; but when there are puppies to be cared for they group themselves around the entrance to their lair. At dawn, one of the adults, generally the same individual each day, gets up, stretches and starts licking its companions until a couple of them are awake. Gradually the whole pack is roused from sleep and for some minutes they tumble about on the ground, nibbling one another playfully, wagging their tails and uttering little whining yelps—just like ordinary domestic dogs greeting their owner. The

Facing page : After killing their prey the hunting dogs dismember the carcase but leave almost half of it untouched. The meat is hastily swallowed on the spot but later regurgitated for the benefit of the young and the adults that have remained behind to guard them.

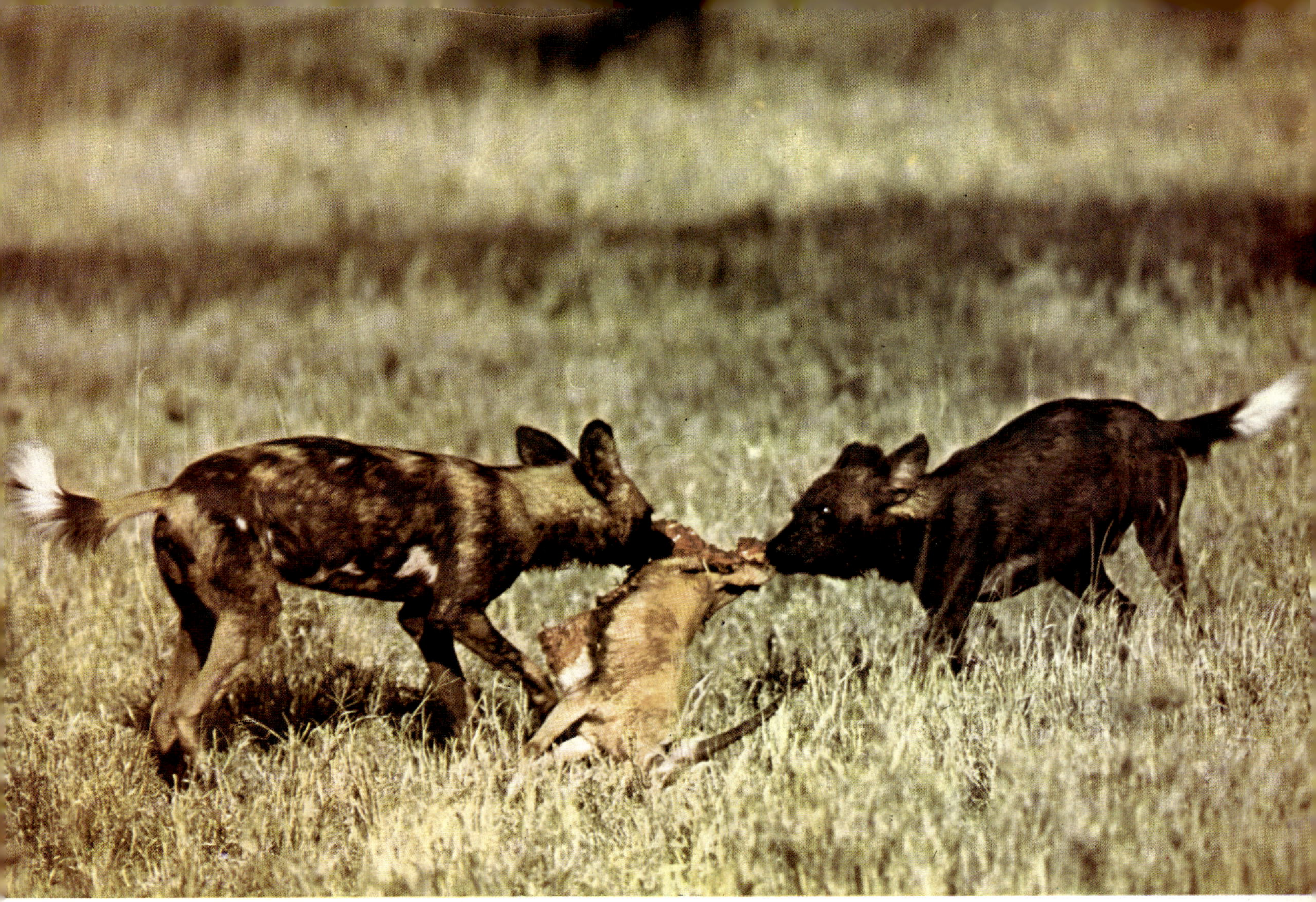

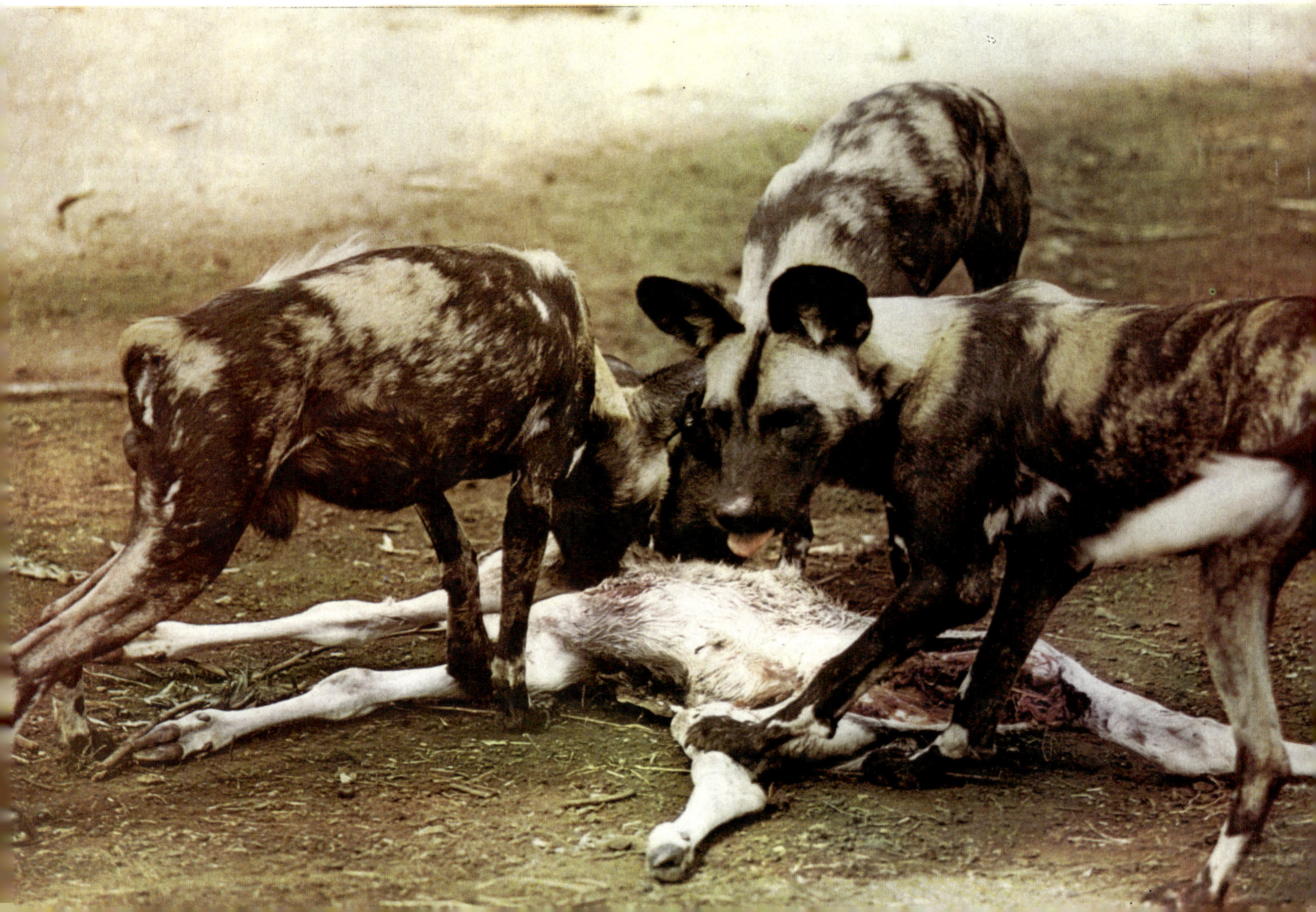

tumult reaches its climax as the entire pack breaks into concerted barking, prior to setting out on the morning's hunting expedition. This ritual being completed, the dogs move off behind one animal which seems to be accepted as leader and guide. At first their progress is slow and cautious, some of the dogs pausing at times to sniff the ground but quickly leaping forward to catch up the others. Soon they all become livelier, breaking into a determined trot and occasionally, if their view is impeded by tall grass, jumping high into the air to take their bearings and perhaps a rapid look at the grazing herds that form the object of their exercise.

Unlike the Big Cats, these canine hunters make little effort to conceal their presence as they advance steadily towards their prey. Silent stalking and ambush play hardly any part in their hunting techniques, and success and failure do not depend on surprise. The victims are fully aware of their predators' intentions and although the gazelles and other small grass-eating animals show visible signs of anxiety and nervousness they are rarely provoked to panic. There is a deceptive air of calm surrounding this stage of the proceedings. The dogs have already selected the group of herbivores to be attacked and have perhaps pin-pointed an individual which is grazing apart from the others. They have now slackened their pace and are moving unhurriedly towards their target, tails down and necks outstretched. This leisurely approach is a deliberate tactic, designed to lull the fears and suspicions of the herd rather than stampede them, as would happen were they to rush noisily into the fray. It is not until the predators are some 200 yards from their goal that the serious business of attack and pursuit begins.

The African hunting dog is capable of keeping up a speed of around 30 miles per hour for a considerable distance but does not normally fritter away energy in this way. It is the leader of the pack that fixes the average speed and this will be a comfortable 20–25 miles per hour in the opening stages of the hunt. The moment the selected victim shows signs of fatigue the rate is stepped up until the entire pack is surging forward at top speed. This final stage of the pursuit may last anything up to twelve minutes, but three to five minutes is generally time enough for the dogs to achieve their objective, after a chase of up to two miles.

Butchers of the wild

We watched one such typical hunting expedition in the Serengeti. While one of the females mounted guard over the puppies, the other dogs set off in pursuit of a group of Grant's gazelles—four females and two males, one of them a young animal. For a while the antelopes held their own against the silent, closely bunched pack of dogs, but after about half a mile the older male, clearly tiring, fell far behind his companions. By so doing the creature signed its death warrant, for the moment it occurred the dogs—as if an unseen hand had suddenly unleashed them—bounded forward with an extraordinary burst of acceleration. The unfortunate gazelle zigzagged frantically in an attempt to shake off the murderous pack, but it was utterly hopeless. Clouds of dust

The naturalist Baron Hugo van Lawick has made detailed surveys of the behaviour of African hunting dogs in the Serengeti National Park, revealing that there is more to these reputedly merciless killers than meets the eye.

marked the points where the dogs were now fanning out in a semi-circle to cut off their victim's escape route.

The gazelle was now completely worn out after a chase that had covered almost three miles. Short of breath, sides heaving, mouth wide open and tongue hanging out, it was practically on its knees but prevented from falling by one of the dogs which now sank sharp fangs into its body. Another dog launched itself full-tilt at the gazelle's shins and this second assault toppled the beast. The pack closed in and began to tear it apart yet somehow the animal summoned up enough strength to get to its feet again. The devilish pack would not loosen its grip. The gazelle's belly was ripped open, its body gashed and bleeding, its legs broken. It must have taken almost twelve minutes to die.

There are few more disagreeable or disgusting sights than a pack of hunting dogs dismembering a live victim, and it is hardly surprising that those who have watched it are tempted to condemn these carnivores as the cruellest and most vicious of killers. Without trying to defend their actions, it is only fair to state that in most cases death occurs much more rapidly than in the above-cited example. Some authors suggest that by the time the final assault is delivered the victim is so stunned and exhausted that it is incapable of feeling very much. Of course this assertion cannot be proved but neither can it be confirmed that the victims of African hunting dogs show any more intense signs of agony than those of other carnivores. Indeed they often seem almost resigned to their fate, offering little resistance.

Baron van Lawick has spent much time watching and taking photographs of hunting dogs chasing and bringing down larger animals such as adult gnus and zebras, and has thrown interesting light on the methods normally employed by the predators in such situations. When the dogs have succeeded in describing a circle around a zebra, their behaviour is rather like that of wolves attacking a solitary horse. A number of dogs deliver simultaneous assaults from several directions, each fastening on a different portion of the animal's anatomy. One may go for the muzzle, thus immobilising the head and the front legs, others will concentrate on the shins, the flanks and the rump, disembowelling the animal, tearing out the internal organs and beginning to eat their victim alive. Even though death probably follows quite rapidly, it is a horrible spectacle—very much the kind of sight to inflame the wrath and hatred of early settlers from Europe, who showed little mercy to the savage carnivores.

Compared with members of the feline family, hunting dogs have fairly poor natural weapons. The Big Cats, thanks to their sharp claws, have no difficulty in holding down a large herbivore and ripping its throat open with their fangs. But these wild dogs have no means of immobilising a victim in this manner and must rely exclusively on their powerful jaws. Keeping out of range of hooves and horns, they snap away at an animal's flesh until it collapses from its wounds. Wolves use exactly the same tactics with animals of medium size and weight such as roe-deer, fallow-deer and sheep, which after prolonged torment also die in the end from sheer exhaustion and loss of blood.

For these reasons the criterion of hunting dogs in selecting a

No two hunting dogs are exactly alike, which makes the task of identification that much easier. Their coats may be blotched in brown, black or yellow, with prominent white markings almost always on the legs and always on the tail.

potential prey is the animal's weight. In a herd of gnus they will automatically pick out either a newly-born, sick or young animal. Van Lawick describes an occasion when, in the course of a survey, he came across a pack of hunting dogs in hot pursuit of a herd of about 60 brindled gnus. Surprisingly, the latter did not begin to run until the dogs were less than 100 yards away, then galloped off, as might have been expected, in the opposite direction to the point of attack. But without any warning the leader of the herd suddenly wheeled about and careered off in a new direction, more or less at a right angle to the pursuing pack. The other gnus followed obediently while the hunting dogs, momentarily caught off guard, stopped in their tracks to gaze after the retreating herd. It was several seconds before they recovered their composure and resumed the chase, focussing their attention on one young animal which was bringing up the rear. The chosen victim made a desperate but futile attempt to quicken its stride and indeed forced the pack to chase it for some two miles before acknowledging defeat. Six of the dogs were needed to bring the animal down and although it had sufficient energy left to struggle to its feet and to let fly with a few feeble kicks, the outcome was inevitable.

The lottery of life and death

Some observers have claimed – quite mistakenly – that these wild dogs of the savannah hunt in relays, basing this belief on the fact that some individuals clearly outpace and outdistance their companions. Van Lawick and Estes, who between them were able to record 112 kills in Ngorongoro and three in the Serengeti, point out that there is no need for the dogs to adopt such a technique for if they do not possess exceptional speed they certainly have sufficient endurance to outrun and bring down any kind of antelope, particularly the two species of gazelle that provide their main source of food.

Unlike cheetahs, hunting dogs do not carry out a long and meticulous survey prior to selecting their prey. Nevertheless, it is invariably the weakest member of the herd that will be their victim. Proving itself unable to stay the course after a sustained pursuit it will, simply by lagging behind, give itself away. The dogs, and particularly the one designated as guide, will be quick to detect the telltale symptoms of fatigue and the entire pack will immediately converge on the unlucky creature. No matter what evading tactics it tries – running in large circles or along a zigzag course – they will not relax their effort or change their objective. Even if another animal, clearly easier to capture, crosses their path, they will not normally be diverted. If they do abandon the chase it will probably be because they have misjudged the staying powers of their intended victim and are forced to cede it victory.

Yet hunting methods sometimes have to be modified to suit the circumstances and there may be special reasons for their breaking off a pursuit of one victim in favour of another. Van Lawick and Estes once saw a young eight-month-old gnu which had been driven almost to a standstill by a pack of hunting dogs

Facing page (above) : Three hunting dogs tear a piece of meat into manageable pieces, so that it can be fairly shared by the entire pack when they return home. (*Below*) larger packs of dogs, requiring a greater amount of food, often tackle herbivores such as gnus and zebras in addition to gazelles. Here a pack converges on a zebra that has been killed after an exhausting chase.

Hunting dogs will pursue a zebra across the savannah until it shows signs of tiring. They can then overtake and surround it, laying it open to attacks from every side. Some go for the muzzle, others for the shins, flanks and rump, literally tearing the animal apart while it is still alive.

Facing page : Lycaon pictus always hunts in packs and is rarely surprised by another predator. But it must keep a constant lookout for hyenas, jackals and lions which are likely to be lurking in the vicinity when the puppies are very young and still vulnerable. Sight and smell also play a vital part in the daily quest for food.

in full cry and which was lashing out with its hooves as its assailants tried to sink their teeth into its rump and flanks. Suddenly a Thomson's gazelle, in an advanced state of pregnancy, strayed imprudently close. The leader of the canine pack wheeled on her and in less than fifteen seconds had caught her. The rest of the pack relinquished their hold on the wounded gnu and just as eagerly accepted the substitute and unexpected prey.

Among the adult animals often selected as victims – doubtless because they are relatively easy to capture – the male territorial 'Tommies' are especial favourites. These proud creatures are invariably the last of a herd to take to their heels, hesitating fatally from a disinclination either to abandon their own little kingdom or to poach on a neighbour's domains. They end up by racing round their territory in ever decreasing circles, giving no

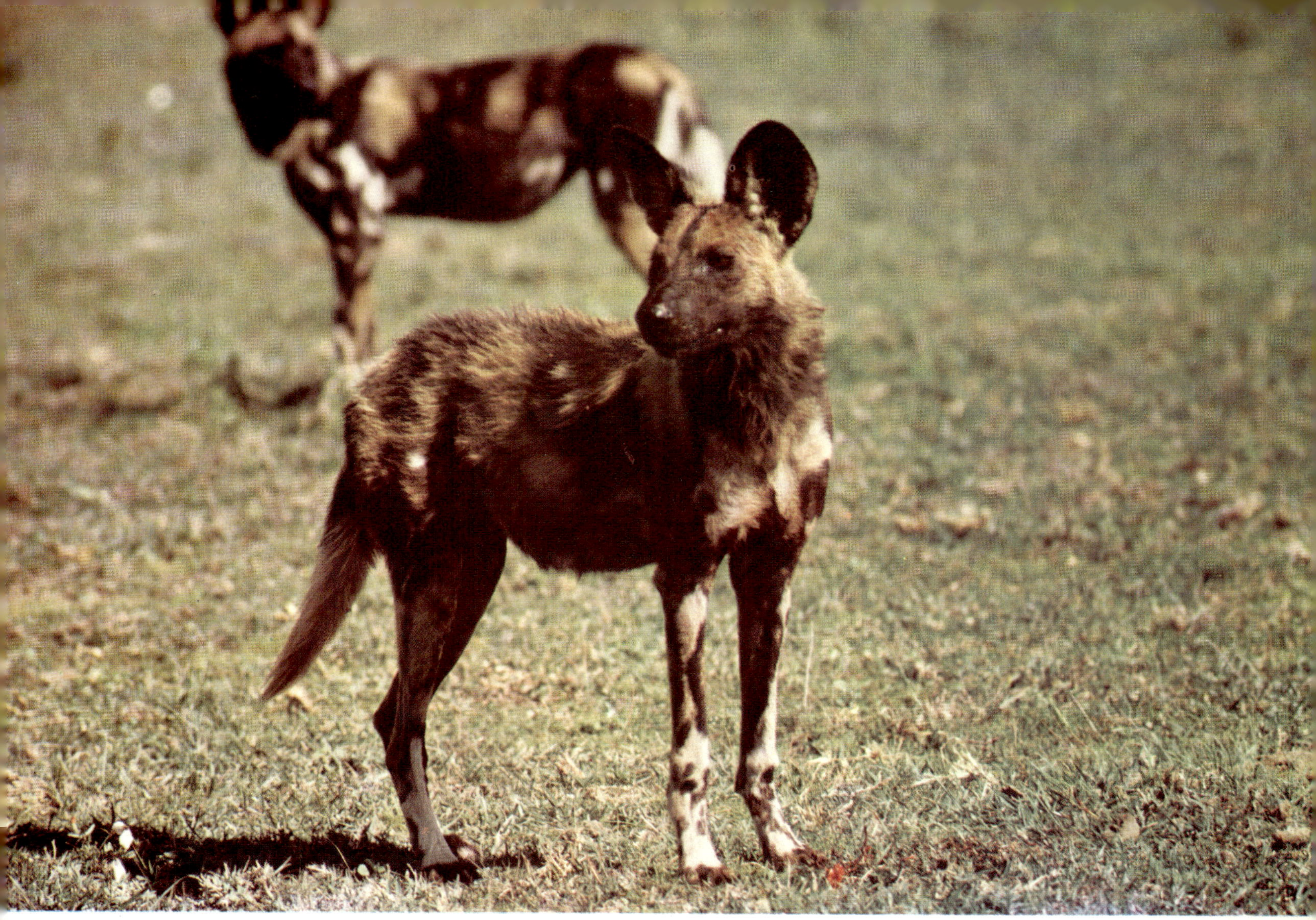

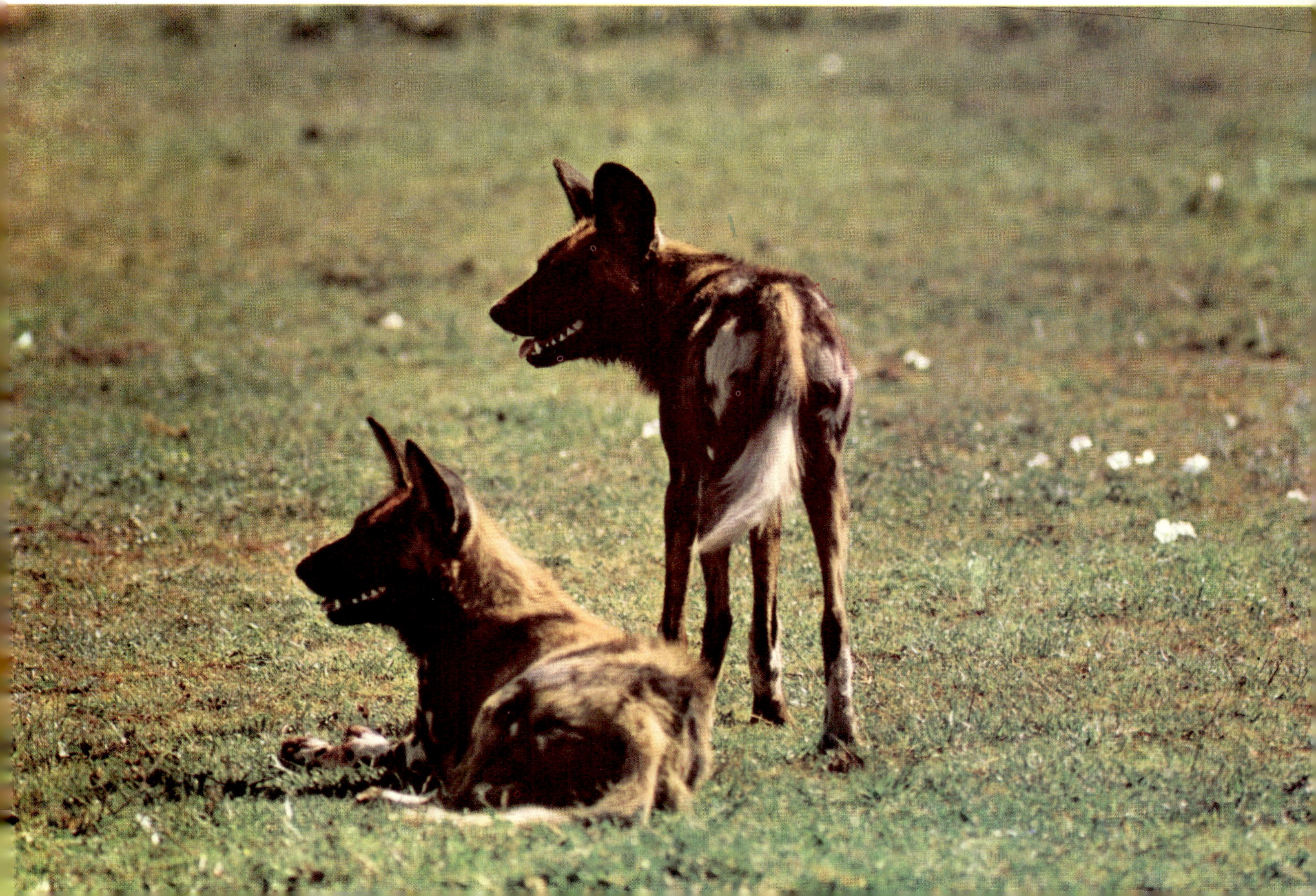

problems to the predators, who proceed to outflank the animal in the customary semi-circular formation and make it impossible for it to get away in any direction.

The smaller herbivores—including the Thomson's gazelles—are usually killed amazingly quickly. Once they are brought to the ground the whole pack will leap in and dismember the beast in a matter of seconds. Grant's gazelles, being larger and stronger, may put up a more spirited defence, but they too cease to struggle once they are down. Gnus, however, are less passive. Adults rarely submit to the attacks of hunting dogs without putting up a courageous fight, often banding together to defend their young or their handicapped companions. Dr Kühme points out that in the Serengeti, when a pack of hunting dogs attack newly-born gnus, the adult males responsible for the protection of the herd charge the predators as soon as they sight them. The dogs avoid these onslaughts fairly easily and try to slip one by one into the heart of the herd where the females and their offspring are assembled. As long as the male guardians succeed in keeping the main body of the herd together the dogs are powerless. Sooner or later, however, a terrified calf, perhaps accompanied by its mother, will find itself isolated from its companions. This is the moment for the dogs to launch their concerted attack. Although the female may fight bravely for a time to keep the assailants at bay, she will eventually be forced to abandon her calf and save her own life by rejoining the herd.

December and January are the months when the female gnus of the Serengeti normally give birth and it is at this season that the mortality rate among newborn calves is especially heavy. At other times the hunting dogs are just as happy to go for young animals that are already weaned.

Even when they are young, in fact almost as soon as they can move at speed, gnus put up a much more serious defence against predatory hunting dogs than do gazelles of a similar age, only submitting to their fate when strength or stratagem have failed. They will sometimes use quite cunning methods to avoid capture. A team of naturalists in Ngorongoro saw a young gnu make a number of successful efforts to escape such attacks by wading into a pool until only its head could be seen above the surface of the water. The dogs were taken completely aback by this unaccustomed ruse and were unable to devise any means of catching their prey, which finally escaped scot-free.

Zebras, even when they are accompanied by young, do not seem to be unduly perturbed by hunting dogs, although attacks are quite frequent and the predators do not take much trouble to conceal their presence. Like all carnivores, the dogs specialise in certain types of prey but in their case there does not appear to be any uniform preference. One pack will chase only gazelles, another will choose the larger herbivores. Much depends on the size of the pack. The one studied by Hugo van Lawick was a fairly large group, consisting of eleven adults and eight young; other packs, perhaps in different terrain, are even larger, some numbering more than forty individuals. It is these more sizeable packs that generally concentrate their efforts on the larger ungulates, having the double advantage of numbers and com-

bined strength, but also requiring a greater amount of food to satisfy their basic needs.

In Ngorongoro Crater Thomson's gazelles stand at the top of the hunting dogs' list of victims—constituting some 53 per cent of the total. Next come Grant's gazelles, furnishing a further 25 per cent of the prey, followed by baby gnus of less than one month (21 per cent). Kongonis make up the remaining one per cent of hunted animals.

According to information collected by different naturalists, the average amount of meat consumed in one day by a hunting dog is about 6½ lb. Allowing for a 45 per cent wastage—namely the bones, skin, stomach contents and other non-edible parts—it is obvious that a pack comprising ten or fifteen animals cannot make do with a single small gazelle daily. Since they will want to conserve energy by attacking one animal rather than several, it stands to reason that the choice should fall on a fairly large and heavy herbivore.

On the savannah and open steppe—their usual habitats—hunting dogs have to face the competition of other carnivores, including lions, cheetahs, leopards and hyenas—all of them interested in more or less the same kind of prey. Lions, in addition to attacking the larger herbivores, are partial to gazelles as well, and often save themselves the trouble of catching them

This herd of gnus shows no visible sign of fear although surrounded by hunting dogs. Evidently the latter have satisfied their immediate food requirement and can be counted on not to attack for the pure joy of killing.

by simply stealing the meat from under the dogs' noses when the latter have effected the kill.

The spotted hyena is the most formidable rival and enemy of the hunting dog in Ngorongoro Crater. Not only will it compete for the same kind of prey but it will often compel a pack of dogs to abandon the remains, devouring them before the pack has time to regroup and fight back. In one unusual incident two hunting dogs managed to outdistance their companions and to kill a gnu. A couple of hyenas had kept pace with them and snatched the animal from the jaws of the dogs, now worn out after a five-mile chase and left no alternative but to watch the thieves from a safe distance. But the moment the rest of the pack arrived they reclaimed their prey and sent the hyenas scurrying off. Fifteen minutes later a couple of lions decided the issue by carrying off the half-eaten carcase and putting both the dogs and the hyenas to flight.

The rivalry between the different species of carnivores is well illustrated by another strange episode in which eighteen hyenas were observed, apparently in pursuit of a pack of hunting dogs. Yet behind them, and hot on their heels, were two cheetahs. All of them were in fact after a single Thomson's gazelle which, to the discomfiture of the entire company, used its superior speed to escape from every one of its pursuers.

Co-operation for survival

Many people who have studied animal behaviour suggest that a domestic dog treats its owner as both parent and chief. As soon as its master comes near the dog adopts infantile postures of submission, indicating its readiness either to play or obey. It will seek human comfort and protection and, should the need arise, will spring to its master's defence. Even an older dog will revert to puppy-like behaviour, lying on its back and waving its paws in the air. In the case of hunting dogs the same kind of master-servant relationship exists within the pack.

On one occasion we saw a wild dog returning to the lair, at the entrance of which the other dogs were resting. Whatever the reason for its absence, the pack was initially hostile, seeing but not scenting the newcomer, since the wind was in the wrong direction. They were clearly on the defensive, necks extended and tails lowered, but when the stranger was a few yards away it quickened its pace, raising its tail, with the distinctive white tuft, like a flag. The change in attitude was remarkable. Now that they recognised their lost companion they threw themselves on him with all the usual signs of affection and welcome.

The white-tipped tail clearly acts as a signal of recognition and probably as a rallying point for the pack after a chase. It is vital that no individual should wander off alone for isolation would place it at the mercy of its enemies. Even if a solitary dog were able to kill a prey it would stand no chance of beating off a raid by scavenging hyenas. In such emergencies the hunting dog has a way of whining softly, the sound carrying some considerable distance. Once we came across a lost dog in the Serengeti. It was crying like a child, a poignant picture of

Facing page : Adult hunting dogs belonging to the same pack display many signs of mutual affection, licking one another, wagging their tails and generally behaving much in the manner of domestic dogs. The muzzle-licking gestures are also indicative of hunger and are designed to provoke the regurgitation and exchange of food.

complete despair and desolation.

Hunting dogs possess special scent glands which give out a highly disagreeable and persistent smell. This too is a valuable method of keeping in touch with one another and signalling their location when the wind is in the right direction.

No single feature of behaviour contributes so powerfully to the stability and survival of the pack as the sharing of food. Other ritual actions are of secondary importance. The procedure of regurgitation ensures that everyone gets enough to eat and can participate in the group's activities, whether it be standing guard at the lair or going hunting. Those that capture the prey hastily ingest large pieces of meat on the spot and then go back to feed the others, the latter indicating their hunger signs by licking the lips and nibbling the muzzles of the hunters.

This type of behaviour is indeed remarkable in the world of carnivores, though the mutual exchange of food is practised by certain insects. It ensures that the entire community receives all the protein necessary for normal development. In other animal societies, notably those of lions, the leaders, with their heavy burden of defending the pride, need to be stronger than their companions. They therefore take it as a right to relieve other members of the group of their share of food, and when this is in short supply the stronger warriors and huntresses will survive while the weaker members will be sacrificed. But in the hunting dog community every member of the group is capable of performing a number of functions – hunting, protecting the family and rearing the young. All are therefore of equal importance and all must be cared for.

Dr Kühme believes that this sharing of duties by all members of a pack shows that there is no system of hierarchy within the group. His own surveys have failed to reveal any form of social priority and the work of Richard Estes and Hugo van Lawick confirms his findings. Much indeed still remains to be discovered about the communal behaviour of these strange animals. For example, we do not know the precise relationship between the members of the same pack nor what links may exist between different packs of dogs occupying the same territory. Little is so far known about the way such territorial boundaries are fixed nor how the area itself is defended. There is insufficient information about the breeding pattern and sexual behaviour of males in packs where, as in the majority of cases, the number of females is severely limited. Doubtless in time these questions will be resolved.

For the time being all that can be said with certainty is that when the dogs are resting either during the heat of the day or at night, they settle down in no particular place or order. Nor does any individual seem to take precedence when it comes to hunting. In all the packs observed the different members behave like children in a large family. Even the adult males beg for a share of food like the young, and have been seen to suck the teats of the females and huddle up against them like puppies. If any individual can claim to be accorded special respect and treatment it is probably the bitch with special responsibilities towards her young, and she indeed sometimes appears to accept such attentions rather haughtily as her proper due.

The slaughter of the hunting dogs

The white man has a dismal record of savagery against wild animals, especially in Africa, and the hunting dog has been one of his favourite victims. The dog's own brutal methods of killing provided the excuse for a massacre which almost took on the guise of a crusade. Wrongfully accused of killing far more prey than they needed for food, hunting dogs were systematically slaughtered even in areas where they were supposed to be protected, as in the Kruger National Park, between the years 1900–1930.

Fortunately, naturalists have now established that hunting dogs kill only in order to satisfy their basic needs and that, far from harming the herds of grass-eating animals, they contribute to the health and stability of the herbivore population by weeding out sick and aged animals. Nowadays the dogs, like other predators, have been granted the right to protection in national parks and reserves. Although their numbers have not reached their former proportions – packs of thirty or forty are still very rare – the steady decrease has been halted.

Those who have studied the behaviour of hunting dogs point out that there is a marked disproportion between the numbers of the two sexes. In the days when they were hunted in the Kruger National Park six males were killed for every four females; and in most litters there are more males than females (out of ten puppies observed by van Lawick in Ngorongoro, only one was a bitch). But this does not seem to be a handicap – on the contrary, the shortage of females is a natural regulating mechanism which keeps the numbers steady while preventing them from becoming excessive. All the same, it is an additional reason for making sure that the species is left to roam freely in the future.

Among the favourite prey of African hunting dogs are Grant's gazelles – a fine specimen of which is seen here – and the smaller Thomson's gazelles.

Three typical representatives of the family Canidae. (*Above, left*) the bat-eared fox; (*above, right*) an African hunting dog; (*left*) the European red fox. The first belongs to the subfamily Otocyoninae, the second to the Simocyoninae, and the third to the Caninae.

FAMILY: Canidae

The Canidae are typical carnivores, but they are less completely flesh-eaters than the Felidae, and many of them, including the familiar domestic species, tend to be omnivorous. Their teeth, while in other respects retaining the fundamental features of those of carnivores, are more adapted to cope with a broadly based diet. Thus the molars, which possess cutting edges, have flat-surfaced crowns for chewing. The upper carnassial, however, is exclusively a cutting instrument. Although there is some variation among the species, the usual number of teeth is 42 – three incisors, one canine and four premolars in each half of either jaw, two molars in each half of the upper jaw, three in each half of the lower.

The Canidae are medium-sized animals with fairly long tails and slender, pointed muzzles. The cranial cavity of the skull is capacious and the brain well convoluted, with highly developed olfactory centres; these provide the members of the dog family with an extremely keen sense of smell, enabling them to detect prey and enemies or to identify their companions with remarkable accuracy. To these ends they are also furnished with a number of tegumentary glands, sometimes enclosed in an anal sac, secreting substances which give each animal a characteristic odour and which play an important role in the demarcation of territorial boundaries and in the erotic stimulation of the opposite sex.

All the Canidae are excellent runners and their chosen hunting method is generally to chase their victims until the latter drop from exhaustion. They owe this faculty of speed to their limb structure. The long, slender legs terminate in four toes (a rudimentary fifth one is found on the fore paws of some species but in others, such as the African hunting dog, this has completely disappeared). They are normally digitigrade – walking on the toes – and have elastic pads at the bases of the toes. The short nails are blunt and non-retractile, therefore not used as claws. The major natural weapon of these animals is the mouth, with powerful jaws and strong teeth for grasping and tearing the flesh of live prey. The stomach of all Canidae is simple and the intestinal tract relatively short. In contrast to other families of the order Carnivora, however, they have a caecum, a blind tube or pouch leading off from the intestine.

The gestation period of Canidae varies between 50 and 84 days, and the number of young in a litter also differs greatly, from one to sixteen. The young of all species are reared with care and affection until they are able to feed and defend themselves unaided.

Being highly adaptable, the Canidae have populated almost every region of the world, except for Australasia and Oceania, where they were introduced by white settlers. There are three subfamilies – Caninae, Simocyoninae and Otocyoninae.

The Caninae consist of dogs and dog-like animals with long muzzles and a fifth toe on the front paws. They are very numerous, including, among others, dogs, wolves, jackals and coyotes (genus *Canis*), foxes (*Vulpes*), Arctic foxes (*Alopex*) and fennec or desert foxes (*Fennecus*).

The Simocyoninae comprise animals with fairly short muzzles, the fifth toe of the fore feet being rudimentary or absent. The three genera are the bush dog of South America (genus *Speothos*), the African hunting dog (*Lycaon*) and the dhole of southern Asia (genus *Cuon*).

There is only one representative of the subfamily Otocyoninae – the bat-eared fox which is found in South and East Africa (*Otocyon megalotis*). This creature is noteworthy for its aberrant dentition. It may have 46 or even 50 teeth – with six or eight molars in the upper jaw and eight or ten in the lower jaw. Neither the upper nor lower carnassial is differentiated from the other teeth.

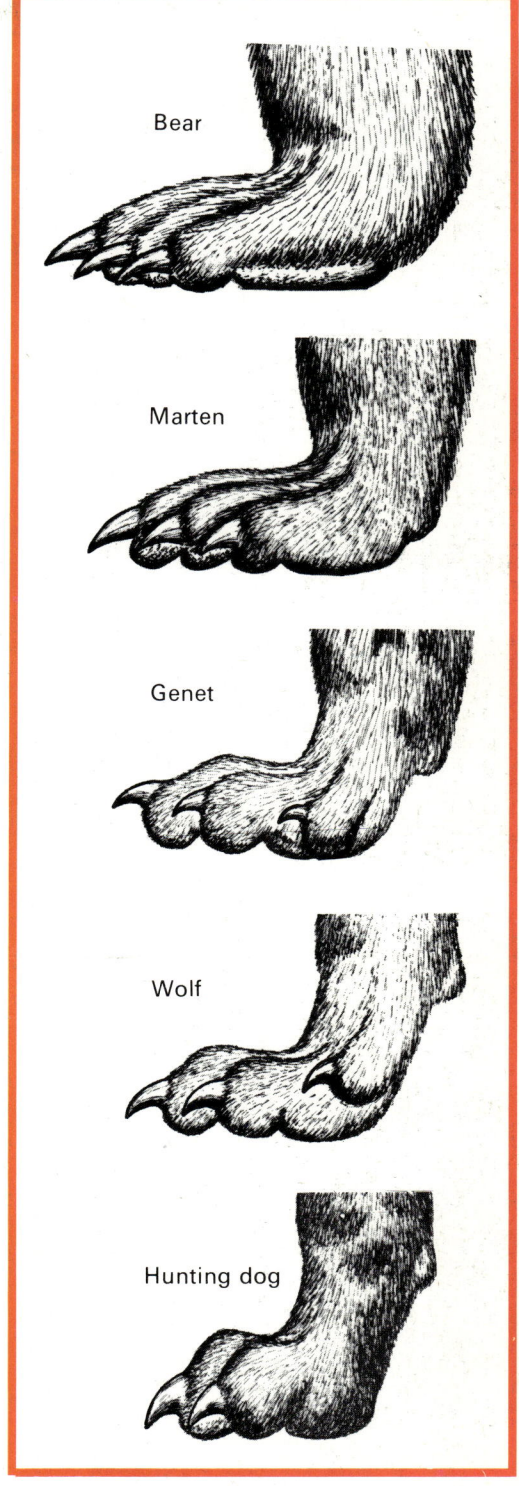

Bear

Marten

Genet

Wolf

Hunting dog

The carnivores have different methods of walking. Between a typical plantigrade, such as the bear – with its ponderous gait – and the fleet-footed wolf, a digitigrade, there are intermediate stages. The marten, for example, is semi-plantigrade and the genet semi-digitigrade. In the faster runners a reduction in size of the great toe provides easier contact with the ground, and in the hunting dog this toe has disappeared.

CHAPTER 10

Spotted hyenas and jackals

If popular opinion accords highest place in the animal kingdom to the lordly lion, the lowest rungs of the social ladder are generally reserved for the scavengers of the wild, particularly the despised hyena. For this unpopularity nature is largely to blame, having endowed the creature with singularly bad looks. Its coat is a nondescript grey or buff colour with brown spots, its head is massive and ugly, the muzzle blunt and thickset, the neck powerful but somehow out of proportion, the large eyes round and lack-lustre, the hindquarters low, the stomach distended, the stance heavy and ungainly. All in all, it is a most unattractive animal, with not a single appealing physical feature.

The local African population has always nursed a special loathing for the carrion-eating spotted hyena, and these feelings of contempt and hatred were dutifully echoed by the first European settlers of the black continent. The native legends helped to stimulate their prejudices. The animals were reputed to dig up the corpses of humans buried on village outskirts, and some tribes regarded them as forces of the occult. Their sloping rump and clumsy gait indicated that they were the devil's own steeds.

Thus the hyena suffered from a double disadvantage—an evil appearance and an evil reputation. But naturalists rightly refuse to judge an animal by human standards. Accusing a hyena of being ugly, frightening or wicked is scientifically valueless. In fact, its very 'ugliness' turns out to be a perfect adaptation to a particular mode of life. The coarse, hairy, wrinkled coat, so unpleasant to the touch, has a natural faculty, as it dries, of shedding all kinds of impurities that tend to lodge there. This self-grooming procedure is a tremendous asset to a creature which literally plunges its head into the stomach of its victim in order to devour the internal organs. Furthermore, the blunt shape

Facing page : Jackals, similar in appearance to foxes and some breeds of domestic dog, are familiar predators and scavengers of the East African plains and savannahs.

Geographical distribution of the spotted hyena.

and brutal appearance of the head may be attributed to the size and development of the jaws, whose carnassials are the largest and sharpest of all terrestrial carnivores. This highly efficient mechanism for mastication is operated by powerful muscles and enables the hyena to break the most solid of bones.

Thanks to its exceptionally fine nocturnal vision, the hyena poses the greatest threat to other animals at night, for most herbivores can only see clearly during the day.

The fore limbs are better developed than the hind legs but if this gives the hyena an awkward, ungainly posture and gait, it helps to concentrate the animal's real strength and momentum in its forequarters, invaluable when it comes to dragging a heavy carcase back to the lair.

The rather curious genital development of the hyena has also given rise to a number of legends – for example, that the creature is really a hermaphrodite, that it changes its sex every year, and that the male need play no part in the breeding process. None of this is true. Hyenas mate and reproduce like all other carnivores. The female's clitoris, however, is well developed, resembling the male's penis, while adjacent projections are not unlike a scrotum. Sexing of individuals by touch is only possible after the male's testicles have descended.

Hyenas of Ngorongoro

The animal that popular legend and opinion have painted in the blackest colours is especially vilified because it is a scavenger and carrion eater. The fact that the noble lion is also prone to such habits is conveniently overlooked, as is the even more important fact that this activity is a part of nature's pattern, the disposal of dead bodies being a necessary stage in the unending cycle of food and energy.

In the public mind the hyena is always regarded as the carrion eater extraordinary, an image fostered by photographs taken at dawn showing a group of lions feeding on a gnu or a zebra, while a band of hungry hyenas wait to converge on the remains. The captions to such pictures generally state that the lions have been responsible for the kill. In fact this is not necessarily true, and the preceding hours of darkness may well have concealed a very different sequence of events.

For some time zoologists working in the Serengeti National Park wondered how the large numbers of spotted hyenas that roamed the plains managed to stay alive on the meagre remains of carcases abandoned by lions and other carnivores. Dr Hans Kruuk, the Dutch naturalist, who admitted quite freely that he too had always considered hyenas to be exclusive carrion eaters, decided to look more deeply into the question by investigating the habits of the hyenas living in Ngorongoro Crater.

The main difficulty that faced him was that hyenas, being nocturnal creatures, rarely show themselves by day, retiring to their lairs in the heart of impenetrable thickets, in the crevices of rocks or in inaccessible natural burrows. Dr Kruuk therefore decided to adopt a marking procedure which, if it failed to provide exact figures, would at least give a fairly accurate idea of the

---

SPOTTED HYENA
(*Crocuta crocuta*)

Class: Mammalia
Order: Carnivora
Family: Hyaenidae
Length of head and body: 52–64 inches (130–160 cm)
Length of tail: 12–13 inches (30–32 cm)
Height to shoulder: 27–36 inches (67–90 cm)
Weight: 120–185 lb (55–85 kg)
Diet: meat (freshly killed, also occasionally carrion)
Gestation: 90–100 days
Number of young: 1–2, sometimes 3
Longevity: 25 years

Adults
Massive head, long neck, powerful jaws, rounded muzzle and short, rounded ears. The short, hairy coat is greyish or yellowish, with dark brown spots. The hair is longer on the neck and back, though not forming a true mane. The hindquarters are sloping, the front legs being longer than the back legs. The tail is short and hairy.

Young
Coat uniformly dark grey or black at birth. At five months the colour becomes lighter on head and shoulders and the first spots appear on the neck region. The spots on the legs are only seen when the animal is fully developed, at about two years.

size of the local hyena population. This method would have the additional advantage of aiding identification in order to study social and territorial behaviour. A special type of rifle was used, with small plastic syringes containing anaesthetic instead of normal bullets. Dr Kruuk selected and anaesthetised 50 animals, cutting a small notch in their ears, then releasing them. He estimated that there were about 420 hyenas in the area.

By following the tracks of the animals he had marked, Dr Kruuk noted that they were not in the habit of straying far from one place, that they were gregarious by nature – their packs numbering from ten to forty individuals – and that they normally made their headquarters in a large hole or burrow, usually one that had been abandoned by another animal. It is in such a lair that one to three young are born, after a gestation period of approximately three months. The coat is completely black at birth and acquires its permanent adult colour after several months. Full maturity comes within two years.

In contrast to the social organisation of most other carnivores, the hyena pack gives the impression of being a matriarchy, simply because the females outnumber the males. In fact the females are less dominant than might appear. They tend to remain close to their burrows, whereas the males and young animals show more independence, often belonging to more than one pack. Despite this tendency to roam, the packs are fairly stable, with perfect mutual understanding among the constituent members and short shrift for outsiders trying to infringe on territorial rights.

The boundaries of a territory are marked out, as is the case with other animals, by means of scent posts, the characteristic odour serving as a warning to all those not belonging to the pack that intruders are unacceptable.

The unrecognised hunter

Since the hair of an animal is not digestible and is excreted, naturalists are able to determine the nature of a mammalian predator's diet by examining its droppings. This is particularly useful in the case of the hyena, a nocturnal animal whose hunting habits cannot easily be observed.

From an examination of 188 samples of waste matter in Ngorongoro, 86 per cent were found to contain hairs from Thomson's gazelles, and 83 per cent hairs emanating from gnus. There were traces of zebra hair in 46 per cent of samples. The figures obtained from a similar survey in the Serengeti were more variable – 54 per cent gnus, 53 per cent Thomson's gazelles and 30 per cent zebras. The discrepancy probably arises from the fact that the herbivores of Ngorongoro Crater are sedentary by nature, while those of the Serengeti are migratory. The conclusion must be that the hyena will feast on whatever prey happens to be available and that when the larger herbivores depart on their annual travels it will turn quite contentedly to the smaller sedentary antelopes. Thus it ranks as one of the great opportunists of the wild; and in any competitive environment it is the animal that depends least on specialisation – the one with the most varied diet – that has the best chance of survival.

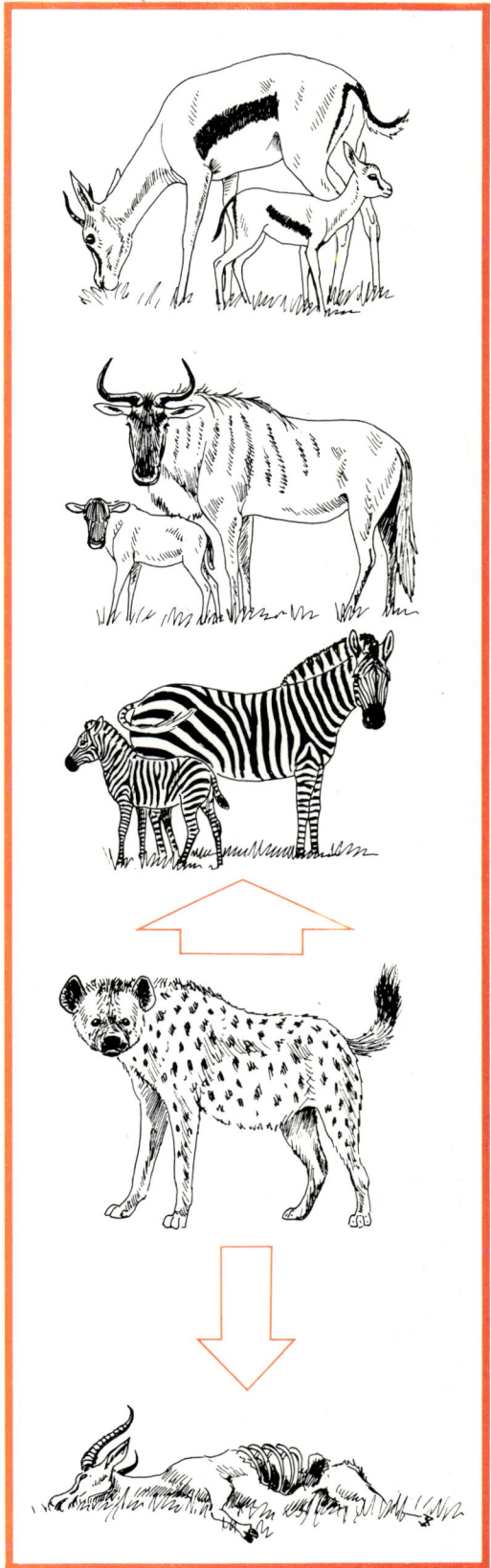

The spotted hyena, though occasionally feeding on carrion, is not primarily a scavenger, as popularly believed. It also kills and feeds on adult and young gazelles, gnus and zebras.

A female spotted hyena rests with two of her young at the entrance to the burrow where the latter have been born and reared. The cubs still have a uniformly dark brown coat colour but by the time they are two years old will look exactly like the adults.

The publication of Dr Kruuk's findings exploded the deeply rooted idea, favoured by popular authors, that hyenas almost invariably eat carrion and that they will only hunt small animals if there is no other way to appease their hunger. This is now known to be inaccurate.

The Dutch zoologist, in the course of his several years of research on the subject, watched hyenas feeding on 1,052 different occasions. He concluded that during the daytime, 34 per cent of their meals consisted of prey killed by other animals – enough, one might think, to justify their reputation of being scavengers; but at night four out of every five victims had been killed outright. Only 11 per cent of the prey consumed had definitely been killed by other carnivores and the balance could not positively be identified one way or the other.

These statistics show clearly that, contrary to popular belief, hyenas are predators first and foremost and carrion eaters only when occasion demands or opportunity offers. Their hunting methods vary. In the rare instances when they venture out by day they are either alone or in pairs, confining their attentions to small or very young animals, especially newborn gnus that have not yet gained the use of their legs or sometimes those that have not even fully emerged from their mother's womb. With this type of individual hunting, the kill is usually effected by the hyena sinking its fangs into the victim's throat, shaking it violently and breaking its neck. But these methods are by no means always successful, chiefly because they do not come naturally to the predator. In fact the percentage of properly executed daytime attacks is pretty low—Dr Kruuk reported that only four out of twenty-one attempts resulted in kills. Far more frequent—and effective—are the group hunting expeditions by night. If the victim is a large herbivore, the entire pack immediately joins in the attack but when the designated prey is of more modest proportions, such as a gnu, the assault may be spearheaded by the leader of the pack which, as soon as it is within range, leaps repeatedly at the creature's hindquarters, inflicting deep bites on the rump. As the chase continues, the other members of the pack take a more active part in the assaults. In the event of the hunted animal leading its pursuers across the bounds of their own territory and into that of another pack, there may be a bitter fight between the rival groups, with victory usually going to the pack defending its home ground.

These night actions are generally successful simply because the attacked herbivores are unable to make as good use of their eyesight—and hence their speed—as during the day. A pack of hyenas in full cry can reach a speed of 35–40 miles per hour at night whereas their prey are lucky to exceed 25 miles per hour in such conditions. Dr Kruuk watched eleven such escapades, of which eight ended in the death of the hunted animal.

When zebras are attacked by a pack of hyenas they gallop off as fast as possible, but the male responsible for protecting the herd hangs back, kicking out at its assailants and doing all it can to keep them at a distance. Unfortunately, the females and young, lacking the male's guidance, often fail to take advantage of the respite afforded by his brave rearguard action. Sooner or later one of the hyenas manages to outflank the defender, exposing the females to direct attack. One of them is singled out and subjected to a flurry of vicious bites on legs and flanks, with a view to slowing her down and separating her from the rest of the herd. Should she be incapable of throwing off this initial assault she is doomed, for the rest of the pack are soon on the scene and under their combined assaults she cannot expect to survive for more than ten minutes.

The actual kill is unpleasantly bloodthirsty, even more savage in some ways than that of hunting dogs, with the prey not dying so mercifully fast. Because they do not have retractile claws, hyenas cannot grip or overthrow their victims in the feline manner and have to rely solely on their powerful jaws, tearing out chunks of flesh from the living animal with their razor-sharp teeth. Larger prey, offering more resistance, are the slowest to die.

The spotted hyena spends the day in burrows or rock crevices which are more spacious than their entrances would indicate. The diagram below shows a typical layout, the figures showing the places where the cubs may be deposited. The passages are several yards long.

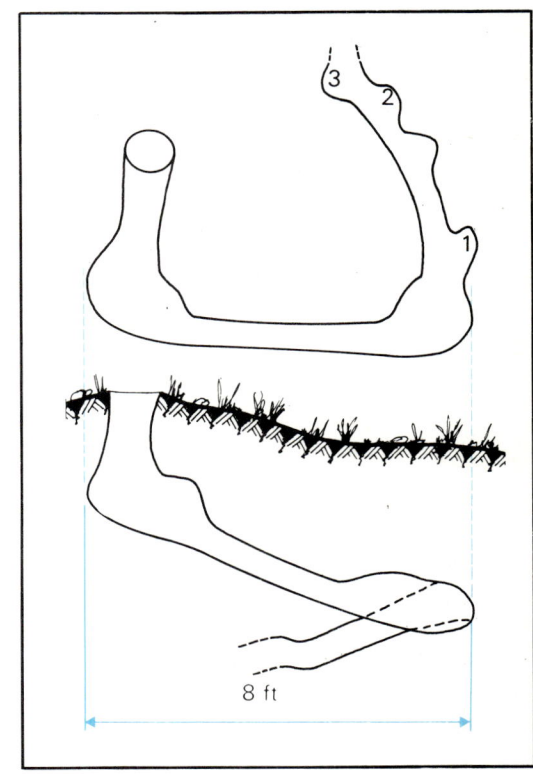

8 ft

Following pages : The massive neck, powerful jaws, coarse coat and sloping hindquarters of the spotted hyena combine to give the animal a highly unpleasing appearance. This has helped to spread the fantastic and sinister legends that have made it one of the most unpopular creatures of the wild.

Jackals feed on small mammals (including rodents and newborn gazelles), fledglings, insects, eggs and fruit. They sometimes resort to eating carrion killed by other carnivores but are themselves predators.

Who was there first?

Let us go back for a moment to that familiar spectacle of a lion devouring its prey, surrounded by a pack of hyenas awaiting their chance to pounce. If, as has now been confirmed, hyenas are capable of killing their own prey, why do they hover about like this for another animal's leavings?

Dr Kruuk's researches suggest in fact that we may well have confused their roles and that the true situation may be the exact reverse of what it seems. On one occasion, for example, he saw a pack of hyenas killing a zebra. Suddenly there was the sound of distant roaring—clearly that of a lion. The hyenas too recognised the noise and silently slipped away into the darkness. A few moments later a lioness sprang out of the bushes, loped over to the body of the zebra and quietly began to feed on it. Time passed and the hyenas slowly regrouped, nosing their way forward once more, no longer silent but barking furiously. One of them, bolder than its fellows, tried to snatch the prey from under the lioness's jaws but she angrily sent it flying with a single blow of her paw. The hyenas accepted this temporary setback but soon returned to the attack, snapping away at the lioness from every side until she was compelled to abandon the zebra's remains. Yet they were still out of luck. No sooner had the lioness beaten a retreat than two male lions with black manes took her place, claiming the spoils and forcing the hyenas to withdraw once more. This time they did not try to fight back but contented themselves with waiting patiently until the lions had gorged themselves to repletion.

To confirm the validity of his findings Dr Kruuk recorded the characteristically strange sounds made by hyenas immediately after killing their prey. One night he set up his tape recorder on the floor of the crater. Predictably, the black-maned lions put in an appearance, clearly puzzled at their failure to find a freshly killed prey. The naturalist was satisfied that in Ngorongoro, at any rate, lions almost never did their own hunting and that it was they, not the hyenas, who were the scavengers.

Commandos of the plains

In the Serengeti National Park, where marking experiments on about 100 animals were carried out—similar to those in Ngorongoro—the hyenas were also observed to live in seemingly female-dominated packs, but there were significant differences between the two groups.

In Ngorongoro, with its sedentary population of lions, hyenas killed the majority of their prey themselves, but in the Serengeti, where the lions are migratory or semi-nomadic and kill on their own account, the hyena population is faced with stiffer competition. The continual coming and going of the herbivores compels the hyena packs to undertake long expeditions—sometimes up to 100 miles—and usually lasting several days, in order to find their food. Invariably, however, they return to their starting point when the hunt is over.

These hyenas, being adaptable creatures like the rest of their kind, rely to a large extent on chance encounters, and operate both as predators and scavengers in about equal measure. During the dry season in the Serengeti, when the plains are practically devoid of wildlife, they cannot be too particular about their next meal and the residue of prey killed by lions unavoidably forms a large proportion of their daily diet. If they have to resort to first-hand hunting they invariably choose weak or newly-born animals that are unable to put up a strong defence. Only as a last resort will they tackle a healthy, adult herbivore. In these parts therefore they often live at the lion's expense whereas in Ngorongoro the positions are reversed.

To sum up, the spotted hyena is far from being the cowardly, evil-minded scavenger of legend and popular science. The animal is in fact a highly efficient and determined predator, with considerable influence on the population levels of the herds of ungulates sharing its habitat, and indeed on the ecological balance of the entire African continent. It is an intelligent animal, remarkably adaptable and, because of this, well equipped to survive, especially when conditions are adverse, as in times of severe drought and shortage of food. It is only because of the devoted work of naturalists studying the animal in its everyday surroundings that such a reassessment and partial exoneration of a misjudged animal has been possible.

The marauding jackal

Among the many animals of the East African savannahs and plains there is one that, in its appearance if not in its behaviour, seems relatively familiar – the jackal. It is typically dog-like, with a well-proportioned body, straight triangular ears, a long narrow muzzle and the kind of expression in its eyes that we are used to seeing in the domestic dog. However far removed some of the Canidae appear to be from what we suppose to be the prototype, there is no doubting the family links of this animal, the tropical equivalent of its more northerly cousin of temperate climes – the wolf. In fact many authorities put forward the claims of both animals to be the ancestor of the modern dog. Others accept the view of Konrad Lorenz that domestic canine species may have some of the characteristics of both wolves and jackals – perhaps suggesting a joint ancestry.

The African jackals are territorial animals, controlling a stretch of ground that may measure a couple of square miles, and marking the boundaries in the customary manner by means of scent posts, in their case the urine of both males and females. This clearly delineated territory is defended against the intrusion of others of their species but – again as in the case of most animals – without resorting to serious life-and-death combat. A show of determined aggressiveness is normally sufficient to establish territorial rights and privileges.

In the Serengeti National Park three distinct species of jackal are found, though this is the only place in Africa where they are all living together – even to the point of gathering round the remains of a zebra or a gnu already half eaten by lions. The shyest, laziest and most cautious of the three is the side-

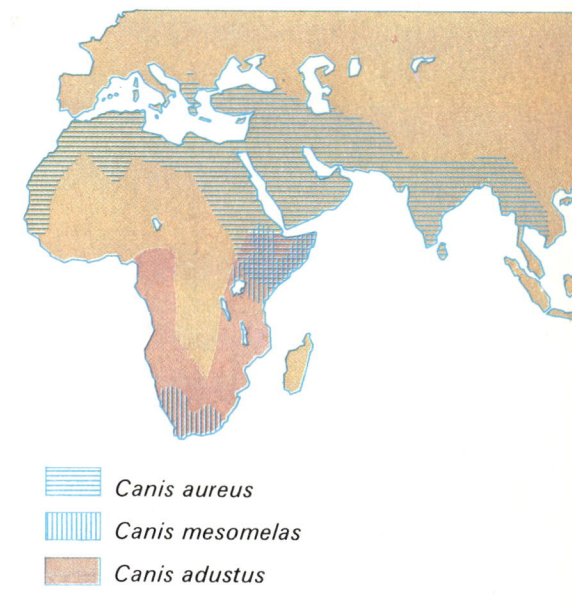

	Canis aureus
	Canis mesomelas
	Canis adustus

Geographical distribution of the common jackal (*Canis aureus*), black-backed jackal (*Canis mesomelas*) and side-striped jackal (*Canis adustus*). The common jackal has a widespread distribution in Africa, Europe and Asia to the Far East.

BLACK-BACKED JACKAL
(Canis mesomelas)

Class: Mammalia
Order: Carnivora
Family: Canidae
Length of head and body: 36–42 inches (90–105 cm)
Length of tail: 13–14 inches (32–35 cm)
Height to shoulder: up to 18 inches (45 cm)
Weight: 22 lb (10 kg)
Diet: meat (small mammals, carrion, birds), eggs, fruit
Gestation: 60–63 days
Number of young: 3–5, sometimes up to 8
Longevity: 13 years

Adults
Long, pointed ears, not widely separated at the base; expressive yellow eyes. The face and much of the body are reddish, the belly and front of the legs are much lighter. The tufted tip of the tail is black. The dark, white-tipped hairs of the back – which give the animal its common name – look silvery from a distance.

Young
The young look much like the adults but their bodies are rounder.

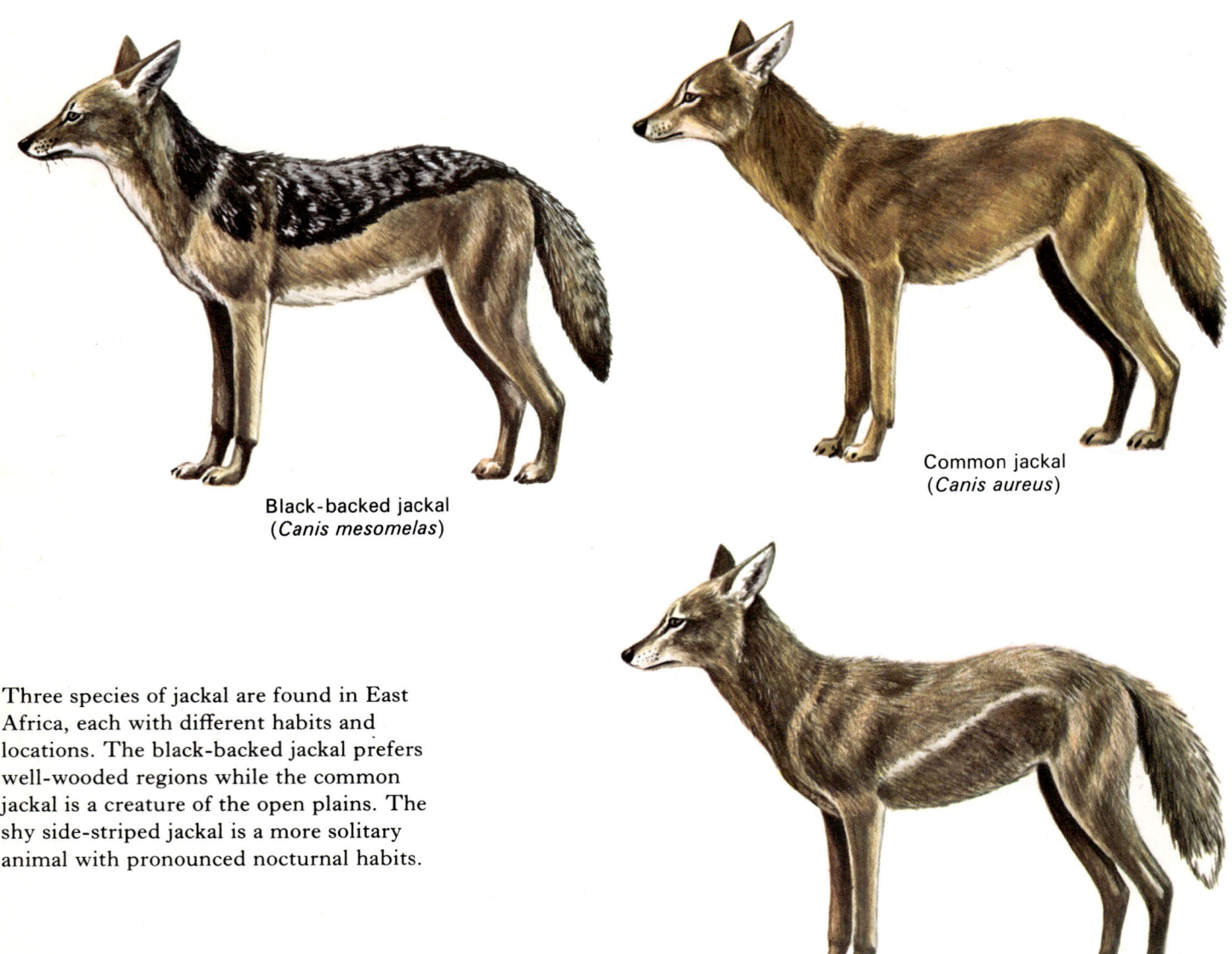

Black-backed jackal
(*Canis mesomelas*)

Common jackal
(*Canis aureus*)

Side-striped jackal
(*Canis adustus*)

Three species of jackal are found in East Africa, each with different habits and locations. The black-backed jackal prefers well-wooded regions while the common jackal is a creature of the open plains. The shy side-striped jackal is a more solitary animal with pronounced nocturnal habits.

Facing page : Hungry jackals attack and disperse the vultures that have led them to the scene of a recent kill.

striped jackal (*Canis adustus*), a solitary creature—like the other species, nocturnal in habit—distinguished by a pale band on the greyish flanks and a tuft of white hair on the tail. It is somewhat larger than the black-backed jackal (*Canis mesomelas*), which, as its name suggests, has a white-flecked black 'saddle', the rest of the body being reddish-brown. The common jackal (*Canis aureus*)—found also in Europe and Asia—has a greyish-red or greyish-yellow coat.

The fact that jackals are often seen feeding on the carcases of herbivores does not signify that they, any more than hyenas, are exclusively carrion eaters. Like the latter, with whom they share an undeserved reputation for cowardice, they are very adaptable creatures, ready to take whatever food presents itself. Although they resort quite frequently to carrion and will eagerly devour the placentas shed by female ungulates after giving birth, they are just as fond of fresh meat and will, if necessary, do their own killing. Both male and female participate in the hunt on such occasions, their favourite prey being a gazelle a few days old. While one of them harasses the mother until she is forced to charge and leave her baby unprotected, the other waits for just such a contingency in order to carry the youngster off.

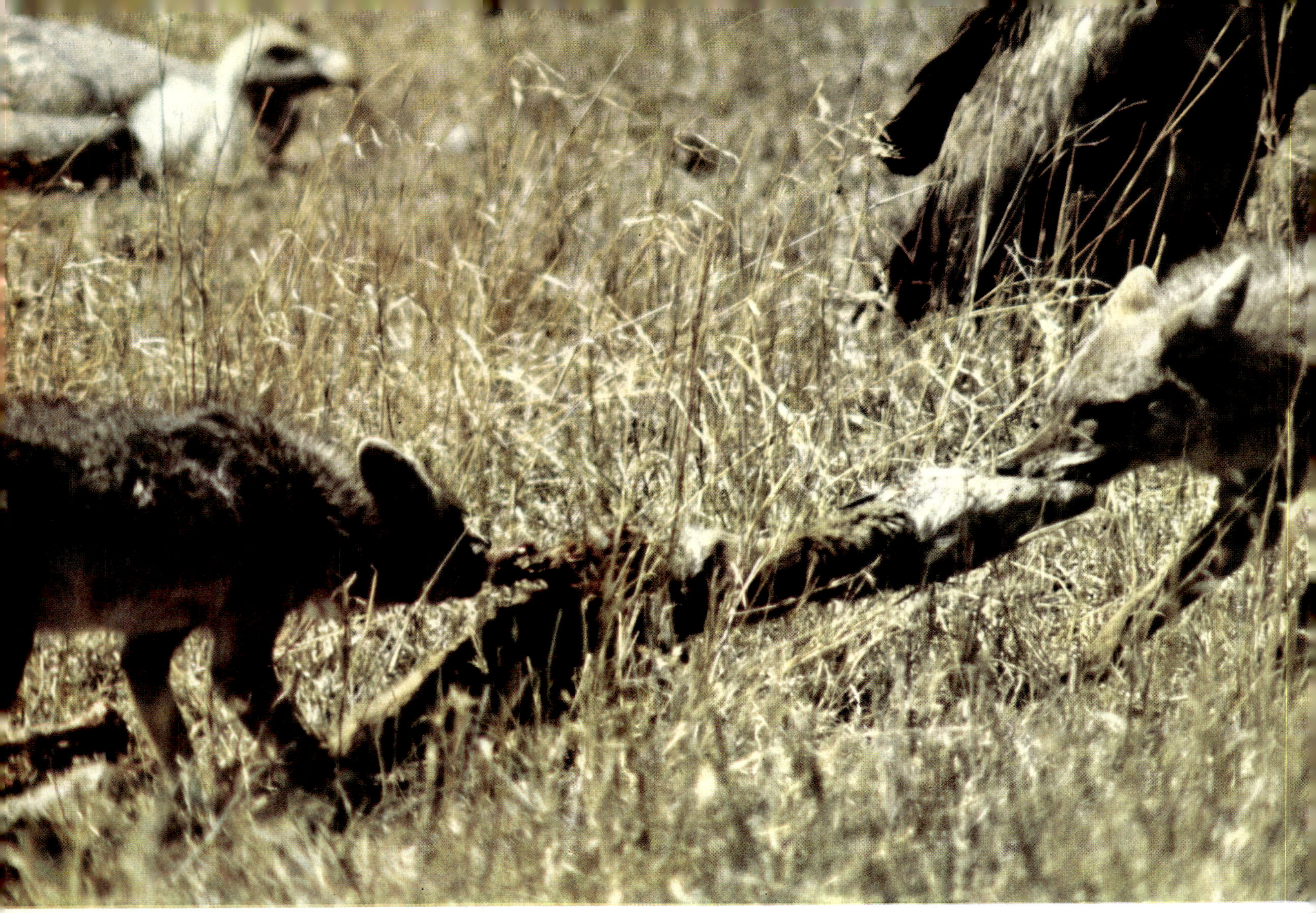
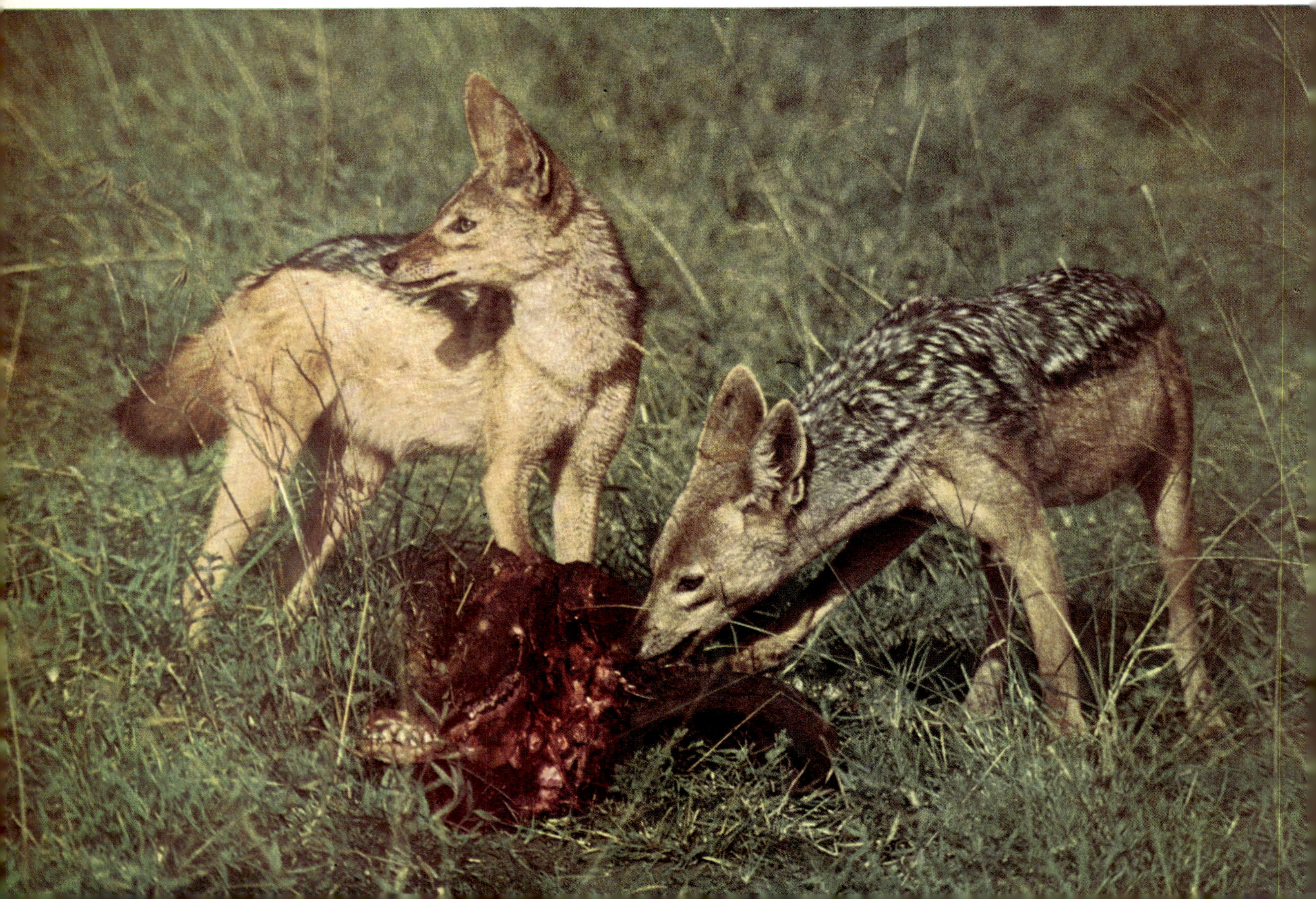

But they will not always confine themselves to defenceless creatures and are often successful in killing adult animals, provided these are of modest size and weight. To supplement their food supply they will catch rodents, fledglings and insects. So jackals have another feature in common with domestic dogs in that they enjoy a widely varied diet.

The more abundant black-backed and common jackals differ in their methods of killing. The former goes outright for the throat of its victim while the latter attacks the flanks and belly, rather in the manner of African hunting dogs. Hunting in pairs brings a greater measure of success than solitary prowling. One survey in the Serengeti revealed that jackals hunting singly killed not more than 16 per cent of the prey they attacked, while two together achieved a success figure of 77 per cent.

Black-backed jackals prefer to inhabit districts with bushes and shrubs, where they can more easily capture small ungulates, whereas common jackals have a liking for the more densely vegetated regions with plenty of insect life. But on the open plain the latter are efficient hunters of gazelles. As for choice of lair, the black-backed species often burrow into anthills, the common jackals preferring to resort to a burrow once, but no longer, occupied by a warthog, hyena or aardvark, which they then proceed to enlarge and tidy up. But whatever their hideout, the females of both species give birth to their litters here, after a gestation period of about nine weeks. The puppies, numbering from two to six, are born blind, opening their eyes when they are nine days old.

Like all members of the Canidae, the mother remains behind to look after her young while the male goes out hunting. When he returns to the lair the whole family scamper out to meet him, tails waving excitedly. He then regurgitates a portion of the food he has ingested during the hunt and the female carefully chews it up again into manageable pieces to divide among the puppies. In due course the latter are ready to leave their place of refuge and to accompany the adults on their expeditions, gradually learning the demanding hunting techniques. At eight months they are more or less self-sufficient but they remain under parental guidance until they are about a year old.

Jackals never hunt in packs—only singly or in pairs. Small family groups of adults and young are common, but larger assemblies of fifteen to twenty animals are comparatively rare, and usually occasioned by the presence of carrion, sighted for them by vultures. Jackals follow the movements of these great birds very attentively and when led to the site of a kill, fight among themselves quite fiercely for the best morsels. One observer watched two jackals quarreling over the remains of a gazelle abandoned by a cheetah. The victor greedily devoured the prey while the defeated rival waited patiently in the wings for any remaining scraps.

Jackals resemble domestic dogs not only in outward appearance but also in their way of barking. But although the sounds are superficially alike the jackal's yelps are louder and more piercing. On the empty plains the strident howls can be curiously chilling—appropriate to the deaths they so often herald.

Facing page : Jackals, although hunters when the occasion demands, will willingly supplement their diet with carrion, especially when the herds of herbivores leave the plains in the dry season.

COMMON JACKAL
(*Canis aureus*)

Class: Mammalia
Order: Carnivora
Family: Canidae
Length of head and body: 32–40 inches (80–100 cm)
Length of tail: 8–12 inches (20–30 cm)
Height to shoulder: 18–20 inches (45–50 cm)
Weight: 22 lb (10 kg)
Diet: meat (small animals, carrion)
Gestation: 60–63 days
Number of young: 2–7
Longevity: 16 years

Adults
The head, lengthened by the pointed muzzle, has fairly long ears; the eyes are yellow. The reddish-grey coat has greyish-yellow markings on the upper part of the body. The tail has a black tip but is not pointed.

Young
The young are born blind and the eyes open at nine days old.

FAMILY: Hyaenidae

In prehistoric times the huge cave hyena of the family Hyaenidae contended with man himself for food and shelter. It belonged to the genus *Crocuta* and was similar to the modern spotted hyena. We do not know what sort of reputation it enjoyed, but its descendants were to be credited with all manner of dire attributes. Popular legend, for example, has accused the hyena of desecrating tombs and graves, of enticing men into the depths of the forest by imitating their voices and there devouring them, of rounding on a pursuer and transfixing him with an unearthly gaze so that he is hypnotised and incapable of further movement. When a hyena dies, so it has been alleged, its eyes promptly turn to stone.

Such fantastic stories have naturally helped to enhance the hyena's sinister reputation, nor has this been improved, as we have seen, by the animal's rather repulsive appearance. There is a visible disproportion between the fore and hind limbs, the coat is coarse, the neck is often covered by longer hairs, the muzzle is short and squat, the expression of the eyes dull—altogether a most unpleasant-looking creature. Added to all this, the animal also has a highly disagreeable smell, due to a solid, fatty yellowish substance secreted by small glands, enclosed in a sac, situated between the base of the tail and the anus.

The unusual structure of the hyena's external genital organs—the female's displaying certain characteristics superficially resembling those of the male—has also contributed to the treasury of legend and folklore surrounding the animal. Even the popular belief that it is a carrion eater pure and simple proves, on closer investigation, to be unfounded. Although hyenas admittedly devour meat killed by other carnivores, they also hunt their own prey and boast highly efficient techniques for the purpose. In fact they are extremely well adapted to play the role both of predator and scavenger, as the occasion and the surroundings demand.

In their guise of carrion eaters they are greatly assisted by powerful jaws and differentiated teeth. The cheek teeth—six or eight premolars and two molars in each jaw—are admirably designed for breaking and crushing the largest and toughest bones that may have resisted all the efforts of other scavengers and predators. The upper carnassials are large, with three sharp cusps especially valuable for ripping up food. Although the teeth structure varies according to genus, the basic formula is:

$$I: \frac{3}{3} \qquad C: \frac{1}{1} \qquad PM: \frac{3-4}{3} \qquad M: \frac{1}{1}$$

The hyena's solid, well-muscled body, supported by long legs, stands it in good stead as a runner, enabling it to chase prey for a considerable distance, a hunting method which, combined with keen eyesight, is especially effective at night. There are four toes on each paw and the animal is a typical digitigrade.

The divided auditory bulla and the structure of the teeth show that the Hyaenidae are closely related to the Viverridae. In fact paleontologists believe they have found in Asia the remains of an ancestor of both families, midway between a hyena and a civet, a creature of the *Ictitherium* type.

Hyenas, animals of Asia and Africa, are divided into two subfamilies. The Hyaeninae include all the typical hyenas, such as the striped hyena (*Hyaena hyaena*), the brown hyena (*Hyaena brunnea*) and the spotted hyena (*Crocuta crocuta*). All have eight premolars in the upper jaw. The subfamily Protelinae is represented by only one species—the aardwolf (*Proteles cristatus*). This is a small hyena-like creature with weak jaws and widely spaced cheek teeth (only six premolars in the upper jaw). Unlike other hyenas it has five toes on the fore paws, four on the hind paws. It is essentially an insect eater, from southern Africa.

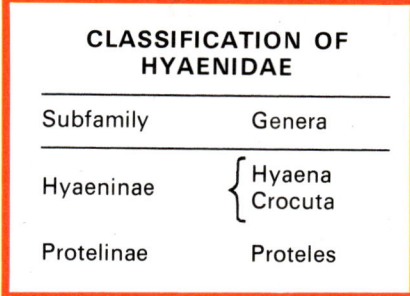

CLASSIFICATION OF HYAENIDAE

Subfamily	Genera
Hyaeninae	Hyaena Crocuta
Protelinae	Proteles

Facing page (above) : The spotted hyena (*Crocuta crocuta*) is a member of the Hyaeninae, one of the two subfamilies of the Hyaenidae. (*Below*) the aardwolf (*Proteles cristatus*) is the sole representative of the other subfamily, the Protelinae.

Following page : Two white-backed vultures perch on a dead tree on the East African savannah. They are one of several species of African vulture, all of which are carrion eaters, though with slightly varying habits.

CHAPTER 11

Vultures, the scavengers of the air

The African savannah is a paradise for the huge carrion-eating vultures. Nowhere in the world do they congregate in such vast numbers as on the high plains used as regular grazing grounds by the herds of wild ungulates. Anyone who has driven along any of the roads that furrow the national parks and reserves of East Africa will surely be familiar with the sight of these great ugly birds perched on a dead tree, never far from the site of a kill.

Vultures habitually feed on dead animals and because of this craving are regarded by most people as despicable, repellent creatures. But the truth is that without their sinister activities the air of the African plains and savannahs would be unbreathable. To a certain extent the hyenas and jackals help to keep these regions hygienic, but it is the vultures that most effectively complete the work of the land scavengers. Once these birds have had their fill, little is left of an animal carcase but bare bones – picked clean of every scrap of flesh – and nothing remains for them but to turn white under the blazing sun.

Yet with the best will in the world – and even bearing in mind the importance of the role they play in preserving the natural balance of the environment – it is difficult not to feel an instinctive disgust for these grotesque, unlovely birds. Most of them are large, with plumage of a fairly nondescript brownish hue, and the long neck, like the head, is thinly covered with down or with rose- and violet-coloured excrescences, known as caruncles.

They are certainly ugly creatures when seen at close quarters, but in flight they are extremely graceful. Their wing movements are almost imperceptible as they soar and glide, carried by the rising columns of air (thermal currents). But the most astonishing feature of these birds of prey is the way in which they manage to locate carrion lying on the plains, which may be hidden a great

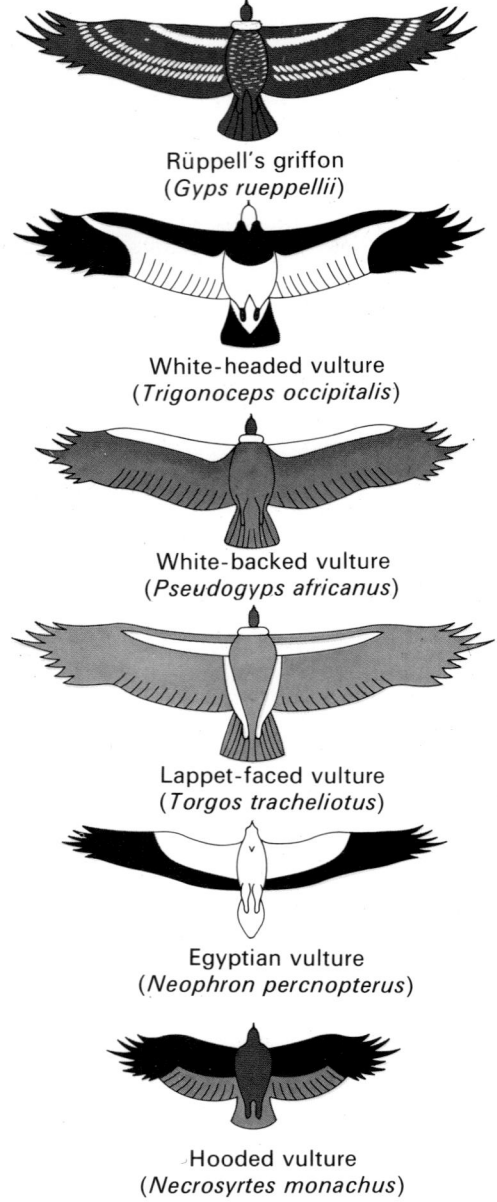

Rüppell's griffon
(*Gyps rueppellii*)

White-headed vulture
(*Trigonoceps occipitalis*)

White-backed vulture
(*Pseudogyps africanus*)

Lappet-faced vulture
(*Torgos tracheliotus*)

Egyptian vulture
(*Neophron percnopterus*)

Hooded vulture
(*Necrosyrtes monachus*)

Silhouettes of African vultures with spread wings.

distance away and often in the most unlikely surroundings.

On a cool, fresh autumn morning some naturalists were taking photographs of a cheetah which had just captured and killed a Thomson's gazelle and was dragging its victim towards an acacia tree. All of a sudden, a vulture almost literally fell out of the sky and alighted on the ground some fifteen yards away. The party's ornithologist easily identified it as a hooded vulture (*Necrosyrtes monachus*), characterised by its modest size, chestnut-coloured plumage, bald, reddish head and strong, hooked beak. The small cap of down on the back of the head – from which the species probably derives its name – was an additional identifying feature. Thirty seconds later a noisy beating of wings caused the naturalists to look up, in time to see another much larger vulture plummeting towards them, wings spread wide, and coming in to land not far from the first one. This was a Rüppell's griffon (*Gyps rueppellii*), most handsome of all African vultures. At the tip of each brownish-black flight feather is a white spot, and a ruff of white down encircles the base of the neck. Other small white feathers are present on the head.

Within minutes there was a great flock of new arrivals – all the birds coming to rest only a few yards from the cheetah. In this noisy, bustling throng the ornithologist was able to pick out African white-backed vultures (*Pseudogyps africanus*) – with white marks on their rumps; Egyptian vultures (*Neophron percnopterus*) – with a flight pattern reminiscent of gulls; a lappet-faced vulture (*Torgos tracheliotus*) – its head naked and its neck covered with fleshy pink caruncles, similar to the wattles of turkeys, situated above and behind each ear; and two white-headed vultures (*Trigonoceps occipitalis*) – especially noticeable for their scarlet beaks and downy collars adorning head and neck.

While the cheetah was still busily ripping open the belly of its victim, more than fifty raptors, belonging to these six different species, had gathered. The hooded vultures tried to snatch a few pieces of meat away from the feline but the latter had no intention of abandoning its trophy so easily and lashed out at the birds, causing them to take wing. This action evidently inspired the respect of the other birds and the cheetah was permitted to finish its meal without further disturbance. Immediately it had wandered off, however, the waiting vultures hurled themselves at the gazelle's carcase and a grisly scene ensued. It began with all the birds pecking away greedily and indiscriminately for a minute or two; then two Rüppell's griffons, with half-opened wings, separated themselves from the seething mass, soared up some distance and swooped down again into the middle of the circle of birds, emitting harsh cries, stretching their necks and inflating their feathered ruffs. This display clearly made a deep impression on the ravenous horde, which gave way at once to their aggressive rivals. The latter helped themselves to the tastiest morsels, then allowed the others to return for their pickings.

It took the entire assembly of vultures about ten minutes of quarrelsome activity to strip the carcase clean. Those that were replete took themselves off, after a series of heavy, rhythmically fluttering leaps. The smaller birds – the Egyptian and hooded vultures – remained behind to collect any overlooked scraps.

Most species of vulture allow themselves to be carried upwards in mid-morning on thermal currents to considerable altitudes. Their far-ranging vision enables them to scan the plains below for carrion and they then plummet down on the site of the kill. Other animals, scavengers and predators alike, follow their flight with interest. Hyenas, lions and other carnivores are thus guided to places where the carcases of dead herbivores are likely to provide them with an unearned feast. Naturalists also take advantage of the vultures' keen eyesight to help locate these sites, where a number of different species will normally collect and where their behaviour may be studied.

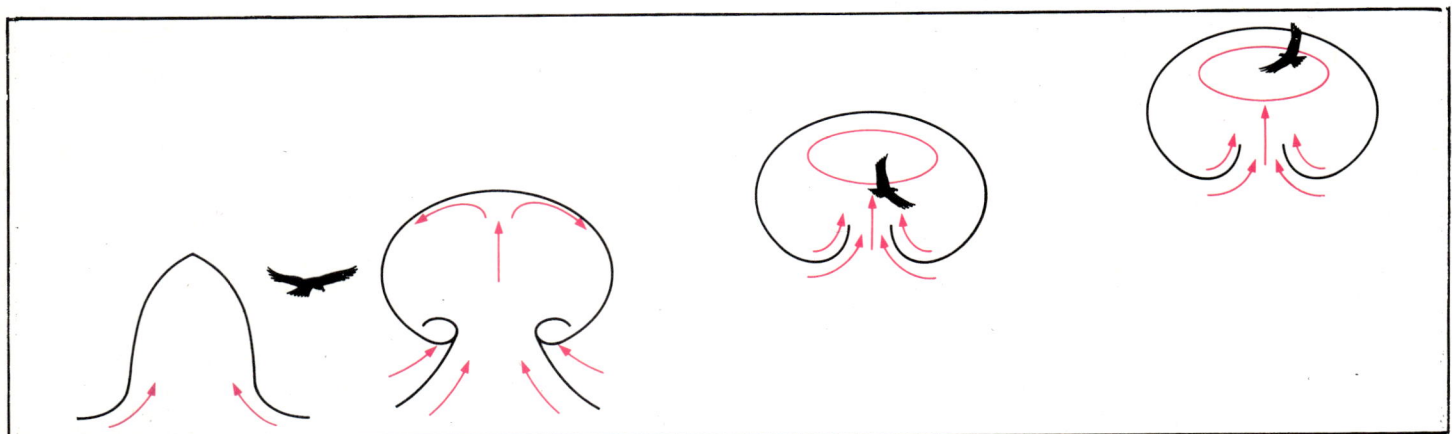

These diagrams show how vultures are borne aloft on warm air currents, enabling them to soar and glide without using their wings to any great extent.

Not all vultures have exactly the same food needs and habits, so they do not really compete with one another very seriously, as the above-mentioned spectacle might seem to imply, for the choicest particles of carrion. In fact they tend to complement one another, each species having its own preference for a different part of the anatomy. Thus the lappet-faced and white-headed vultures go first for the skin and the tough muscle and nerve fibres, which are easily ripped to pieces by their particularly sharp beaks. The Rüppell's griffon and white-backed vulture, on the other hand, prefer the tenderer internal organs, which they extract from the abdominal cavity by literally plunging their heads inside, either through the natural body openings or through holes that they themselves peck with their beaks. The Egyptian and hooded vultures seem to be content with the chunks of flesh torn away but tossed aside by the larger species.

Yet if there is a reasonably fair division of the spoils, there is still a well-observed and literal 'pecking order'. From many observations of both European and African birds of prey it seems that precedence is almost always given to the hungriest raptors. In the instance recorded above it is more than likely that the two Rüppell's vultures that were successful in scattering their companions had not eaten for some time.

All these normal behaviour patterns may of course be modified according to the individual and the circumstances, acute hunger being one of the more influential factors. Like all other animals, vultures display marked signs of aggressiveness when they are undernourished, such symptoms diminishing gradually as they succeed in satisfying their hunger pangs.

Recent studies support the view that the vulture's astonishing faculty for locating prey from a great height and distance is due solely to its remarkable powers of vision. As soon as the sun begins to warm the air of the savannah—about two hours after it has risen—the vultures, which until then have remained perched on the branches of the acacias or on rocky outcrops, take to the sky. They allow themselves to be borne upwards on the thermal currents and soon manage to reach a considerable height (depending on atmospheric conditions)—sometimes as much as 7,000 feet. Around mid-morning the sky may be darkened by thousands of glider-like forms as the birds soar and swoop above the open plains, scrutinising the ground below for the least sign of movement and keeping a close watch on one another at the same time.

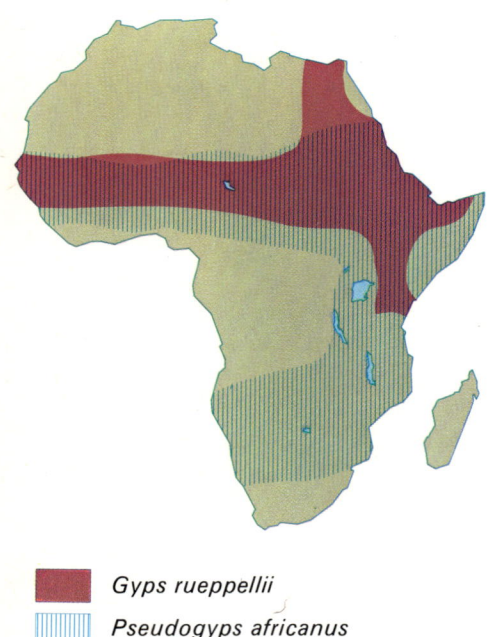

■ *Gyps rueppellii*

▮ *Pseudogyps africanus*

Geographical distribution of Rüppell's griffon and white-backed vulture.

The moment one of their number spots an animal's dead body and dives straight downwards to investigate, the birds in the vicinity immediately follow suit, legs outstretched and wings flapping violently. The message soon passes from one to another and it is not long before they are all crowding round the scene of the kill. When they have finished their meal, only the skeleton is left, cleaned of flesh.

The amazing eyesight of these birds of prey is also of value to other animals, particularly hyenas which, when they decide to go searching for food during the daytime, follow the course of the vultures' flight and hasten towards the spot where the birds alight. Lions too use the vultures as aerial scouts. A splendid male was once seen following a flock of vultures until it was guided to the place where a cheetah was about to feed on its prey. Certainly these great African raptors have no equal when it comes to scouring the plains in broad daylight and no deed of violence can be committed on the savannah without their knowledge and prompt action. Thus naturalists and guides also learn to rely on these birds to lead them to the sites where the feline predators are likely to be found, still feasting on their victim.

Rüppell's griffon

The griffons are huge vultures which are normally found in mountain regions and Rüppell's griffon is among the largest of African birds of prey. It has a wide range of distribution—from the plains and mountains of Eritrea and Sudan southwards to Tanzania and westwards to Guinea.

These birds are gregarious by nature and their colonies may sometimes be made up of as many as a hundred pairs, roosting for the night and nesting at the appropriate season on the ledges of steep rocks or cliffs. Shortly after the rising sun begins to illuminate the plains below, they take to the air. Their flight is leisurely and unhurried and they generally permit other species to guide them to the site of carrion, rarely being the first birds of prey to arrive on the scene.

The breeding season of these raptors varies according to the region which they inhabit, but although mating may take place at any time of the year, it is more usually during the traditionally dry months. At these times the male and female may be seen describing wide circles in the sky around the prominence or peak which they have selected as both home and nuptial territory. The nest will be built among the rocks, sometimes being fairly elaborately constructed but more often simply consisting of a collection of leaves and branches, sealed together with the birds' excrement. The female lays a single egg—exceptionally two—the shell being white to pale green in colour, sometimes flecked with brown.

The precise length of the incubation period is not known, nor has it been established whether one or both birds share the incubation. The chicks, however, remain in the safety of the nest for a minimum period of three months, during which time they are fed on regurgitated carrion. They are then virtually self-sufficient and ready to embark on short flights, though it may be some time before they are permitted by the parents to stray too far.

RÜPPELL'S GRIFFON
(*Gyps rueppellii*)

Size of wing: 24—26 inches (61—65 cm)
Tail: 11—12 inches (27·5—30 cm)
Tarsus: about 4½ inches (11—11·5 cm)
Wingspan: up to 96 inches (240 cm)
Weight: 14—20 lb (6·4—9 kg)
Food: carrion
Number of eggs: 1—2

Adults
Head and neck covered by thin, dirty-white down; white ruff. Feathers of back, either light or dark, look like scales. Remiges and rectrices brown to black. Cere and naked skin of face grey to blue-grey. Eye yellowish-brown, extremities grey.

Young
Brown down on head and neck. Body more uniform in colour and darker than that of adults.

WHITE-BACKED VULTURE
(*Pseudogyps africanus*)

Size of wing: 22—24 inches (55—60 cm)
Tail: 10—11 inches (24—27·5 cm)
Tarsus: 4—4½ inches (9—11·8 cm)
Wingspan: 89—90 inches (222·5—224 cm)
Weight: 11—17 lb (5—7·7 kg)
Food: carrion
Number of eggs: one
Incubation: 45 days

Adults
Smaller than Rüppell's griffon, without scaly appearance on back. White rump. Tail feathers dark, lower wing coverts light. Eyes dark brown. Black skin of head and neck covered with white or yellowish down. Limbs dark.

Young
Darker coloured body than adults; head and wings spotted. Head plentifully covered with down.

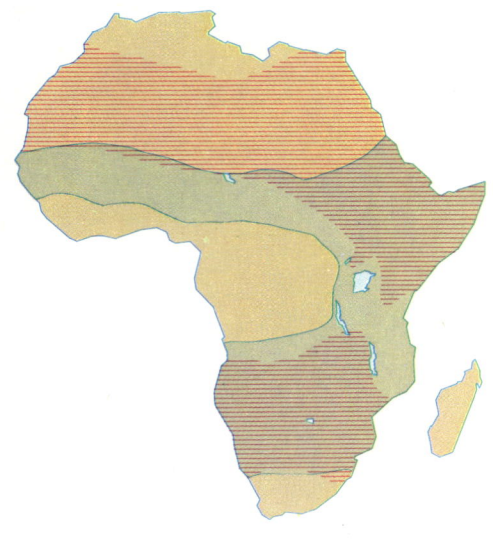

Geographical distribution of lappet-faced vulture and white-headed vulture.

LAPPET-FACED VULTURE
(*Torgos tracheliotus*)

Size of wing: 28½–32 inches (71·5–79·5 cm)
Tail: 14–15 inches (34–38 cm)
Tarsus: 5–6 inches (12·4–14·6 cm)
Wingspan: 103–104 inches (258–262 cm)
Weight: 29½ lb (13·5 kg)

Adults
Head and neck bright red. Ruff of short brown feathers at base of neck. Back and tail coverts dark, remiges and rectrices brown to black. Long feathers of stomach have chestnut-brown centre and lighter edges. Upper part of legs covered with white down. Males smaller than females.

Young
Head and neck dull, thighs brown.

WHITE-HEADED VULTURE
(*Trigonoceps occipitalis*)

Size of wing: 24½–25½ inches (61–64 cm)
Tail: 10½–12 inches (26·5–29·5 cm)
Tarsus: 4–4½ inches (9·5–11 cm)
Wingspan: 81–88 inches (202–220 cm)
Weight: about 10½ lb (4·8 kg)
Incubation: 43 days

Adults
White tuft on top of head. Naked skin of face and neck ranges from whitish to salmon-red. Back dark brown, but some remiges are lighter coloured. Tail black. Lower neck and lower body and legs white. A dark band crosses the chest.

Young
The whole body is dark, with the exception of the upper part of the head and of the lower neck, which are white.

The white-backed vulture

The white-backed vulture is slightly smaller than Rüppell's griffon, an inhabitant of the African plains and savannahs and even found occasionally in desert regions, although never far from a convenient watercourse.

These birds may live either singly or in small colonies, in the former case not wandering too far from the places where others of the species are gathered. They normally roost in trees bordering rivers but may sometimes come to rest on a large isolated tree in the middle of the empty plain. At daybreak they leave their refuge and soar upwards to make a thorough reconnaissance of the surrounding terrain. Although the initial take-off stage seems to require a considerable effort their flight becomes easy and unencumbered as altitude is gained. Like Rüppell's griffons they too allow smaller raptors to guide them to the site of a kill. As dusk falls they wing their way back to the trees where they intend to roost for the night.

During the mating season these vultures fly in great sweeping circles, the beat of their wings more leisurely than usual, male and female so close together that their wingtips are almost touching. The nest is built in the higher branches of a tree and as many as six nests may sometimes be seen in a single tree. Along a 130-mile stretch of the Uebi Shebeli river—a typically well-watered and wooded region—observers counted 250 vultures' nests.

One egg only is laid, generally during the dry season, and while the chick is reared by its mother, the male is responsible for the food supply, returning from his forays to alight by the side of the nest and regurgitate whatever he has found. The female bird will later take care of all the fledgling's needs, not straying too far from the nest.

The lappet-faced vulture

Although this vulture has been seen in the Pyrenees and has ventured as far east as the Dead Sea to breed, its usual habitats are the bare plains and deserts of the African continent. Although it is by nature a solitary creature, usually found alone, it is not uncommon in some areas to see it congregating in large numbers, often more than a hundred perching on the same tree.

The lappet-faced vulture is one of the world's largest birds of prey and it uses its remarkable strength and massive beak to great advantage when it comes to dividing up the spoils. Unlike the vultures already described, it frequently captures and kills small animals by itself and may also supplement its diet in other ways. It will, for example, attack flamingos—whose eggs it finds particularly succulent—and will also make a meal of grasshoppers or ants. Moreover, it will not hesitate to rob smaller raptors of their prey. On one occasion observers in the Serengeti saw the bird steal a half-devoured gazelle from a Verreaux's eagle.

The nest of these vultures is a large flat construction, carpeted with pieces of skin, hairs and grass, and sealed with excrement. It is generally built in the lower branches of a thorny tree, the sharp spines providing a certain protection against possible

predators. In those areas where suitable trees are few and far between, the nest may be constructed on rocky ledges that are equally inaccessible to other animals or birds.

The female lays a single large egg, which is a whitish colour, flecked with brown. Both birds take turns to incubate the egg, the one that is brooding literally flattening its body against the ground so that it is effectively camouflaged.

Should the chick be endangered in any way, it will remain quite motionless, open its beak wide and feign dead. At four months old it is sufficiently self-reliant to leave the nest but will prudently refrain from wandering far away.

The white-headed vulture

This widely distributed species is very rarely seen grouped with other vultures around carrion but is frequently found near watercourses, being obliged to drink at regular intervals. Even here, however, the birds usually gather singly or in pairs for, like the lappet-faced vultures, they have solitary habits.

In many respects this bird fails to conform to the typical image of the vulture swooping vertically down from the sky towards the carcase of a zebra or a giraffe. For although it will not turn down the chance of feasting on one of the larger herbivores, it has a

The six species of African vulture shown here complement one another's food habits. While the Rüppell's griffon (1) and white-backed vulture (3) plunge their heads into the abdominal cavity of their victim to devour the internal organs, the lappet-faced vulture (4) and white-headed vulture (2) use their great strength to rip off the skin and eat the muscles and tendons. The Egyptian vulture (5) and hooded vulture (6), being smaller, have to make do with scraps of flesh torn off and tossed aside by the others.

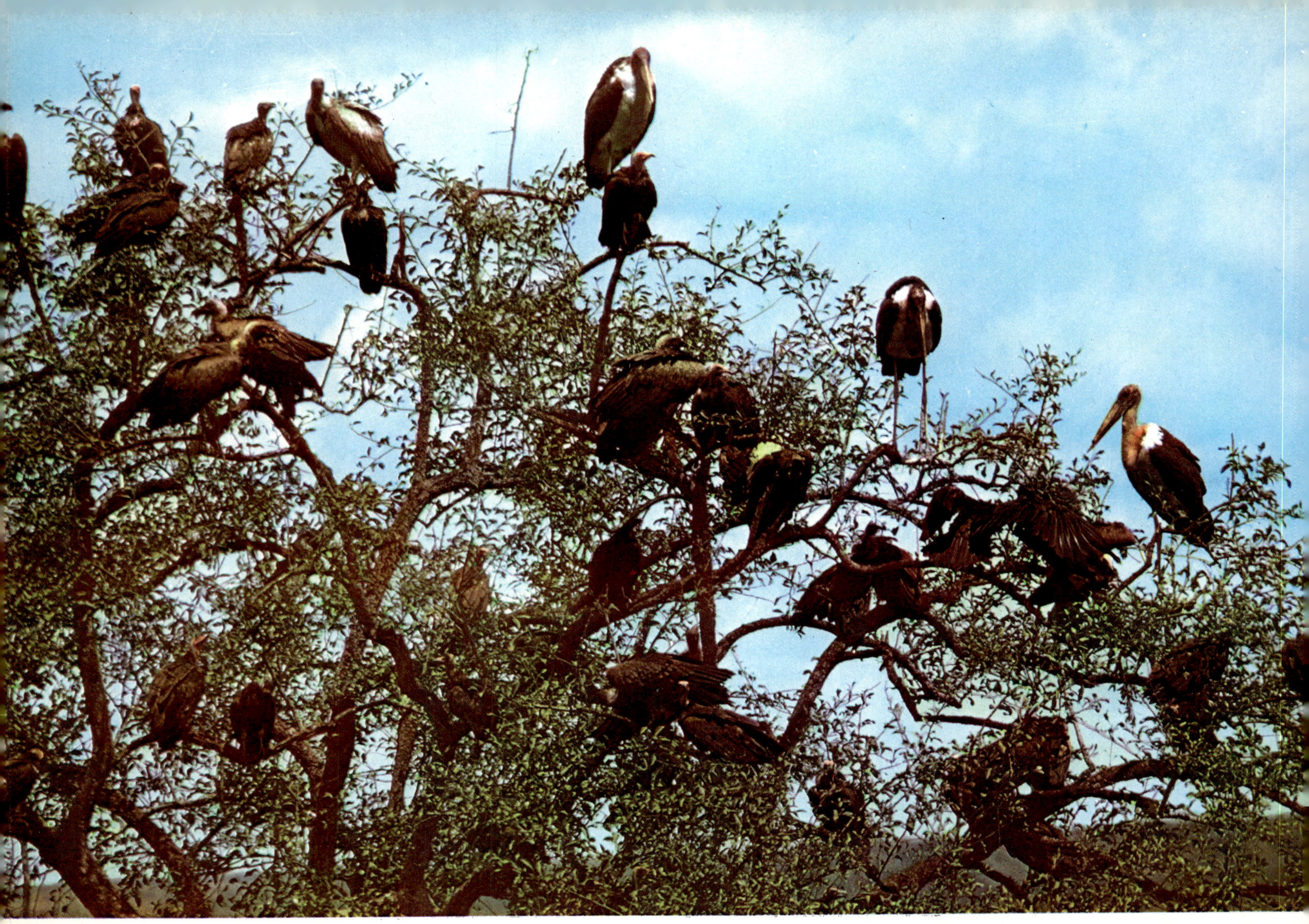

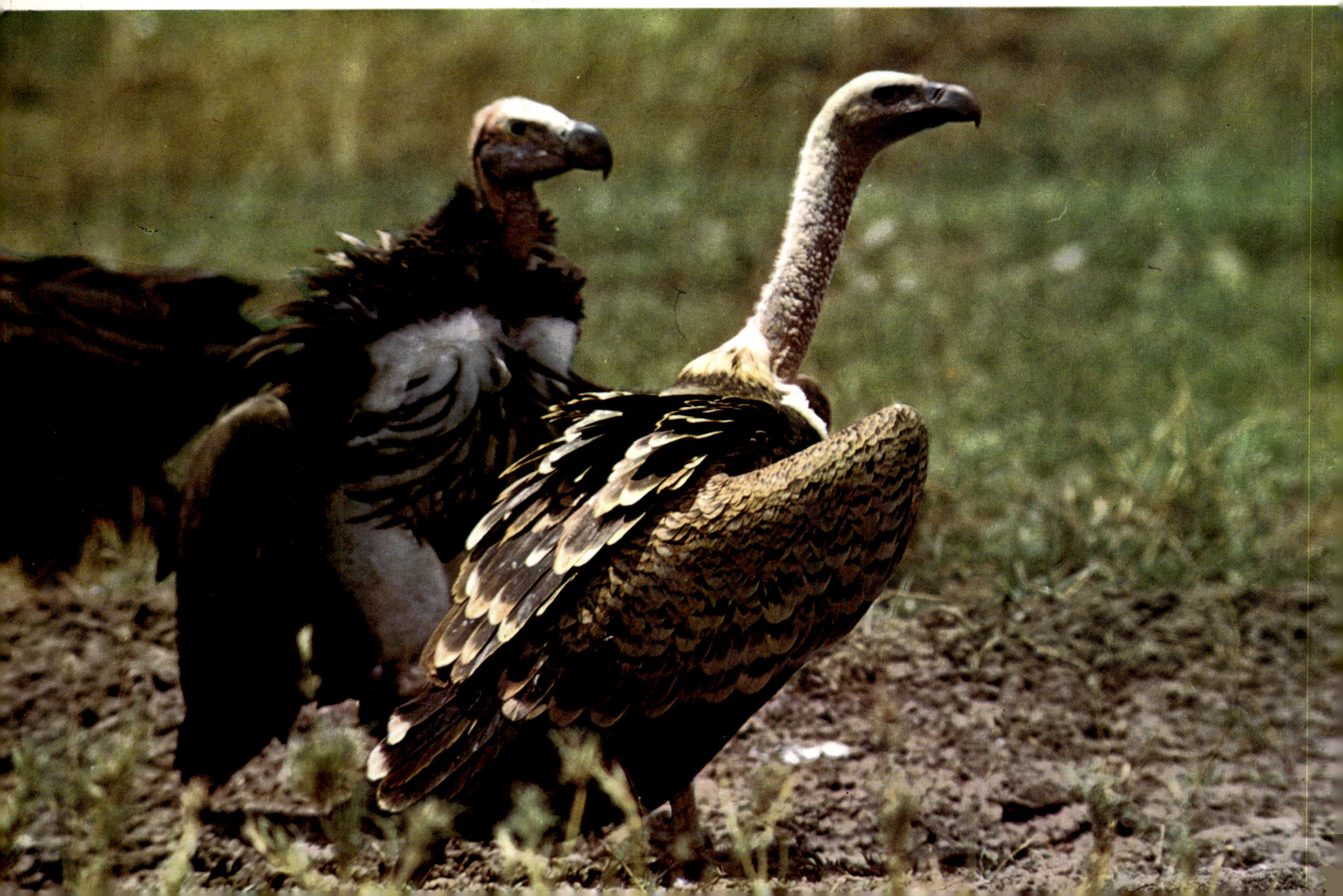

decided preference for animals of more modest size. What is more, it is an expert hunter, capable of killing a young antelope, a flamingo or a guinea fowl, and extending the range of its diet to include lizards and flying ants. Thus it leads a double life of both scavenger and predator.

The nest of the white-headed vulture is a large construction of branches, the base of which is lined with hairs and grass. It may be as much as a foot and a half in diameter and situated either in an acacia or a baobab. The female lays one egg and the chick is hatched after an incubation period of six weeks.

The hooded vulture

The well-known hooded vulture, a little larger than a crow, is the only African species that ventures into wet forest regions. Its familiarity stems from the fact that, like the Egyptian vulture, it makes frequent contact with humans, often flocking in populated areas, especially in the neighbourhood of village markets, slaughterhouses and rubbish dumps. African villagers recognise that this bird's function is that of a scavenger and cleaner, and for this reason they do not hunt or otherwise harm it. In fact it may often be seen hopping in and out of the rows of huts, undisturbed by the comings and goings of the local inhabitants.

This bird is rarely seen on the savannah and never in colonies. Because of its small size it takes to the sky very early in the morning, not relying for uplift on thermal currents. Thus it is often the first to alight on carrion, though this is not a marked advantage because its beak is not powerful enough to tear open the skin of a dead herbivore. Only when the larger vultures arrive, ripping flesh and muscles with their huge hooked beaks, does the small hooded vulture stand a chance of picking up a few morsels. These are not always sufficient to satisfy the bird's appetite and the diet may be supplemented by grasshoppers, ants and other insects; and in village regions, where scraps of meat may be collected without fear of competition, the birds will often alight on cultivated fields, hopeful of feasting on worms and other creatures that may be thrown up in the ploughed furrows.

The hooded vulture is commonly found all over Africa, with the exception of the Congo and the extreme north and south of the continent. During the mating season, although the courtship display is not very remarkable, the male sometimes makes short swooping descents on his mate, while she turns clumsily in small circles, with claws outstretched.

The nest is built in a tree, generally a baobab, but the vulture may also take advantage of an old nest constructed by another bird, recovering it with leaves, hair and other miscellaneous materials. The female lays one egg, which is incubated in turn by both parents, although the male will be employed for the greater part of the time in finding food for himself and his mate.

At three months the chick is able to hop about on the branches near the nest and two weeks later will fly for the first time. A month after that it is fully independent, but even when it has quitted the family circle the parents will remain together and return to the same nest each year to breed.

The raptors, or birds of prey, are generally furnished with strong, curved, pointed claws, which they use to capture, kill and carry off their victims. A typical example (*above*) is the eagle. But the claws of those vultures that feed on carrion (*below*) are blunter and unsuitable for gripping objects. Nevertheless they help the birds to move about more easily on the ground.

Facing page (above): These vultures have flocked with marabou storks on the branches of a tree not far from a slaughterhouse, waiting to feed on the unwanted remains of the dead beasts. (*Below*) the larger vultures, such as these lappet-faced and Rüppell's vultures, are seldom seen in colonies and prefer to feed on carcases found on the savannahs and plains.

Following pages: Completing the work begun by hyenas, jackals and other scavengers, vultures strip animal carcasses bare and thus perform a useful sanitary function.

These photographs by Hugo van Lawick show the ingenious method used by Egyptian vultures to break open ostrich eggs. The birds search for stones or pieces of rock of suitable size, carry them in their beaks to the place where the egg is lying and then drop them repeatedly until the shell cracks. This type of behaviour is extremely rare in the animal kingdom, ranking them with Darwin finches, sea otters and indeed man himself as creatures making use of artificial objects in order to carry out a particular task. In the case of these vultures it is not possible to say with certainty whether this represents an example of acquired behaviour – as taught by and copied from parents – or whether it is purely instinctive and no more 'intelligent' than hunting for food or building a nest.

The Egyptian vulture

This small vulture, particularly when young, bears a strong resemblance to the hooded vulture, being about the same size, with a similar slender, pointed beak and possessing a number of other comparable anatomical features. When it hops about on the ground, the Egyptian vulture, with its yellow face, whitish plumage and uncertain gait, reminds one of a huge and somewhat clumsy hen; but when it opens its black-bordered wings and spreads its tail for flight, soaring upwards in great gliding circles on the warm air currents, it has all the elegant and effortless grace of a typical water bird.

This species, like the hooded vulture, is relatively tame and sociable—also frequently found in villages and swooping on refuse dumps for scraps of food. It too offers what amounts to a sanitation service in return for its immunity. Thanks to the birds' enormous appetites, many slaughterhouses built on the fringes of savannah country prove to be more than ordinarily spick and span. The small vultures perch patiently on the acacia branches and then hurl themselves greedily on the unwanted portions of newly-slaughtered animals. When they have finished their meal and taken their departure, the place looks spotless.

The Egyptian vulture also makes a meal of many species of small animals as well as snails and assorted insects. It has a special liking for the eggs of flamingos, scooping them up in its beak and dropping them on the ground to break them.

The fondness of Egyptian vultures for birds' eggs has led to investigations which have revealed a level of ingenuity found in very few animal species. For these birds show a faculty for making use of an alien object in order to achieve a desired purpose. Baron Hugo van Lawick has demonstrated that in certain regions of East Africa—including Ngorongoro Crater and the Serengeti National Park—the vultures actually pick up stones in their beaks and drop them on ostrich eggs until the shells are cracked. This is a most interesting phenomenon, though it cannot be positively said whether this behaviour is 'acquired' as a result of learning from parental example and imitating the action, or whether it is inborn—as instinctive as the nest-building process.

Egyptian vultures flock together for the night, roosting and nesting on cliffs, in ruins or in occupied buildings. But at breeding time they tend to pair off—well away from their companions—both sexes helping to construct the nest. This will probably be used in consecutive seasons by the same pair of birds. At the end of the spring the female lays from one to three eggs—normally two—which are incubated alternately by both parents and which hatch six weeks later. At three months the chicks are fully developed and ready to leave the nest.

The Egyptian vulture has a wide distribution range that stretches from Africa across Asia as far as India. The most northerly communities are migratory, flying up to North Africa when the winter ends and then by way of the Straits of Gibraltar into southern Europe around the end of February or beginning of March. After spending the summer in the Mediterranean countries they return to Africa during September and October.

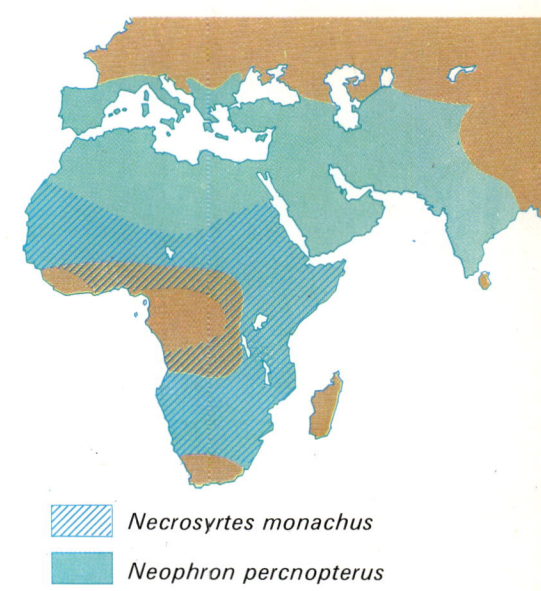

Necrosyrtes monachus

Neophron percnopterus

Geographical distribution of hooded vulture and Egyptian vulture.

HOODED VULTURE
(*Necrosyrtes monachus*)

Size of wing: 18–20 inches (45·5–49 cm)
Tail: 9–10 inches (22–25 cm)
Tarsus: 3½–4 inches (8·2–9·4 cm)
Wingspan: 63–68 inches (157–169 cm)
Weight: 4½ lb (2 kg)
Food: carrion and insects
Number of eggs: one
Incubation: 16 days

Adults
The reddish head and neck are covered with fine grey down. Rest of body chocolate-brown with white down on chest and thighs. Iris brown, blue-grey or grey.

Young
Similar to adults, but darker down on chest and thighs.

EGYPTIAN VULTURE
(*Neophron percnopterus*)

Size of wing: 18½–21 inches (46–53 cm)
Tail: 9–10½ inches (22–26·6 cm)
Tarsus: 3–3½ inches (7·5–8·8 cm)
Wingspan: 64–66 inches (160–165 cm)
Weight: 3½–4¾ lb (1·5–2·2 kg)
Food: omnivorous (carrion, eggs, insects, snails, small crustaceans)
Number of eggs: 1–2, sometimes 3
Incubation: 42 days

Adults
Plumage creamy-white with a few black feathers. Eyes red, facial skin and extremities yellow or orange. Beak dark brown.

Young
Dark-coloured body. Tail feathers greyish, light yellow at tips. Iris chestnut-brown. Naked skin of face blackish.

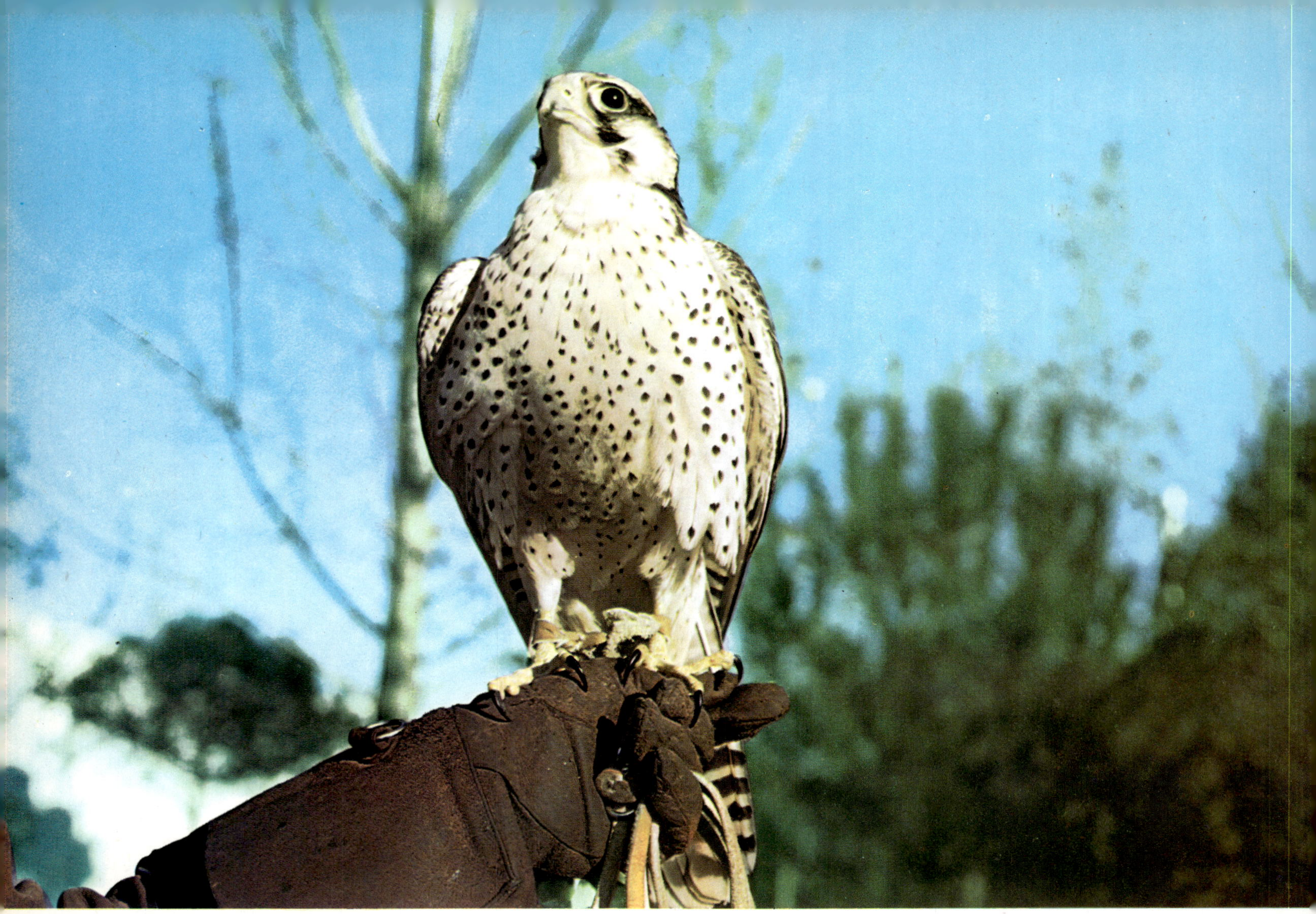

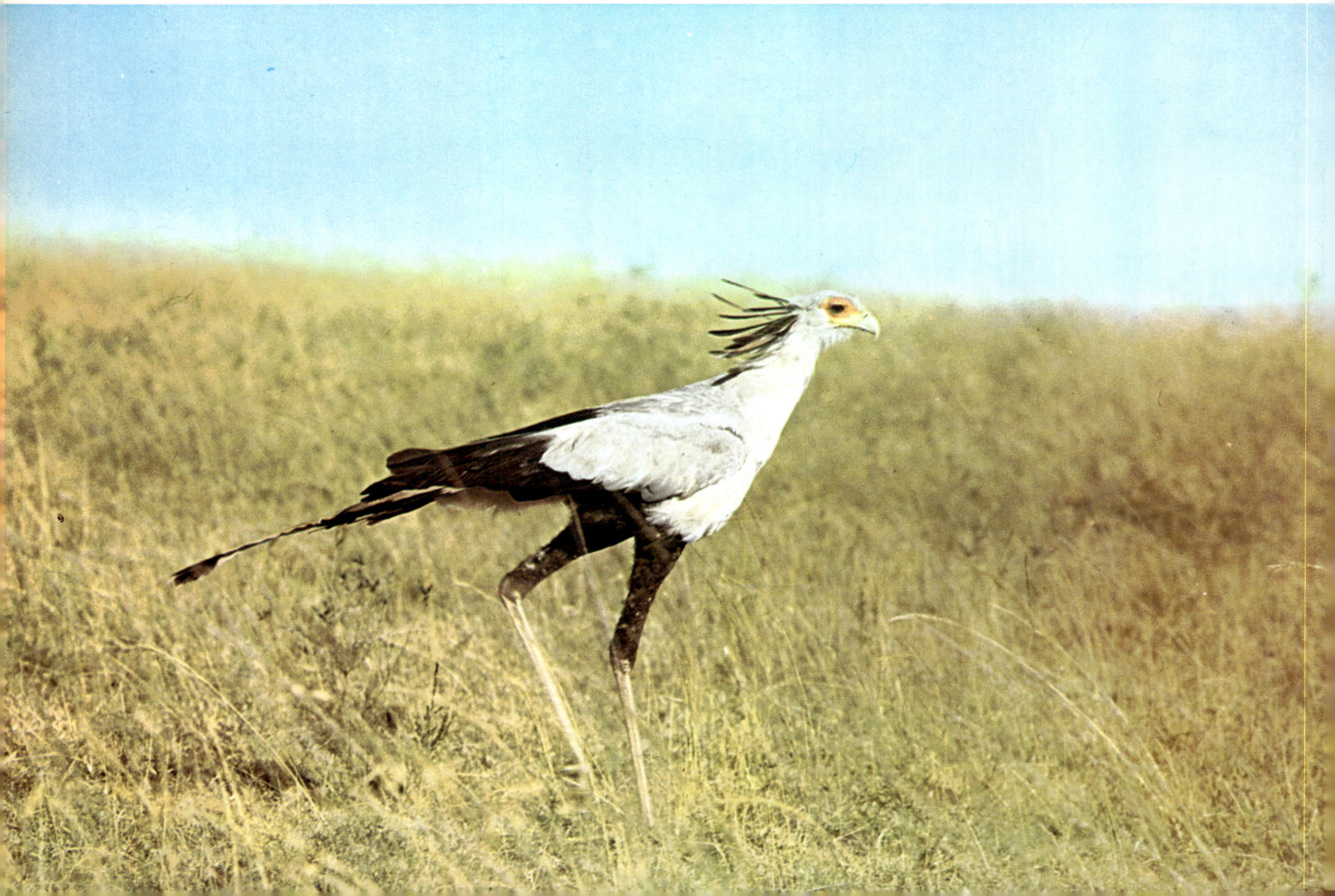

ORDER: Falconiformes

The order Falconiformes derives its name from the diurnal birds of prey of the genus *Falco* – the true falcons – but it also comprises other genera. The latter, however, with only a few exceptions, have not succeeded in attaining the true falcons' very high level of performance either in flying or in capturing and killing prey.

The Falconiformes are nevertheless all raptors, relying almost exclusively on a meat diet – some of them eating carrion, others catching their own prey usually in the form of small animals, and in such cases concentrating their attacks on the more vulnerable species. All are physically strong and resilient – the predators well equipped for hunting, the scavengers capable of enduring any privations imposed upon them by their restricted fare.

In bone and body structure the Falconiformes present a homogeneous pattern, the body being thickset and powerful, the limbs sturdy, the head fairly large and more or less rounded. The beak is strong and hooked, with a thick wax-like growth at the base (the cere) through which the nostrils open. The cere may be variously coloured but usually matches the feet.

The beak has a sharp cutting edge suitable for tearing flesh but its shape differs from one species to another. Thus the true hunters, such as the goshawks, sparrowhawks and falcons, have tougher, thicker beaks than those of vultures which, for the most part, are scavengers. Falcons' beaks, used for breaking the necks of their prey, have tremendous punching force, the upper section having two lateral projections corresponding with two indentations in the lower part.

Other raptors kill their victims by strangulation and do not begin to devour them until all bodily movement has ceased. For this purpose they make use of their feet, ideally formed for seizing prey, with four very large, strong toes, each of which is armed with relatively long, pointed claws.

Among all the birds that make up the order only the vultures, which feed entirely or largely on carrion, have blunted claws that generally lack sufficient strength and flexibility for gripping prey.

All birds that depend on their ability to fly in order to survive – the vast majority of species – must keep their plumage in good condition. The birds of prey most adept as hunters invariably possess the glossiest feathers. Compare the drab-coloured carrion-eating vultures, for example, with the handsome peregrine falcons which capture their prey on the wing.

The contour feathers, generally sparsely distributed, have downy tufts at their base. The remiges (flight or wing feathers) and rectrices (steering or tail feathers) are relatively large; there are usually 10 primary remiges, 13–16 secondary remiges and 12–14 rectrices. The leg feathers come down over the thighs like a pair of tights or breeches and in some species extend farther to cover the shanks or tarsi.

Because they are mainly predators, raptors are concerned not to be too conspicuous and for that reason their feathers are generally not very gaudy or spectacular. The colours most frequently found are harmonious blends of grey, yellow, brown and sometimes blue. The black pigment, melanin, is highly resistant both to variations of light and to wear, so that its distribution on the plumage of birds of prey provides some hint of the bird's strength. The primary remiges, which are subjected to continual use, are very often tipped with dark colours. Furthermore, the alternation of dark colours and lighter areas helps to lend the wings elasticity as well as camouflaging the bird among rocks and thickets.

The process of moulting is critical for all birds of prey, which must be capable of flying at all times. In them, the replacement of the feathers (which generally occurs once a year) is a gradual procedure, with only a few wing and tail feathers being shed at the same time. Since both parts are always

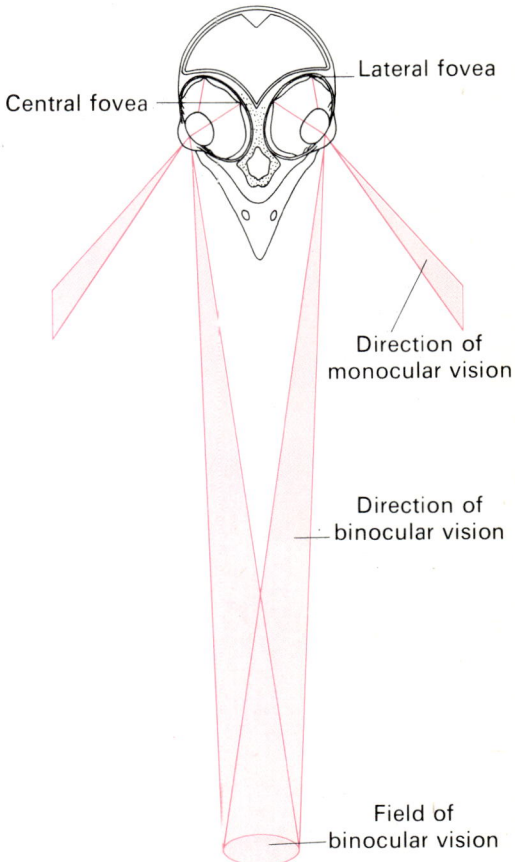

The diurnal raptors possess very remarkable vision, thanks largely to their special eye structure. Each eye has two foveas – points of extreme sharpness of vision – whereas most birds have only one. Thus although the eyes are at the sides of the head, as in other birds, they have three distinct fields of vision – two lateral zones of monocular vision and one central zone of binocular vision, enabling them to distinguish contours and gauge distances.

Facing page: Two representatives of the order of Falconiformes. (*Above*) a gyrfalcon (*Falco rusticolus*), member of the family Falconidae; (*below*) a secretary bird (*Sagittarius serpentarius*), only member of the Sagittariidae.

King vulture
(*Sarcoramphus papa*)

Gyrfalcon
(*Falco rusticolus*)

Golden eagle
(*Aquila chrysaetos*)

Four typical representatives of the families making up the order of Falconiformes. (*Above*) the king vulture (Cathartidae), the gyrfalcon (Falconidae) and the golden eagle (Accipitridae). The secretary bird (*facing page*) represents the Sagittariidae.

furnished with feathers, flight capacity never becomes impeded.

All the Falconiformes possess exceptionally acute vision (the falcon, for example, can detect the movement of a bustard at a distance of some five miles), the size of the eyes and of the optic lobes of the brain being large in relation to the cranial cavity and the body as a whole. The iris of falcons is dark and unremarkable but that of many other species is very beautiful, ranging in colour from bright yellow or amber to green, brown or scarlet. What is even more significant is the fact that, unlike most other birds, these raptors have simultaneous monocular and binocular vision. Each eye independently covers a lateral field of vision of about 130° so that a very wide area can be surveyed; and where the two lateral zones overlap, contours and distances can be gauged with great accuracy. To reinforce this wide-ranging capacity the retina of each eye has two small pits or foveas (points of maximum visual sharpness), providing vision keener than that of any mammal.

Compared with the sense of sight, neither those of hearing nor smell are especially highly developed. The auditory ducts have openings at the sides of the head, usually bordered by small feathers. More effective than the sense of smell are the delicate hair-like feathers situated on the fringes of the cere and the eyes that act as organs of touch.

The cries of these birds of prey are seldom heard, although they all emit distinct signals to indicate alarm, anxiety, hostility and even hunger. Although they are inclined to be noisy during the mating season, and sometimes at night, it would seem that gesture and aerobatics play the most important role in their nuptial displays and assertions of territorial rights.

The Falconiformes are admirably built for flying, with hollow bones and subcutaneous air-sacs to lighten the body weight as well as aiding respiration. Their liquid intake is minimal and a small quantity of food, easily digested and assimilated, builds up great reserves of energy.

The pattern of flight and method of hunting seem to condition the shape of the wings. The woodland raptors, which may have to change direction suddenly, have fairly short, broad, rounded wings with supple remiges; the falcons, hunters of the open plains, have narrower, more pointed wings; and the vultures and eagles, the most powerful fliers, support themselves on large, elongated wings with remiges that are separated at the tips. The largest and longest tail feathers are those of the goshawks and sparrowhawks, which track their prey through the undergrowth.

The Falconiformes are found all over the world, except for Antarctica. Some migrate for the winter, including the osprey or fish hawk, whose fishing grounds—especially in eastern and northern Europe—become iced up. Normally shunning one another's company, raptors at migration time tend to flock together.

Almost all species breed during early spring. Sometimes both sexes build a nest but certain species of falcon lay eggs straight onto a rock ledge or the ground. Mating is usually preceded by nuptial displays by the male, any intruders being expelled with angry, piercing cries.

The eggs are generally round with strong, pitted shells, the colours ranging through every shade of white, grey, yellow, brown and green, sometimes with darker spots.

The larger species normally lay either one or two eggs, the smaller birds up to four or five. At time of hatching the chicks are covered with a thick layer of white or greyish down. The feathers proper develop later and the parent birds are most assiduous in caring for their broods. The carrion eaters provide their chicks with predigested and regurgitated pulp, while the predators feed their fledglings on small pieces of meat, the male bringing the food back to the nest and the female distributing it fairly.

This division of labour may account for the difference in size between the sexes. The females, responsible for tearing up the meat and for defending the nest, are generally a third heavier than the males, whose lighter and more agile bodies stand them in good stead for chasing and catching prey.

Even when they are old enough to leave the nest the fledglings are fed and protected for some time by the adults, but once capable of flying and finding their own food they are left to lead independent lives.

The food of the various raptors may consist of carrion, of vertebrates and invertebrates, and sometimes even excrement and vegetable matter. The digestive process, beginning in the crop, is efficient. Non-edible substances, such as bones, hair, feathers and tendons, are regurgitated in the form of compact little pellets when digestion is complete.

The Falconiformes have often been regarded as harmful birds, whose numbers should be restricted by indiscriminate hunting and other means. Recent surveys, however, have shown that, on the contrary, such birds of prey have a beneficial effect, keeping the species on which they feed stable and healthy by disposing of weak and sickly individuals, and rivalling the Corvidae (crows, magpies and jays) in destroying the genuinely harmful rodents. In many countries, therefore, these raptors are protected.

According to the most recent classification, the Falconiformes comprise five families. The Cathartidae are the vultures of the New World, including, alongside the condors, the king vulture (*Sarcoramphus papa*). The Accipitridae are made up of the Old World vultures and many predatory birds such as eagles, hawks, kites and buzzards. The Falconidae are represented by the true falcons, some of which have been trained by man for hunting since ancient times. The Pandionidae have for their sole family representative the osprey or fish hawk (*Pandion haliaetus*), a long-tailed bird found on the banks of lakes and rivers in many parts of the world, which lives on fish. Unlike other raptors, which have toes of uneven length, the osprey has toes of equal length, the outer one being reversible. The Sagittariidae also have one member—the curious secretary bird, which feeds on snakes and other reptiles, small mammals, fledglings and eggs.

Secretary bird
(*Sagittarius serpentarius*)

FAMILY: Accipitridae

Many ornithologists consider that the Accipitridae should be regarded as a subfamily of the Falconidae rather than a separate family, pointing out that the birds comprising this group do not possess characteristic differences that would justify their being classified on their own. This view, however, is not universally accepted and other authorities are of the opinion that there are sufficiently important variations in anatomy and behaviour between the Accipitridae and the Falconidae for both to be regarded as distinct families.

Having decided to accept the latter viewpoint and to classify them separately, one is immediately faced with additional difficulties concerning systematics within the family itself—that is, to sort out the various subfamilies in a satisfactory manner. The family is often divided into eight subfamilies, but although some of them, such as the Aegypiinae or Old World vultures, are clearly homogeneous, most of them tend to merge and overlap, with species that display intermediate features. One of the most practical attempts to solve the problem is the classification adopted by Brown and Amadon, who hold that it is incorrect to speak of subfamilies at all and that it is better to group together those species that possess common characteristics, these groups more or less coinciding with the traditional subdivisions.

Even if one accepts this procedure—as we do here—some confusion is bound to arise over the use of common names. Species named eagle, hawk and kite, for example, are to be found in several groups. This cannot be avoided, whichever system of classification is adopted.

The Elaninae are the white-tailed kites, small raptors distributed all over the globe. They feed on moderate-sized animals, including rodents, reptiles and insects. Dawn and dusk are their favourite times for hunting, though they may also venture out in broad daylight. They are not highly specialised birds, lack a strong territorial instinct and sometimes live in colonies. Male and female collaborate in building the nest in a tree, often making use of an old nest belonging to another species, and lining it with grass and new leaves. Two to five eggs are laid and both birds share the responsibility for incubation, which lasts about a month.

The Perninae include the swallow-tailed kites of both the Old and New Worlds. A typical representative of this group, despite its common name, is the honey buzzard (*Pernis apivorus*) from Europe and Asia—a bird that spends the winter in Africa and is so named because of its habit of raiding the nests of bees and wasps for their larvae. All the Perninae have tarsi that lack feathers but are covered with strong scales. The claws are relatively weak. Small mammals and insects form the basis of their diet, the movements of prey being scrutinised either from the air, after soaring to a considerable height, or from nearer the ground, perched high on a tree. Although these raptors may form flocks of several hundreds when migrating, nest-building activity generally takes place privately. The nest—either a new one or one already existing—is situated in a tree and the two or three eggs take 30–35 days to incubate.

The true kites belong to the Milvinae and include some species that hunt and eat small animals and others that are scavengers. They build their nests in trees or among rocks, several pairs sometimes forming colonies. The black kite (*Milvus migrans*) is—as its specific name suggests—one of the migratory birds of prey. All species of this group are strong fliers.

The large group of Accipitrinae includes many predatory hawks, distributed all over the world and found in well-wooded regions where they prey on small mammals and birds. Typical members of the group are the various species of goshawk and sparrowhawk. All have fairly short wings and a long tail, enabling them to fly rapidly and make sudden changes of direction as they pursue their prey over terrain studded with thick undergrowth. They

Black-shouldered kite
(*Elanus caeruleus*)

The birds shown above and on the facing page are both members of the family Accipitridae. The black-shouldered kite is one of the Elaninae, the northern goshawk one of the Accipitrinae—regarded by most authorities as groups rather than subfamilies.

use the element of surprise, relying on acute vision to locate prey and on strong claws to deliver the finishing touches. There is a noticeable size variation between the sexes, the female sometimes being twice as big as the male.

Hawks, buzzards and eagles make up the largest and most highly differentiated of these groups – the Buteoninae. Most of them are accomplished hunters, but some are content to feed exclusively on carrion should the need arise. Small mammals form the main feature of the diet of the majority of these raptors, some hunting fairly large terrestrial birds and others preying on water birds. Their feet are perfectly adapted for predatory activities, the toes being relatively short, though strong, and the claws exceptionally tough. There is little difference in size between the sexes.

The birds belonging to this group are distributed the world over and have succeeded in adapting themselves to contrasting environments. In their anatomical structure and behaviour they naturally display a wide range of variations. Among them, for example, are the majestic golden eagle (*Aquila chrysaetos*), once widely distributed over the entire northern hemisphere but now, unfortunately, quite rare, and the bald eagle (*Haliaeetus leucocephalus*), the national bird of the United States of America. The species of the genus *Buteo* include both buzzards and hawks (the former term is more commonly used in the British Isles, the latter in North America). Best known is the common buzzard (*Buteo buteo*), which is similar to the American red-tailed hawk (*Buteo jamaicensis*).

It may be as much as three years before these birds attain sexual maturity and the females lay very few eggs. This is doubtless why, despite restrictions placed on hunting and precautions taken to protect the nests and fledglings, many species are threatened with extinction.

More homogeneous and thus easier to classify are the Aegypiinae or Old World vultures, inhabitants of the warmer countries of Europe, the larger part of Africa and the dry regions of Asia. They are mostly very large birds, some of them weighing more than 20 lb and having a wingspan of 8–9 feet. Their diet is highly specialised, consisting almost entirely of dead animals, though some species are predators as well as scavengers.

The carrion eaters locate their prey by virtue of their exceptional eyesight. Their beaks are long and powerful, suitable for ripping the skin off the carcases on which they feed. Some dismember the animal by plunging head and neck deep into the stomach cavity, this being rendered easier by the fact that these parts of the anatomy are either naked or only sparsely covered with down. The claws are comparatively poorly developed and for this reason they have little difficulty in moving about on the ground. Their wings are very large and sometimes rounded – especially well adapted for their characteristic soaring flight. Vultures are in fact capable of spending hours on end aloft, gliding lazily above the plains, without showing any perceptible wing movements.

The Circinae or harriers are mostly species of the genus *Circus*. They have long legs and wings and may be distinguished by a curious ruff-like formation of feathers encircling the head which gives them the facial appearance of owls. It is probable that, like these latter nocturnal birds of prey, the harriers rely to a large extent on the sense of hearing to locate their victims, which consist of small mammals, birds and reptiles. These are literally harried as the raptors fly backwards and forwards over open ground. The birds nest on bare ground and the females lay three to six eggs.

The Circaetinae are the serpent eagles, hunters of snakes and other reptiles as well as amphibians. One striking representative of this group is the Bateleur eagle (*Terathopius ecaudatus*), though this African bird's diet also extends to small mammals and carrion. They are all fairly large birds, with broad wings and thick plumage which helps to protect them from the bites of poisonous snakes. The tarsi are not covered with feathers but with hard scales. The serpent eagles normally build their nests in trees and the females lay one or two eggs.

Northern goshawk
(*Accipiter gentilis*)

Following page : A fine example of a golden eagle, one of the most handsome and powerful predators of the Accipitridae. It has a particularly strong, hooked beak and the characteristic far-ranging and acute vision of the family to which it belongs.